Models for Technical Writing

Philip C. Kolin

University of Southern Mississippi

Janeen L. Kolin

Forrest County General Hospital

St. Martin's Press New York

To Eric and Kristin
Julie and Loretta
Tony and Mary
and J. C.

Library of Congress Catalog Number: 83–61618

Manufactured in the United States of America.
98765
fedcba
For information, write St. Martin's Press, Inc.,
175 Fifth Avenue, New York, NY 10010
cover design: Darby Downey
text design: North 7 Atelier Ltd.
ISBN: 0–312–53587–2

ACKNOWLEDGMENTS

"Influence of Time of Herbicide Application on Control of Kudzu." Reprinted by permission of Weed Science Society of America.

"Kudzu." Copyright © 1963 by James Dickey. Reprinted from *Helmets* by permission of Wesleyan University Press. This poem first appeared in the *New Yorker.*

"Word Processors for Attorneys." Reprinted by permission from *Trial Magazine,* the Association of Trial Lawyers.

"Memo about Word Processing Needs at a Corporation." *Purdue University Business Writing Samples: Letters, Memos, and Reports.* Ed. Jeanne W. Halpern and Agnes M. Lokke. West Lafayette, IN: Purdue University, 1983.

"Apple II Super Serial Card Instructions." Copyright 1984, Apple Computer, Inc. Used by permission of Apple Computer, Inc., 20525 Mariani, Cupertino, CA 95014.

"Exploring Technical Aspects of Formatting Floppies." Republished with permission from *Word Processing & Information Systems.* Copyright 1980 by Geyer-McAllister Publications, Inc., New York.

"The Freedom 200." Copyright by Computer Design Publishing Company, © January 1984. All rights reserved. Reprinted by permission.

"Shock Absorbers." Herbert E. Ellinger, *Automechanics,* 2nd Ed., © 1977, pp. 482–484. Reprinted by permission of Prentice-Hall, Inc., Englewood Cliffs, N.J.

Acknowledgments and copyrights continue at the back of the book on page 465, which constitutes an extension of the copyright page.

Preface

Models for Technical Writing offers more than 65 pieces of writing actually prepared for the world of work, most of them appearing in their entirety and in their original formats. Our purpose in assembling these models is to give instructors a wide range of interesting, teachable, and realistic examples that they can use to introduce students to the variety of technical documents written on the job and to the rhetorical skills necessary to prepare such documents. Drawn from many scientific and technological fields, the models include letters, memorandums, instructions, investigative reports, feasibility studies, proposals, progress reports, opinion surveys, news releases, product brochures, and technical advertising. We have made a special effort to include many examples of kinds of on-the-job writing that are not readily available to technical writing instructors.

The emphasis throughout *Models* is on audience analysis. We selected models that would show students how a writer or a group of writers effectively analyzed an audience. Some of the models were intended for one reader with a single purpose, while others were directed to a number of readers with multiple purposes. Each model, therefore, reveals how the writer(s) selected technical details, language, visuals, organization, and format to meet the needs of the audience.

Each model is introduced by a headnote giving students background information about the audience; the audience's and writer's purpose; the type of organization in which or for which the model was written; especially appropriate uses of language, visuals, or format; and the types of information the writer had to gather before drafting the model. Thus alerted to the rhetorical situation, students will be better prepared to read and evaluate the model and to see how it solved a problem or accomplished a specific task. We believe that from this kind of analysis students can more effectively organize and write their own technical documents.

Two sets of questions follow each model. The first set asks students to discuss the effectiveness of the writer's language and style, visuals, organizational strategy, and format—the elements previewed in the headnotes. Through these questions, students have an opportunity in class to explore why and how the writer met the reader's needs. The second set of questions, "Suggestions for Writing," generates two types of writing assignments. One type asks students to respond appropriately in writing to the model (e.g., answer a letter, provide a subsequent report, rewrite a model or parts of it for a different audience). The other type requires students to adapt the techniques found in the model to their own writing for their major or on the job.

While the headnotes and questions focus on the specific models, detailed introductions to each chapter concentrate on the process of writing, giving students the background they need to evaluate the models as well as their own writing. These introductions explain how to gather, arrange, and present information, emphasizing the use of rhetorical techniques and the importance of revision in the writing process. In keeping with our emphasis on audience, each introduction begins with a series of questions that readers are likely to ask about a particular piece of technical writing (e.g., an internal proposal) or a specific rhetorical technique (e.g., classification). By knowing at the outset the kinds of questions readers will ask about a document or technique, students will be encouraged to see each model—and ultimately their own work—through a reader's eyes.

Models for Technical Writing is organized into three sections. The first section, comprising Chapter 1, provides carefully selected and ordered models for audience analysis. Unique to technical writing textbooks, this chapter presents two different sets of models (one on kudzu, a plant used to prevent soil erosion, and another on word processors). Through these models, students are introduced to various audience types (experts, decision makers, operators, consumers) and to the different purposes these audiences have in reading. Instructors may assign one or both sets of models, but in either case students will see how a single topic is adapted in a variety of ways depending on an audience's purpose, background, and interests.

Section Two, comprising Chapters 2–5, explains and illustrates the basic rhetorical techniques used in technical writing—physical description, classification, partition, definition, and process explanation. These four chapters lay the groundwork for the students' application of the techniques in later chapters. Because rhetorical techniques are best discussed as a part of the writing in which they occur, we have included a number of pieces of on-the-job writing in these early chapters as well as examples selected from technical journals and reference works.

The third section, "Applications," which includes Chapters 6–12, is rightfully the longest part of *Models for Technical Writing.* Chapter 6 focuses on instructions and is organized around the formats instructions most frequently follow—visuals, numbered steps, decision trees, and troubleshooting charts. Chapter 7 is devoted to investigative reports, including laboratory, field, and opinion survey reports. The kinds of data gathering and evaluating emphasized in this chapter will assist students in writing the longer technical documents discussed in Chapters 8 and 9. Proposals, the subject of Chapter 8, offers both internal and external proposals for students to analyze and imitate. Chapter 9 discusses feasibility studies and includes studies written for industry and government, while Chapter 10 covers both progress reports and performance reports. With Chapter 11, "Technical Publicity Announcements," *Models for Technical Writing* parts company with other technical writing books. This unique chapter provides information about and examples of news releases, product brochures, and technical advertising. Chapter 12, "Correspondence," the last chapter of *Models,* includes model letters of application (with résumés), inquiry, response, complaint, adjustment, and sales letters in addition to a variety of memos. To give instructors greater flexibility, we have included a number of examples of each type of correspondence.

Including real models from business, industry, and government is one of the major strengths of *Models for Technical Writing*. For the privilege of reprinting a few of these models, we agreed to change some names, dates, places, and costs. We think that these changes are a small price to pay for using the documents and in no way affect their overall realism or teaching effectiveness.

Many individuals assisted us in gathering models and provided suggestions for writing about them. We are grateful to the following individuals from the University of Southern Mississippi—Ralph Bisland, Newton Fawcett, David Goff, Col. Joe Kinnan, James Larsen, Ronald Marquardt, Mackie Odom, David Roberts, and Mark Winchell. A very special thanks goes to Laura Weaver (University of Evansville), J. Madison Davis (the Behrend College of Penn State), and Colby Kullman (University of Mississippi), and Walter J. Stephenson and James Thomas, U.S. Air Force, retired.

We are grateful to the many individuals in business, industry, and government who supplied us with models or helped us to locate them. Our thanks go to our many friends at the Leaf River Forest Products Company—Gene Flanders, Jerry Perkins, Vic Mattson, and Rob Eddy; John Moore of the PQ Corporation; Ted von Sprocken; Jerry Golden; Harold Cufaude; Llewellyn Toulmin; Robert Pindroh and Kenneth Wattenberger of the Lockheed Employees' Recreation Club; Richard Councill of the Family Lines Railroad; Michael Aun of the U.S. Bureau of Prisons; Malcolm Broom and Elton Raby of the U.S.D.A.

We also want to express our gratitude to our editors at St. Martin's Press for their assistance—Nancy Perry, Marilyn Moller, and Mark Gallaher.

Our sincere thanks also go to the exemplary reviewers whom St. Martin's asked to evaluate our manuscript. Their sage advice gave us much guidance—Carol Barnum, Southern Technical Institute; Virginia A. Book, University of Nebraska—Lincoln; William Coggin, Bowling Green State University; Michael Connaughton, St. Cloud State University; Eileen Evans, Western Michigan University; Jone Rymer Goldstein, Wayne State University; John S. Harris, Brigham Young University; Nell Ann Pickett, Hinds Junior College; and Jack Selzer, Pennsylvania State University—University Park.

Our typists—Anne Sykes and Beverly Broome—deserve kudos for their efficiency and speed.

To our children, Eric H. Kolin and Kristin Julie Kolin, go our love for their patience and enduring interest in our work.

Finally, we are eternally grateful to and for each other.

Philip C. Kolin
Janeen L. Kolin

Contents

part 1

Audience Analysis

chapter 1

Models for Audience Analysis

Know your audience. No advice is more basic or valuable for the technical writer. The most effective technical writing suits the experience and meets the needs of its audience. The following description of the squirrel-cage motor, for example, is a good piece of technical writing—functional, lucid, usefully detailed.

> Figure 1–1 illustrates a representative performance of a standard squirrel-cage induction motor.
>
> The two basic components of a squirrel-cage induction motor are the stator (stationary part) and the rotor (rotating part) reminiscent of a squirrel cage.
>
> A stator consists of a laminated iron core within and around the inner rim of which are distributed insulated windings. These primary windings are three in number for three-phase current and two for two-phase current. The arrangement of primary windings is controlled by the line voltage, number of phases, and number of poles. The power source is connected to the primary windings.
>
> The rotor consists of a laminated iron core with bar windings in various shaped slots around the periphery of the core. The bar windings at a design resistance specific to various NEMA design are interconnected (short-circuited) with rings. The rotor mounted on a shaft is in turn mounted in bearings. There are no direct power connections to the secondary windings. The current in the rotor is induced, therefore the name induction motor.

The writer has made a number of appropriate choices for the intended audience—engineers, particularly those with an interest in air-conditioning design—selecting the right level of scientific detail and including necessary technical vocabulary. Although appropriate for engineers, this description would be inappropriate for consumers trying to select which brand of air conditioner to buy. More than likely, consumers would not know the meaning of words like *stator* or *windings*. Certainly, they do not need to know about NEMA designs or differ-

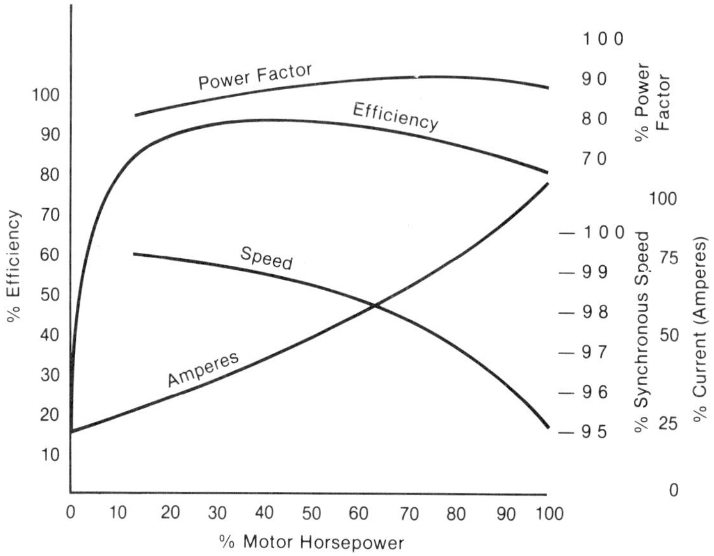

Figure 1.1 Typical performance, standard squirrel-cage induction motor (medium size).

ences between primary and secondary windings to make an informed decision. An article in a consumer magazine (e.g., *Consumer Reports*) could answer a nonspecialist audience's questions without providing these technical details. Therefore, the essential point to remember as you start your study of technical writing is that *what constitutes a clear and helpful message can vary greatly from one audience to another, from one reader's needs to another's.*

This chapter will help you to define your audience, assess its needs, and design your written work to meet those needs. The models in this chapter will show you how the same topic is discussed in appropriate ways for a variety of readers with different needs, interests, and backgrounds.

DEFINING YOUR AUDIENCE

Writing is a process by which you communicate what you know about a subject in a way that will be useful and relevant for your audience. In writing for any audience, you must gather appropriate information, plan what to say about that information, prepare some preliminary drafts (when deadlines permit), and revise your work. A key part of the writing process is to analyze your audience: you need to ask questions about your reader as well as about the work you are preparing. This and all the other chapters in *Models for Technical Writing* will give you the types of questions that must be asked—and answered—about a particular audience or type of technical communication.

Research on the writing process indicates that many pieces of writing fail more because they are ineffective for a particular audience than because they are poorly written. Despite their scientific or technological knowledge, some technical writers have difficulty communicating with an audience. These writers usually err by making choices of content, vocabulary, or style that are not appropriate for the targeted audience. One good way of making the right choices for your audience is by *decentering*, the technique of removing yourself from your written work to read it from the targeted audience's viewpoint. By following the suggested guidelines for decentering, you can more effectively prepare written work appropriate for your audience.

Consider your audience's technical expertise. Your goal is to provide the reader with written work that can be understood and used to complete a job. Problems can arise when writers overestimate a reader's technical expertise. Not every reader will know a subject as well as you do. When your reader is not a specialist in your area, supply helpful background information and definitions, omit technical terms, and relate new information to old through analogies, metaphors, and similes. At the same time, a technically advanced audience may require greater documentation of the writer's assertions than a nontechnical audience would.

Be especially careful when you are supplying directions. A set of instructions that is unclear or missing steps can result in lost work time for employees, damage to equipment, or even personal injury. An unclear letter or report can lead to misunderstandings requiring further time and money to correct. Incomplete descriptions of a piece of equipment can delay an order or repairs. Since you will seldom be available for your audience to question what it does not understand, strive to make your written work self-explanatory.

Consider the needs of your audience. A relevant piece of writing will help a reader accomplish a specific task in a specific situation. Since you will not always write to the same reader for the same reason, adjust your level of technical detail to match the particular needs of a given group. A consumer, for example, would find the description of the squirrel-cage motor on page 3 too detailed, whereas an engineer would not. In some instances you will omit or summarize, in nonspecialist's language, the work you do; when a reader is concerned primarily with costs and availability of parts, he or she will not want extensive technical detail. But when your audience needs these details to complete a job, include them.

ANALYZING YOUR AUDIENCE'S NEEDS

In the world of business and industry, the more you know about your audience's particular needs the more able you will be to provide helpful information. A good starting point is to prepare a profile of your audience. You can do this by considering the following sorts of questions about your audience's responsibilities, purpose in consulting your work, perspective, education, and experience.

Keep in mind that your audience may be one individual or a group of individuals, hence your hypothetical "reader" may in reality include many people.

The Reader's Responsibilities

- What is the reader's title in the organization?
- What job does the reader perform?
- Does the reader supervise a particular kind of project or type of work?
- Is the reader the only one to see your work or will it be sent to another reader?
- How will your work affect the reader's own work?
- Does the reader perform tasks similar, or identical, to yours?

The Reader's Purpose in Consulting Your Work

- Does the reader primarily want an explanation of how something is done or made?
- Does the reader want instructions on how to perform a task?
- Is the reader examining your work to make sure some job or task has been documented?
- Is the reader using your work to decide which of many options is the most feasible?
- Is the reader examining your work to give or withhold authorization of a project?

The Reader's Perspective on Your Work

- What sort of evidence or support is the reader most likely to accept?
- Does the reader have any strong preferences (about design, arrangement of information, style) you need to know about?
- Is the reader likely or unlikely to accept your message?
- What does the reader know about you or your organization?

The Reader's Education and Experience

- What is the reader's level of education?
- How many years of professional experience does the reader have?
- How much about your job does the reader know?
- Has the reader ever done your job?
- What are the reader's major professional interests, accomplishments, and skills?

These or similar questions can help you to identify the specific audience you are addressing. In general, a reader of technical writing will fall into one of four groups discussed below.

IDENTIFYING BASIC AUDIENCE TYPES

Most technical writing is written to four basic types of readers: (1) consumers, (2) decision makers, (3) operators, and (4) experts. Throughout *Models for Technical Writing,* audiences will be identified as one or a combination of these types. They can be generally defined as follows.

Consumers

Everyone is at some point a consumer, a buyer or user of goods, services, technical expertise, and advice. The consumer is the nonspecialist—the general reader, the layperson—with no technical knowledge or experience *in your area.* The consumer may be informed about a wide range of subjects but not the one you are writing about. Consumers can be homeowners, do-it-yourselfers, members of a community organization, or computer scientists reading about water pollution. As consumers, they have a great deal of flexibility. Unlike other audiences, consumers often are not required to read a piece of writing. They have all the options of the marketplace. Consumers read to be informed, persuaded, or entertained, to learn about new developments in electronics or health care, to solve a household problem, or to determine if a particular product or service is best for them. Even though such a reader may be a nonspecialist, he or she is still likely to greet your writing with some probing questions. Following are some such questions:

1. Is the topic interesting to me? Is it an attractive product? a useful service? valuable information?
2. Does this information serve any of my needs?
3. Is the time I spend reading worth it?
4. Can I understand it? Are technical terms and concepts clearly explained, and are the explanations of technical material relevant to my purposes?
5. Is the writing lively and easy for me to follow?
6. Are there interesting and clear illustrations, drawings, diagrams, pictures?
7. How does this product or service compare with similar ones?
8. What is the reputation or track record of any company or organization associated with the product?

Decision Makers

Decision makers may be your most demanding and important audience. From the technical writer's point of view, a decision maker is the person who must approve or reject a project. A decision maker can be almost anyone in an organization responsible for planning budgets, setting priorities, or establishing policies that affect an organization. Decision makers include presidents, account executives, portfolio managers, department heads, cost engineers, marketing chiefs, personnel managers, or line supervisors. As these examples show, decision makers are found at all levels of business, industry, and government. Many decision makers are trained in traditional business fields—marketing, personnel management, accounting, business administration—and do not have backgrounds in science or technology. Others, however, are engineers or other specialists who have been promoted into administration because of their managerial talents—e.g., an electrical engineer who directs a company's training program. Decision makers are principally concerned about costs, personnel, productivity, programs, distribution, work loads and stoppages, customers, and contracts. They want "bottom line" information—conclusions and recommendations, not scientific methods and theories—for the practical management decisions they must make. Here are the types of questions decision makers are generally interested in as they read:

1. Is what I am reading about feasible? Will this venture succeed?
2. Is there a need for what is being proposed?
3. What will it cost? Are there tangible details offered about expenditures (for equipment, supplies, transportation, labor)?
4. Is it cost effective?
5. Who is doing it? Will additional employees have to be hired or current ones retrained? What are the writer's or the agency's qualifications for the job?
6. How well is it being done? Are things running on schedule? Are all problems identified or solved? Are all local, state, and federal regulations being met? Are all quotas being achieved? Are union guidelines being followed?
7. What's going to happen next? When is the next phase of the job to be completed? When is the next report due? How long will it take to finish the job?
8. What advantages will result for the organization? a better product than those of our competition? new markets? more efficient procedures? improved corporate relations? a better community image?

Operators

A very large part of your audience will be operators. They may have graduated from vocational and technical schools, may have earned certificates from company or military schools, and may have years of, or very little, on-the-job experience. Within an organization operators may include machinists, repair persons, assembly line workers, technicians, counters, and clerk/typists. They are the individuals who manufacture products, or install, use, maintain, and repair equipment. Basically, operators want instructions on how to do things. Because they perform consistent procedural tasks, they are less interested in theory than in practical "hands on" information about applications and results. When you write for them, here are some of the questions you can expect operators to ask:

1. What step-by-step procedures must I follow to begin, continue, and complete a job?
2. What happens when I perform a certain procedure?
3. How long does the procedure take?
4. What equipment must I install or operate? What does it look like and how does it work?
5. What parts, tools, or power sources do I need?
6. Are alternate parts or procedures useful or possible?
7. What preparations must I make and what precautions must I take in operating this equipment or performing this procedure?
8. What can go wrong, and what must I do if it does?
9. What repair or maintenance work must be made and how often must it be done?

Experts

Experts have formal educations and backgrounds in the sciences and technology. They may read and contribute to professional publications in their fields. Experts may be chemists, physiologists, engineers, architects, geologists, nurses, or any other specialists. Experts work for and with management. They design the equipment that operators run, perform experiments on new products and services, or monitor environmental conditions. Through rigorous scientific investigation, experts solve technical problems that arise in an organization. When something goes wrong, it is their job to retest, reformulate, and redesign. They frequently offer training programs for operators and work with them in identifying and correcting problems in equipment or site conditions. Experts are familiar with technical terminology, formulas, and theories. They will expect you to document your methods and sources of information carefully so they can apply your findings to their own research. Here are some questions members of this audience are likely to ask:

1. What problems need to be solved?
2. What standard procedures are used to perform this type of study or experiment? Are the standard procedures to be followed or changed? If they are to be changed, why?
3. What kinds of scientific or engineering constraints are there in a study of this problem or experiment?
4. How did the production model behave in relation to the experimental one?
5. Does the product meet design and performance specifications?
6. Is there adequate data to support the conclusions reached and was this data presented with appropriate documentation?

7. What changes in equipment, design, procedure, or sampling need to be made to eliminate the problem?
8. How can these conclusions help improve the way products are made or improve the environment?
9. How can these conclusions assist us in further research, development, or training?

Strictly speaking, these four audience types refer less to specific individuals than to the roles individuals assume when they read your work. It is misleading to think that every time you write it will be to a single reader firmly entrenched in one of these audience categories. A reader takes on many roles, depending on his or her purpose in reading your work and the varying degrees of interest and expertise. It is wise, then, to regard the four audience designations as various roles a reader can play. What is especially important to remember when you consider the needs of your prospective audience is that one person may be many audiences, that many audiences may read a single piece of technical writing, and that differences exist among the members of the same audience. Each point deserves further explanation.

One person may be many audiences. In the course of a single day, an individual can be any or all of the four audiences depending on his or her job and reason for reading. The decision maker may be an expert as well as a manager. For example, he or she might have a degree in chemical engineering but may recently have worked as the assistant manager of a plant that manufactures plastic products used in the home furniture industry. When reading a report on the problems connected with distribution and shipping, the individual assumes the role of a decision maker who must make changes in production schedules and delivery routes. When reading an article in a professional engineering journal, this assistant manager is the expert studying the latest approaches to, and experiments in, a complex technological discipline. When shopping for a new ten-speed bicycle for a son or daughter, this same manager-expert may read as a consumer and in fact may consult a copy of *Consumer Reports* to learn which ten-speed bicycle is the most safely constructed. After purchasing the bicycle, this individual may have to read directions like an operator in order to know how to assemble it.

Many audiences may read a single piece of technical writing. If one person can be many audiences, then the converse is also true. One piece of technical writing can be addressed to many different audiences. Take the example of a long report that was commissioned by the Occupational Safety and Health Administration (OSHA) to establish acceptable noise levels in the foundry industry. The author of such a report must anticipate which sections will be read by all audiences and which sections will be of interest to just one audience. All readers study the abstract (see pages 219–220). Scientists who want information on how tests were conducted will read sections on test methods, equipment, and results. Sections on budgets, timetables, and worker disabilities would be of most interest to decision makers, the officials both at the consulting firm that won the government contract to do the study and at OSHA. Many of the specific conclusions of the report might make their way into a set of guidelines for operators in the industry advising them of potential dangers and safety procedures.

Differences exist among the members of the same audience. As the preceding example about the chemical engineer-manager illustrates, members of a single audience do not all have the same background and experience. As we have seen, many decision makers do not have engineering degrees or work experiences; they may have backgrounds in marketing, personnel management, or finance. Knowing a manager's background helps you to determine how much, if any, technical information to include. Two individuals may both be classified as operators, yet one could be more experienced than the other or could have used completely different sets of tools or techniques. For example, one operator may use tools that are metrically calibrated while another reader's equipment is not. You would have to anticipate these problems if you were writing instructions for these two readers.

WRITING FOR SPECIFIC AUDIENCES

As we have stressed, you will be writing for a variety of audiences with differing backgrounds, interests, and needs. You will have to adapt your writing to meet their needs. Carefully consider the following elements of writing as you shape your work for specific audiences: content, format and organization, style, and visuals.

1. Content

Give your audience the information it wants and needs. After assessing your reader's background and need for your work, you should be able to decide which details are necessary. Scientists and engineers reading a long report in their area will expect to find detailed technical information about previous studies and methods used in the report. For the decision maker, who is mainly concerned with the financial feasibility of the project, such information would be inappropriate. Table 1–1 shows the kinds of information about a new automobile model relevant to four separate groups of readers, each with a different need and background.

2. Format and Organization

Format and organization are interrelated. Format refers to the physical appearance of your work on the printed page. It controls how you divide and label

TABLE 1–1. TYPES OF INFORMATION REQUIRED BY READERS ABOUT THE MODEL X-670

CONSUMER (***Customer***)	***DECISION MAKER*** (***Automobile Manufacturer***)	***OPERATOR*** (***Assembly Line Worker***)	**EXPERT** (***Design Engineer***)
EPA mileage rating	cost-profit analysis	new skills/retraining needed for assembly	engineering parameters
warranty and rebates	units produced per month	ease of assembly	drag coefficient
optional equipment available	promotion campaign	types of tools required	design of safety equipment
availability of parts	delivery dates	location of parts	life expectancy of parts
service after the sale	consumer complaints	safety measures to be followed in installing parts	aerodynamic qualities

the parts of your work and how you order and arrange those parts on the page. You help to organize information for the reader by choosing an appropriate format.

Technical writing is not like writing an essay for a freshman composition class, where one paragraph of unbroken prose follows another. Instead, technical writing breaks information down into smaller parts separated and identified in a variety of ways. Your choice of format makes clear for your reader the division and relationship of the various parts of your work. Following are some format devices frequently used by technical writers:

1. headings and subheadings
2. numbered paragraphs and subparagraphs
3. numbered lists
4. bullets, or raised periods, as used below
5. italics and underscoring to highlight important points
6. extra white space, including indentations, to set elements apart

An attractive and helpful format makes your work easier to read and thus predisposes your audience to react to it favorably. Different kinds of technical writing require different formats. Instructions (discussed in Chapter 6) may use numbered steps. Progress reports (discussed in Chapter 10) can be in outline format or include headings to separate paragraphs into different divisions.

Organization refers to the order in which you present material for maximum effectiveness. An effective organization assists the reader in getting the most relevant information quickly and clearly. A variety of organizational patterns can be used in technical writing. Here are a few examples:

- classification (for comparison and contrast of related items)
- cause to effect (for reports moving from problem to solution)
- chronological sequence (for some progress reports)
- spatial (for site descriptions)
- hierarchical (for reports on a sequence of topics from most to least important)

Organizational plans can vary according to the reader's needs. Scientists reading about experiments expect information to be listed under conventional headings pretty much in the following order—abstract, introduction, materials, methods, results, discussion, and conclusion. Proposals and feasibility studies (discussed in Chapters 8 and 9) often print purposes, conclusions, and recommendations at the start of a report to save decision makers from having to read the entire document. For the reader's safety, job effectiveness, and convenience, instructions are frequently organized chronologically into sequentially numbered steps.

3. Style

Style in technical writing basically refers to the complex choices writers make to ensure that their writing is easy for the reader to comprehend. Elements of style include word choice, sentence type and length, paragraph length, and tone (the author's attitude toward the work and reader). There are some basic rules all writers need to follow regardless of whom they are writing for:

1. Be concise and to the point. Cut wordiness.
2. Be clear and coherent. Start each paragraph with a topic sentence that summarizes or explains. Arrange sections, paragraphs, and sentences in an orderly way so the reader can follow them easily.
3. Be precise. Use accurate language.
4. Be concrete. Use words that involve the senses.
5. Be direct. Use the active, rather than the passive, voice when possible. (Active voice: The technician did the job. Passive voice: The job was done by the technician.)

Following these basic rules will help you to achieve clarity, valued by all audiences. But what constitutes clarity for any audience will further depend on the reader's needs, purpose, and level of expertise. Readers have different "sentence understanding" skills. Consequently, a style appropriate for the expert often will not be clear to the layperson.

The following discussion of style is organized around the four potential audiences for your technical writing and identifies how the needs of a particular audience govern your stylistic choices.

Consumers. For consumers you are writing primarily for readers who wish to satisfy a personal curiosity, to gain some technical information or a general understanding of a technical topic, or to get some practical advice. Consumers cannot be expected to understand the complexity characteristic of something written for an expert. Make your writing easy and relaxed for the consumer. Consequently, the following points are important when writing to this audience:

- Use an informal tone.
- Choose common words over technical terms.
- Give definitions of any terms a non-expert is likely to misunderstand.
- Write clear and direct sentences that make the topic easy to understand.
- Use the first or second person.

The sentences below illustrate how information about a tree fungus and a type of computer are appropriately written for consumers. Each sentence shows the writer making the stylistic choices indicated above.

a. A fungus has attacked many of the seedlings at Idlewild Park. You will see the city replacing these trees in the next month to keep our park beautiful.
b. The relational model will allow you to retrieve information quickly and easily. It will give you access to any type of data stored in the data base.

Decision Makers. Writing for decision makers requires a style that is more "informed" about managerial considerations than writing for consumers requires, but less "scientific" than the style used for experts. For decision makers remember that "time is money." They want concise writing, free from distracting jargon or easy-going conversation. Strive to be clear and to the point. Decision makers have a job to do and they want what they read to help them do their job efficiently and well. They are not particularly interested in the technical documentation that experts want, nor do they particularly need to be put at ease as consumers do. However, persuasion (selecting words to obtain approval and authorization) plays a major part in your choice of stylistic elements for decision makers. Basically, this audience will find a relatively formal style most appropriate. Use the following stylistic elements when writing for them:

- Write sentences that are free of clutter and bureaucratic legalese.
- Use the third person, although this should not preclude the possibility of something more personal.
- Use technical terms the audience knows ("specific performance" for attorneys, "ARM's" for accountants).

Rendered in an appropriate style for decision makers, the two sentences about fungus at Idlewild Park and the computer become:

a. It will cost the city approximately $7,500 to replace the fungus-damaged seedlings at Idlewild Park. To that $7,500 add another $4,000, the initial expenditure for the trees last year.

b. The Data Base Management System will initially cost $48,000. The annual maintenance fee of $1,500 includes system updates and modifications.

Operators. Operators will require a more technical style than necessary for the two previous audiences but not as scientific as that used for experts. The primary consideration for the operator-reader is "leave no chance for error." This audience requires that a writer's language be immediately understood and that the tone be helpful and encouraging, not intimidating. When you write for this audience, do the following:

- Make sentences brief.
- Use the second person for instructions and directions, but the third person for descriptions.
- Include jargon and technical language that is a part of an operator's duties.
- Write in the active voice most often for instructions, in the passive voice for process descriptions.

For operators, directions on removing the diseased trees or using a computer command might read as follows:

a. Remove infected trees to truck at once.
CAUTION: Do not let infected trees touch healthy trees.
b. Enter the SQL command as follows:

Select	Domain Name
From	Table Name
Where	Predicate

Experts. For experts you are writing to provide information that will help them solve a problem or satisfy a professional curiosity. Your presentation must be precise, objective, sometimes even dense. Meeting these goals will generally involve a highly formal style.

- Use first as well as third person.
- Use passive constructions (to satisfy the need for objectivity).
- Include technical language (to satisfy the need for precision).
- Use generally longer, more complex sentences (to satisfy the need for complete, more exact information, which would be distorted if expressed exclusively in simple, short sentences).

Experts will require the most scientific presentation, as the following sample sentences show.

a. According to data gathered at the Idlewild site, trees in the 9 and 10 d.b.h. classes were infected with fusiform cankers (*Cronartium fusiforme*); the danger of infestation to neighboring regions is about 40 percent.

b. The relational data base management system is implemented by a back-end/processor using associative memory addressing and responds to Boolean queries in 23 nanoseconds.

4. Visuals

Visuals play a vital role in technical writing. Notice how often a piece of technical writing—a textbook in your major field, for example—uses them. Look through *Models for Technical Writing* to see how many of the selections depend upon visuals to convey information to an audience. Photographs, drawings, graphs, tables, charts, flow charts, diagrams, schematics, and maps help writers to clarify, organize, emphasize, and summarize important information for the reader. Significant relationships, concepts, and designs are frequently expressed visually. Although visuals are sometimes the reader's primary source of information, most often they are equal partners with your words in helping a reader understand and act on your work. As with style, visuals need to be appropriate for your audience and their purpose. A homeowner reading about a piece of equipment might be helped by a photograph or perhaps even a drawing showing the way the internal parts are arranged. But an engineer who designs the equipment would require and prepare far more elaborate visuals (blueprints) than a photograph or simple diagram. The use of visuals requires great care. Never use them just for decoration. They should play an important functional role in your work. Not only must they be clear and relevant, but you have to know how and where to introduce them in your work. Chapter 2 will give you specific procedures to follow when you use a visual.

A NOTE ABOUT THE MODELS IN THIS CHAPTER

This chapter contains two groups of models. The models in the first group are all concerned with kudzu, a fast-growing plant used for foliage and sometimes for soil conservation. The models in the second group concentrate on the uses of word processors. Each model within a group is addressed to a specific and separate audience. By studying all of the models within a group, you will see how the presentation of a single topic—kudzu or word processors—is adapted to meet the needs of a different audience. As you read each model within a group, pay close attention to the types of choices writers have made about technical detail, organization and format, language, and tone. These choices were made only *after* the writer had successfully analyzed the audience's background and its purpose in reading the work. The materials for audience analysis in this chapter will show you that all effective technical writing starts with an understanding of who the reader is.

A CHECKLIST FOR AUDIENCE ANALYSIS

1. Determine your audience's technical expertise.
2. Estimate how much knowledge about the topic your audience has.
3. Consider your audience's purpose in reading your work.
4. Find out the specific job responsibilities of your audience.
5. Try to imagine your audience's perspective (organizational views, personal tastes, likelihood to agree with you).
6. Categorize your audience (consumer, decision maker, operator, expert).
7. Decide what kind of information, style, and visuals your audience requires.

KUDZU

This very brief article appeared in the EPA Journal, *published by the Office of Public Awareness of the Environmental Protection Agency (EPA). The chief aim of the* Journal *is to publicize the effectiveness of the EPA and to promote its activities. Written by an EPA staffer, the article on kudzu is intended for nonspecialists (consumers) who consult the* Journal *for information about environmental issues—air and water pollution, pesticides, solid waste disposal—and the EPA's role in responding to them. The article provides a varied, lively history of kudzu—how and why it was introduced in the United States, what happened to it here, and what its current status is.*

Living with the Kudzu

Charles Pierce

As one of the world's most vigorous plants rests through the winter months 1
its networks of tangled brown vines keep their grip on the red earth and nearby trees and telephone poles along thousands of miles of Southern roadways.

With the return of spring temperatures and more direct sunshine the Kudzu, 2
a vine imported from the Orient, will once again begin its astounding growth.

The introduction of this plant provides a classic illustration of the need 3
for careful study of an import's overall environmental impact.

The Kudzu was brought to the U.S. a century ago and planted in the Florida 4
panhandle. From there it has gradually spread over about one million acres of land in the Southeastern States.

At first the plant, which can grow as much as one foot a day and 100 5
feet in a year, was regarded as a savior of the South. Its advantages included:

The growth of great billowing waves of Kudzu washing over fields and 6
highway embankments helped control soil erosion.

The vine, a member of the legume family, helped fix nitrogen in the soil. 7

Cattle, goats and other livestock would eat this plant. 8

Its deep-probing roots help aerate the soil. 9

For many years it was regarded as a miracle crop that would help restore 10
the South's worn and eroded soil. Conservationists and others joined in a crusade to help promote distribution of this marvel crop.

One of the plant's main boosters was Channing Cope, former long-time 11
agricultural editor of the *Atlanta Constitution,* who proclaimed, "Cotton isn't king here any more. Kudzu is King."

In the 1950's, however, farmers began to become disillusioned with this exotic crop. They noticed after the vine was planted it often began to intrude into neighboring fields where it ruined other crops. 12

Telephone poles were sometimes pulled down by the vine and power companies were forced to spray Kudzu with herbicides periodically to prevent it from climbing high-voltage towers, and causing damage. 13

Good lumber and pulpwood trees were destroyed when the thick foilage on the climbing vine shut off the sunlight needed for growth. 14

Also, substitutes were found which were considered better than Kudzu for pasture and soil erosion control. 15

However, by the time cultivation of the vine had stopped it had leapt fences and run wild. There are parts of Mississippi where highways slice through countryside so completely enveloped with Kudzu that all other vegetation has been killed. A traveler feels he has entered another and somewhat ominous world. 16

James Dickey, the noted Georgia poet, referred to the plant in his poem "Kudzu" as "green, mindless, unkillable ghosts." The poem states: 17

> In Georgia, the legend says, 18
> That you must close your windows
> At night to keep it out of the house,
> The glass is tinged with green,
> even so . . .

In areas where it is necessary to control the plant, along highway rights of way, for example, the Kudzu is mowed. The chemical 2-4-D is also frequently used to check the spread of this plant. 19

Gale A. Buchanan, a former editor of the magazine, *Weeds Today,* and 20 agriculture professor at Auburn University, has conducted many experiments on controlling the plant.

Regarding the future of Kudzu, Buchanan said: "I don't see it taking over much more territory. I think it has probably reached its limits. It is a problem along highways, but not in cultivated fields where cows, for example, can kill it by over-grazing. However, cows can't chase it up telephone poles. 21

"There are chemicals and other control measures that can be used to keep it in check. The level of effort being made now to control it is probably adequate. The lesson we need to learn from the Kudzu experience is how careful we must be about importing new plants into our environment." 22

QUESTIONS FOR DISCUSSION

1. What is the purpose of this article? Why would the EPA publish a story about Kudzu, especially for the intended audience?
2. Describe the article's diction, sentence and paragraph length, and tone (view on the subject). Are these different elements appropriately chosen for the audience? Explain.
3. What functions do the quotations serve? How are they a part of the author's strategy in adapting the material for the audience?
4. Contrast the discussion of 2-4-D in this model with that in the model on the influence of time on herbicide application on pages 30–35. Should the author of "Living with the Kudzu" have given readers more information about the chemical? Explain.
5. What other visuals might have been used in this model? Indicate why your choice(s) are relevant for the audience and the writer's purpose.

SUGGESTIONS FOR WRITING

1. This article describes kudzu to an audience that probably knows little, if anything, about it. Write a similar piece about a topic of your professional or academic specialty for a consumer readership. Here are some suggestions:
   - desalination
   - a security system
   - lasers
   - blood banks
   - solar reactors
   - energy from biomass
   - adjustable rate mortgages
   - electric cars
   - genetic engineering
   - an animal disease
   - holographs
   - biorhythms
2. Write a summary (50 words maximum) of the main points in this article. Be prepared to explain what you deleted and why. For a good example of a summary to follow, look at the abstract for the herbicide application model on page 30.

The following memorandum is a piece of in-house writing. It was prepared by a landscape engineer (an expert) for the business manager of an industrial park (a decision maker). The memorandum contains a feasibility study (discussed in detail in Chapter 9, pages 300–306) in which the author identifies and evaluates appropriate ground covers for a particularly troublesome section of the park. Her reader knows nothing about landscape design or kudzu and is interested only in having the information that will help him to make the wisest financial decision. In making her recommendation, the author uses criteria that are of primary importance to her reader—cost, maintenance, and overall appearance.

Memo about Landscaping an Industrial Park with Kudzu

B. Barnham

GULF COAST INDUSTRIAL PARK

Hasting, Florida 32164

INTEROFFICE MEMORANDUM

TO: F. R. Hendering
Manager

FROM: B. Barnham B.B.
Landscape Engineer

RE: Ground Cover for Data Processing Center

DATE: June 7, 1984

Purpose

Before June 30, as construction of the Data Processing Center nears completion, we need to decide on an appropriate ground cover. I have studied three possible covers and I am writing to ask you to approve what I believe is the best choice: kudzu. 1

Background and Problem

There is no difficulty on three sides of the Center; concrete driveways run along both sides of the building and a parking lot sits behind it. But the area in front of the Center slopes approximately 20 to 30 percent and thus makes our choice of cover difficult. This area is fairly large, approximately 20,000 sq. ft., the equivalent of one half acre. The topography is gullied and covered by a fine clay soil. Both the contour and composition of this bank contribute extensively to soil erosion. 2

Traditional grasses (e.g., Bermuda, bent) used for landscaping elsewhere in the park are not feasible here because they would not solve the erosion problem. An 3 appropriate ground cover must be selected that will (a) stop erosion, (b) stabilize the bank, and (c) be easily maintaincd.

Possible Solutions

Three ground covers might be used to solve the problem: 4

1. Weeping love grass, a long grass that bends like a weeping willow; it lays, or "weeps," over the ground in a cascading fashion.

2. Ivy, a vine with small, shiny leaves that covers the ground.

3. Kudzu, a deep-rooted plant with velvety leaves.

Each of these ground covers was evaluated in terms of the 5
following criteria: adaptability, maintenance, appearance, and installation costs.

Adaptability

1. Weeping love grass is a slow-growing cover requiring at least six months to become established. Since it grows in clumps and bunches, there is the risk of erosion around each clump until it is established. Once established, however, weeping love grass holds the soil effectively, sheds water quickly, and retards erosion.

2. Ivy takes three to four months to become rooted, which is much more efficient than weeping love grass. However, sunny conditions make the ivy, because of its root system, less useful and effective in holding soil.

3. Kudzu is native to the area and starts holding the soil in 30 to 40 days. It is a fast-growing climbing plant, spreading at the rate of six to eight inches a day. Its crowns, or plants, spread quickly in horizontal directions.

Maintenance

1. Weeping love grass on this slope would be difficult to mow and water. Of the three choices it would require the most maintenance. We would have to obtain a plex-mower with triple cutters. Because of the slope, we would run the risk of water runoff and poor fertilization.

2. Ivy, because it is a vine, does not present the mowing problems of weeping love grass. But like weeping love grass, it would require frequent watering.

3. Kudzu is virtually maintenance free except for the need to cut any runners extending into the parking area. Contained by the road embankment at the bottom of the slope and two concrete driveways, kudzu in this area would produce a thick mat that might have to be leveled with a scythe once or twice a year.

Appearance

1. Weeping love grass is light green and would blend in with other lawns in the park. In the winter, however, it turns brown.

2. Ivy, which is green all year round, has palmate leaves (diverging from one central point) and would contribute to the appearance of the park.

3. Kudzu is less attractive than either weeping love grass or ivy. More primitive looking than these other two covers, kudzu has large blue flowers and its leaves have a velvety touch. Like weeping love grass, kudzu dies down to a dull brown in winter.

Installation Costs

1. Weeping love grass would have to be planted from seed, and the seed bed would have to be prepared and controlled in the area for six to eight months, until the plants are up and established.

Seed, 15 lbs./half acre @ $10.00 a lb.	150.00
Preparing the seed bed (40 hrs.)	1200.00
Planting the seed (8 hrs.)	215.00
Water control (6–8 months)	2500.00
	$4065.00

2. Ivy requires the same type of soil preparation but entails additional work and expense for handsetting each plant.

1000 plants @ $.35 plant	350.00
Preparing the seed bed (40 hrs.)	1200.00
Planting the vines (40 hrs.)	1835.00
Water control (4 months)	1250.00
	$4635.00

3. Kudzu costs less than ivy and requires less maintenance than the other two covers. It is therefore cheaper to use on the slope.

1000 crowns (plants) @ $15.00 a hundred	150.00
Preparing the seed bed (40 hrs.)	1200.00
Planting the crowns (40 hrs.)	1835.00
Water control (1 month)	350.00
	$3535.00

Conclusion and Recommendation

Although kudzu is less attractive than either weeping love grass or ivy, it is by far the most effective means of preventing soil erosion on the slope in front of the Data Processing Center. It is also more economical to install and maintain than weeping love grass or ivy. 6

For these reasons, I recommend the use of kudzu on the slope in front of the Center. With your approval, we can begin preparation of the seed bed within the next ten days. 7

QUESTIONS FOR DISCUSSION

1. What kinds of research did Barnham have to do to give her reader the relevant information he had to have about kudzu and the other ground covers?
2. Why does Barnham mention her choice of ground cover both in the first paragraph and in the last section of the report?
3. Based upon Barnham's descriptions of the ground covers, has she accurately assessed the amount of information her reader needs to have about these topics? Explain your answer with examples from this model.
4. How does the author deal with the problem of kudzu's tendency to overgrow? Is this discussion adequate?
5. How does the writer's format help her reader to make a decision easily?

SUGGESTIONS FOR WRITING

1. Write a brief response from Hendering to Barnham about her report. Assume Hendering's viewpoint and try to react to her proposal as he might. Ask questions to try to get more information if you think it necessary. The preceding model, "Living with the Kudzu," might give you ideas for questions needing more attention.
2. Taking relevant facts from this memo, write a brief description of this site for a consumer or for an expert. Explain in your first paragraph why you are providing such a description.

The following selection is an information sheet on kudzu distributed by the agronomy division of a county agent's office. The audience knows about planting techniques, crops, and herbicides but needs precise instructions on how to grow, graze animals on, and control kudzu. Like many other documents available from a county agent's office, the emphasis here is on practical results, getting a job done quickly and accurately with the least chance of error. Consequently, the piece is written for readers whose needs are those of operators.

How to Grow Kudzu

Kudzu saves soil and makes a good grazing crop at the same time. It can be 1
grown on rough, eroded land that cannot be mowed. Kudzu chokes out bushes, briers and weeds. It never gets too dry for well established kudzu to grow.

Land for Kudzu. Kudzu will grow on rough, gullied land, but not on shallow 2
or hardpan soil. Bottom land makes poor growth of kudzu. The sandy areas, deep brown loam soils and well-drained, deep soils are best. Do not set kudzu on high water table land or overflow land. Canal banks are good kudzu areas.

Fertilizing and Setting. Kudzu must be fertilized with a complete fertilizer 3
at first to do well because usually it is set on poor, eroded land. On land not too gullied, make rows 12 feet apart and set crowns 3½ feet apart. Apply 500 pounds of mixed fertilizer per acre at time of setting. It will pay dividends in quicker growth and survival to use barnyard manure if land is poor and low in humus.

Kudzu crowns can be set on edges of gullies and short rows between 4
gullies. Apply a handful of fertilizer around the crown or plant at time of setting. If land is badly eroded, holes can be dug with a post hole digger. Put manure and a complete fertilizer in the holes. Mix with top layer of dirt and set the crowns.

If minerals are used, lime land to a pH of 6.0 to 6.5 and use 500 pounds 5
of a mixed fertilizer such as 0-20-20 or its equivalent. September or October is the best time to apply minerals if crimson clover, ryegrass, or wild winter peas are interplanted with the kudzu. Apply minerals in March if no interplanting is done.

Time to Set. March or April. 6

Rate to Set. Set on 12 foot rows, 3½ feet apart, using 1000 crowns per 7
acre. A cover should be had in two years if cultivated. Where rows can be cultivated in between the kudzu plants, 500 crowns may be set on 25 foot rows, 3½ feet apart. This should cover the land in about three years.

Cultivation. Plants must be cultivated in the first year to get good growth. 8
Weeds and other plants will stop growth until the ground gets covered with kudzu vines. Cultivation can be done quickest with a plow or cultivator, but

if the land is too rough or gullied, it will pay to cultivate with a hoe. Cultivation is not needed after a good cover is on the ground.

Grazing Management. Grazing should be started after a good cover is on the ground, which should be the second or third year. Kudzu is best used for grazing in late July, August, and September. Grazing should be done by the hour in place of leaving cattle on all day. However, if enough land is allowed, livestock can be left on all day. Overgrazing will kill kudzu. It should never be grazed close enough to get all the leaves off and new growth all grazed back. 9

Value of Kudzu Grazing. Kudzu will grow good grazing on special areas if managed right. Dairy cows will increase in milk from summer pasture to kudzu grazing because of the high protein content of the leaves. Beef cows will carry calves longer and calves will gain on the grazing. Hogs will do extra well on kudzu grazing. Overgrazing is the big enemy of kudzu. 10

Plants to Interplant in Kudzu. Since kudzu will furnish grazing only about three months of the year, it will pay to interplant crimson clover and ryegrass or winter peas and ryegrass. Plant 20 pounds of crimson clover and 25 pounds of ryegrass in October after the leaves are grazed off. Or, 40 pounds of wild winter peas and 25 pounds of ryegrass can be sown on top of the kudzu in October. Crimson clover, ryegrass and wild winter peas will reseed if seeds are allowed to mature. 11

Begin grazing in the late fall and continue grazing until April 15, when growth of crimson clover, ryegrass, or winter peas is good. Move livestock in late April to let seed mature. 12

Let established kudzu grow until late July before grazing. 13

Caution. Kudzu will kill trees and will grow out into a forest if set near it. 14

This is a real problem along highway rights of way in forest areas throughout the lower south. It can be controlled by continuous grazing for a couple of years where cattle can reach it. Kudzu can also be controlled with chemicals. Two or three repeated applications with Silvex or 2, 4, 5-T will usually give good control. Follow the label when using all chemicals. 15

Extension Agronomy
Mississippi State, MS 39762

QUESTIONS FOR DISCUSSION

1. How effectively has the writer used information about the audience's background and purpose in this selection? Identify two or three examples that illustrate that the writer knew the specific duties and roles the audience would have in reading this model.
2. The writer's language is informal and technical at the same time. Is it consistently appropriate for the reader's needs? Why do so many of the sentences contain imperative verbs (*plant, apply, begin*) or other words of command (*must, should*)?

3. In what ways has the writer's format helped the audience achieve its purpose for reading the information sheet?
4. Explain how the audience will use the brief introduction and the background details found throughout the model. Would this information be important for any other audience type? Explain.
5. Contrast the discussion of the herbicide in this information sheet with that in the models from the *EPA Journal* (pages 18–20) and from *Weed Science* (pages 30–35). Would a more detailed description of herbicides and control processes be appropriate or inappropriate for readers of these selections? Explain.

SUGGESTIONS FOR WRITING

1. Rewrite the section "Fertilizing and Setting" and "Plants to Interplant with Kudzu" from this model for consumers.
2. Rewrite a section of the piece to make it relevant to a decision maker. For example, the "Land for Kudzu" section can be rewritten to help the owner of a large, varied plot of land decide about its most economic use. Or the sections on grazing can be rewritten considering dairy farmers as an audience of decision makers.

The next model on kudzu was written for experts—soil scientists, biologists, agricultural researchers. It is a field research report that appeared in Weed Science, *a technical journal reporting on scientific experiments for academic researchers. (You will learn more about field research reports in Chapter 7.) The authors' purpose was to study a problem and offer a solution—when was the best time to apply a particular herbicide to control kudzu so as not to harm surrounding crops at a particularly sensitive time in their growth. The authors correctly assume that readers have a great deal of knowledge about kudzu, herbicides, and growth regulators. Readers can apply the results of this study in specific programs instructing farmers when and how to use effective measures against kudzu.*

Influence of Time of Herbicide Application on Control of Kudzu

Ray Dickens and G. A. Buchanan

Abstract. Field experiments were conducted during 1968 to 1970 to determine the influence of time of herbicide application on the control of kudzu [*Pueraria lobata* (Willd.) Ohwi]. Picloram (4-amino-3.5, 6-trichloropicolinic acid) effectively controlled kudzu when applications were made as early as May 22 and as late as October 28. Acceptable control was obtained with rates of picloram 0.34 kg/ha or higher. In contrast, (2,4,5-trichlorophenoxy)acetic acid (2,4,5-T) was most effective when applied during the first half of summer. Essentially no control of kudzu was obtained from 2,4,5-T applied 1 week before the first killing frost. 1

INTRODUCTION

Kudzu is a perennial leguminous vine introduced into the United States from Japan (8). Early promotion for hay, grazing, and erosion control led to its introduction in many localities throughout the southeastern United States. Its rapid, viny growth characteristics and persistence assured its success as a major weed problem. At present, kudzu is a pest on thousands of acres of forest and noncropland in the southeastern United States (1, 2, 7). 2

Control of kudzu with phenoxy herbicides generally has not been successful. Davis and Funderburk (6) reported that 13 applications of a 1:1 mixture of the propylene glycol butyl ether esters of (2,4-dichlorophenoxy)acetic acid (2,4-D) and 2,4,5-T were required to eradicate kudzu. Applications were made over a 5-year period. In the same study, 2,2-dichloropropionic acid (dalapon) and a 3 mixture of dalapon and 2-(2,4-dichlorophenoxy)propionic acid dichlorprop) were ineffective in controlling kudzu.

In 1965 and 1966, several workers (3, 4, 5) reported that picloram was 4
highly effective in controlling kudzu. Brenler (3) concluded that spray formulations were more effective than granular formulations.

Research results over the past 20 years indicate that most perennial-brush 5
species are more susceptible to the phenoxy herbicides in the late spring or early summer when food reserves are generally low (9). Unfortunately, this is also the time in which most crop species such as cotton (*Gossypium hirsutum* L.) and soybean (*Glycine rex* (L.) Merr.) are most sensitive to these herbicides, which may drift considerable distances depending upon application conditions.

The objectives of this series of experiments were to determine the influence 6
of time of herbicide application on control of kudzu and to determine the minimum rate of picloram necessary for control.

MATERIALS AND METHODS

Field-plot experiments were conducted on highway rights of way near Enterprise, 7
Alabama, in 1967, near Evergreen in 1968, and in an abandoned field at Auburn in 1969 to determine the influence of date of herbicide application on control of kudzu. All experimental plots were located in dense stands of kudzu. In 1967, treatments were applied May 24 and August 4 on 93-m^2 plots. Treatments were not replicated in this initial test. In the 1968 experiment, plots were 6.1 by 12.2 m while in the 1969 experiment they were 6.1 by 6.1 m. A completely randomized design was used with two replications in 1968 and three replications in 1969.

An alley at least 120 cm wide was maintained between each plot to prevent 8
new kudzu growth from contaminating experimental plots. These alleys were maintained weed free by application of 2-*sec*-butyl-4,6-dinitro-phenol (dinoseb) in diesel fuel on a biweekly schedule throughout the growing season.

In the 1968 experiment, 2,4,5-T, 2-(2,4,5-trichlorophenoxy)propionic acid 9
(silvex), 3,6-dichloro-*o*-anisic acid (dicamba), 2,3,6-trichlorobenzoic acid (2,3,6-TBA), and picloram were included in the experiment. The 1969 experiment included only picloram, 2,4,5-T, and a picloram + 2,4-D combination.

Herbicides were applied with a hand-carried knapsack sprayer calibrated 10
to deliver 930 L of solution per hectare. Herbicide applications were made May 22, July 9, and August 20 in 1968 and June 6, July 24, September 18, and October 28 in 1969.

Control of kudzu was measured both by individual estimations of percent- 11
age control by the authors and also by counting living kudzu crowns the year following application. Initial control ratings were made 4 to 6 weeks after each application in 1968 and 30 days after each application in 1969. In both experiments, crowns were counted in April following the previous year's application.

Data were subjected to analysis of variance and means were compared 12
using Duncan's Multiple Range Test.

RESULTS AND DISCUSSION

Results of a preliminary experiment in 1967 indicated that August applications 13
of picloram were just as effective in controlling kudzu as were applications made in May (Table 1).

In most instances, late August applications of herbicides in 1968 were more 14
effective in reducing the number of surviving crowns than the earlier applications (Table 2). There was no difference between May 22 and July 9 application dates when results were averaged over all herbicides. The initial control of kudzu foliage did not vary with date of application, but differences in the amount of foliage killed by the various treatments were considerable (data not presented). The most effective treatment both in initially killing and subsequently reducing the number of surviving crowns was picloram at rates as low as 0.28 kg/ha. Silvex, 2,3,6-TBA, and 2,4,5-T were the next most effective treatments.

In 1969, the amount of kudzu foliage killed 30 days after application (Table 15
3) was similar among June 6, July 24 and September 18 application dates for all treatments giving adequate control. Foliage initially killed by the October 28 application could not be evaluated because a killing frost occurred 1 week after application.

The rate of picloram required to kill essentially all kudzu foliage (control 16
rating $\geq$ 90%) in 30 days was between 0.17 and 0.34 kg/ha. Picloram at 0.34 kg/ha was equally as good as 4.48 kg/ha of 2,4,5-T amine in initially controlling foliage. In most instances, the addition of 2,4-D to picloram did not improve these results over picloram alone.

Analysis of variance of the number of surviving kudzu crowns revealed 17
that dates, treatment, and treatment by date interactions were significant (P <

TABLE 1. INFLUENCE OF RATES OF PICLORAM AND DATE OF APPLICATION ON CONTROL OF KUDZU NEAR ENTERPRISE, ALABAMA, 1967 AND 1968

		Kudzu Apparently Killed	
Herbicide	***Rate***	***Sept. 10, 1967***	***May 23, 1968***
	(kg/ha)	(%)	(%)
Picloram[a]	2.24	100	98
Picloram[b]	0.56	100	80
Picloram[b]	1.12	100	95
Picloram[b]	2.24	100	99

[a] Applied May 25, 1967.
[b] Applied August 4, 1967.

TABLE 2. EFFECT OF HERBICIDES AND DATE OF APPLICATION ON KUDZU CONTROL NEAR EVERGREEN, ALABAMA, DURING 1968 AND 1969

		Dates of Application in 1968			
Herbicide	***Rate***	***May 22***	***July 9***	***August 20***[a]	***Average***
	(kg/ha)	Live crowns/100 m^2 in 1969[b]			
2,4,5-T amine	2.24	8	87	30	42 bcde[c]
	4.48	70	74	9	51 bc
Dicamba	2.24	81	94	7	61 b
	4.48	65	59	0	42 bcde
Picloram	0.28	17	8	0	8 ef
	0.56	9	7	0	5 ef
	1.12	1	3	0	1 f
2,3,6-TBA	5.60	67	74	3	48 bcd
	11.20	4	7	0	4 ef
2,4,5-T ester	2.24	35	19	11	22 cdef
Silvex	2.24	24	52	11	30 bcdef
Check	—	229	210	215	218 a
Mean		51 A	58 A	24 B	

[a] Plants on these plots were much reduced in size and vigor compared to other dates of treatment.
[b] Counted April 24, 1969.
[c] Means followed by the same letter in the same case are not different using Duncan's Multiple Range Test ($P < .05$).

TABLE 3. EFFECT OF HERBICIDE AND DATE OF APPLICATION ON KUDZU CONTROL APPROXIMATELY 30 DAYS AFTER APPLICATION AT AUBURN, ALABAMA

		Control of Kudzu 30 Days after Spraying — *Date of Application in 1969*		
Herbicide	***Rate***	***June 6***	***July 24***	***Sept.18***
	(kg/ha)	(%)	(%)	(%)
Picloram	0.11	70 b[a]	52 b	85 b
	0.17	87 c	73 c	80 b
	0.34	96 d	90 d	98 c
	0.56	99 d	95 d	99 c
Picloram + 2,4-D	0.28 + 1.12	98 d	80 cd	90 bc
2,4,5-T amine	4.48	99 d	97 d	97 c
Check		0 a	0 a	0 a
Mean		78	70	78

[a] Means with a column followed by the same letter are not different using Duncan's new Multiple Range Test ($P < .05$).

.01). Picloram at 0.11 or 0.17 kg/ha and picloram + 2,4-D gave the same degree of control regardless of the date applied (Table 4). There was, however, a response to dates from the higher rates of picloram with less crown survival from the earlier applications. Even though the control from the September 18 and October 28 applications of picloram at the two higher rates was statistically less than that obtained from the June 6 application, these late applications did reduce survival of kudzu crowns from 80 to 94% when compared to the untreated areas.

The two early applications of 2,4,5-T were more effective than those in 18
September and October. Survival of crowns on plots treated in June and July showed 4.48 kg/ha of 2,4,5-T gave control equal to that by 0.34 kg/ha of picloram. However, at the two later dates of application, 0.34 kg/ha of picloram was superior to 4.48 kg/ha of 2,4,5-T. Comparisons of treatments made October 28 show that all but the lowest rate of picloram gave control superior to 2,4,5-T.

A possible reason for the difference in reaction between picloram and 2,4,5- 19
T may be differences in their rates of uptake and movement in kudzu. It is possible that a higher percentage of the picloram than 2,4,5-T was absorbed and translocated to the crown and root system of the kudzu prior to the top-growth being killed by frost. Additional studies would have to be conducted to prove or disprove this theory.

The herbicide 2,4,5-T is most effective in controlling kudzu when applied 20
during the first half of the summer. Kudzu was not effectively controlled when applications of 2,4,5-T were made during or after late September.

Kudzu was effectively controlled by picloram at rates as low as 0.34 kg/ 21
ha. Rates below 0.34 kg/ha gave only marginal control. Picloram also tended to give better control when applied during the first half of summer; however, the difference in degree of control is not of practical importance. By using picloram, one could delay spraying kudzu until soybean, cotton and other sensitive crops have matured and thus reduce damage from herbicide drift.

TABLE 4. INFLUENCE OF DATES OF APPLICATION ON THE CONTROL OF KUDZU THE YEAR FOLLOWING HERBICIDE TREATMENT AT AUBURN, ALABAMA

		Dates of Application			
Herbicide	***Rate***	***June 6***	***July 24***	***Sept. 18***	***Oct. 28***
	(kg/ha)	Number of living crowns/100 m^2[a]			
Picloram	0.11	431 a[b] A	237 ab A	194 ab A	334 ab A
Picloram	0.17	237 a A	151 abc A	162 ab A	172 bc A
Picloram	0.34	22 bc A	54 d AB	97 bc B	97 c B
Picloram	0.56	11 c A	18 e AB	43 c B	32 d AB
Picloram + 2,4-D	0.18 + 1.12	54 b A	86 cd A	97 bc A	32 d A
2,4,5-T amine	4.48	75 b A	65 cd A	183 ab B	474 a C
Check		614 a A	625 a A	485 a A	571 a A

[a] Counted April 22, 1970, following herbicide applications in 1969.

[b] Means within a column followed by the same lower case letter or means within a line followed by the same upper case letter are not different by Duncan's Multiple Range Test ($P < .05$).

LITERATURE CITED

1. Brender, E. V. 1960. "Progress Report on Control of Honeysuckle and Kudzu." *Proc. So. Weed Conf.* 13:187–192.
2. Brender, E. V. 1961. "Control of Honeysuckle and Kudzu." *U.S. For. Serv., S. E. For. Exp. Sta. Paper* 120. 9 p.
3. Brender, E. V. and E. L. Moyer. 1965. "Further Progress in the Control of Kudzu." *Down to Earth* 20(4):16–17.
4. Couch, R. W. and D. E. Davis. 1965. "Kudzu Control with Tordon." *Proc. So. Weed Conf.* 18:309. (Abstr.)
5. Couch, R. W. and D. E. Davis. 1966. "Kudzu Control with Tordon." *Down to Earth* 22(3):1, 24.
6. Davis, D. E. and H. H. Funderburk, Jr. 1964. "Eradication of Kudzu." *Weeds* 12:62–63.
7. Klingman, G. C. 1961. *Weed Control: As a Science.* John Wiley and Sons, Inc., New York, N.Y. 421 p.
8. Martin, J. H. and W. H. Leonard. 1949. *Principles of Field Crop Production.* The Macmillan Company, New York, N.Y. 1176 p.
9. Robbins, W. W., A. S. Crafts, and R. N. Raynor. 1952. *Weed Control.* McGraw-Hill Book Co., Inc., N.Y. 195 p.

QUESTIONS FOR DISCUSSION

1. The information in this and similar reports follows a conventional order. What are the major parts of this report and why is each part important for the reader? Explain how the level of technical information in the various sections of this report has been adapted to the education, background, and responsibilities of readers.
2. Describe the authors' style. Are their sentences relatively short and simple (following a subject-verb-object pattern) or are they long and complex? Are they usually in the active or passive voice? Explain how the authors' choices are relevant for their audience and purpose.
3. Characterize the authors' tone or attitude toward their subject. Is it formal or informal? Why? Do they ever refer to themselves directly? Where and how?
4. What kinds of terms, abbreviations, and symbols do the authors use? Are these appropriate for the intended audience? Do the authors supply definitions or comparisons for readers? Why or why not?
5. Explain the function and relevance of the four tables. Would the reader be confused if the tables were omitted?

SUGGESTIONS FOR WRITING

1. What parts of this report would interest another audience or audiences? Select one part and write a brief description (500 to 600 words) for a nonexpert audience. Identify that audience and explain what changes you made in the wording and details selected from the original article to meet that audience's needs.
2. Find a recent article written for specialists in your field. Choose a section of at least 300 words and rewrite it for nonspecialists. Bring the original article to class along with your rewritten copy.

This section concludes with a poem about kudzu written by James Dickey, the author of the novel Deliverance. *The poem includes many of the same details about the plant found in previous models—its origins, color, texture, amazing growth, use as foliage, and life cycle—as well as some description of the problems it causes on a particular farm in Georgia. Certainly Dickey's work is not technical writing, and that's why we include it—to show you the differences in objectivity, use of language, and tone between this poem and the other models in this section.*

Whereas the other models in this section focus on the botanical, agricultural, or financial aspects of kudzu, Dickey uses the plant to comment on religious and philosophical themes such as deception, evil, and the relationship of human beings to nature. Dickey's audience, people of many different backgrounds, read his poem not for factual data or practical instruction but for pleasure and wisdom.

Kudzu

James Dickey

Japan invades. Far Eastern vines
Run from the clay banks they are

Supposed to keep from eroding,
Up telephone poles,
Which rear, half out of leafage,
As though they would shriek,
Like things smothered by their own
Green, mindless, unkillable ghosts.
In Georgia, the legend says
That you must close your windows

At night to keep it out of the house.
The glass is tinged with green, even so,

As the tendrils crawl over the fields.
The night the kudzu has
Your pasture, you sleep like the dead.
Silence has grown Oriental
And you cannot step upon ground:
Your leg plunges somewhere
It should not, if never should be,
Disappears, and waits to be struck

Anywhere between sole and kneecap:
For when the kudzu comes,

The snakes do, and weave themselves
Among its lengthening vines,
Their spade heads resting on leaves,
Growing also, in earthly power
And the huge circumstance of concealment.
One by one the cows stumble in,
Drooling a hot green froth,
And die, seeing the wood of their stalls

Strain to break into leaf.
In your closed house, with the vine

Tapping your window like lightning,
You remember what tactics to use.
In the wrong yellow fog-light of dawn
You herd them in, the hogs,
Head down in their hairy fat,
The meaty troops, to the pasture.
The leaves of the kudzu quake
With the serpents' fear, inside

The meadow ringed with men
Holding sticks, on the country roads.

The hogs disappear in the leaves.
The sound is intense, subhuman,
Nearly human with purposive rage.
There is no terror
Sound from the snakes.
No one can see the desperate, futile
Striking under the leaf heads.
Now and then, the flash of a long

Living vine, a cold belly,
Leaps up, torn apart, then falls
Under the tussling surface.
You have won, and wait for frost,
When, at the merest touch
Of cold, the kudzu turns
Black, withers inward and dies,
Leaving a mass of brown strings
Like the wires of a gigantic switchboard.
You open your windows,

With the lightning restored to the sky
And no leaves rising to bury

You alive inside your frail house,
And you think, in the opened cold,
Of the surface of things and its terrors,
And of the mistaken, mortal
Arrogance of the snakes
As the vines, growing insanely, sent
Great powers into their bodies
And the freedom to strike without warning:

From them, though they killed
Your cattle, such energy also flowed

To you from the knee-high meadow
(It was as though you had
A green sword twined among
The veins of your growing right arm—
Such strength as you would not believe
If you stood alone in a proper
Shaved field among your safe cows—):
Came in through your closed

Leafy windows and almighty sleep
And prospered, till rooted out.

QUESTIONS FOR DISCUSSION

1. Identify the realistic, factual details about kudzu that Dickey includes in his poem.
2. Identify examples of Dickey's language that are emotionally charged, i.e. that appeal to a reader's feelings, impressions, fears. Why would such diction be inappropriate in a memo to a decision maker, in a set of instructions, or in a report on a scientific experiment? For each term you identify, suggest a more objective synonym that would be appropriate for these three forms of technical writing.
3. What kinds of comparisons does Dickey use to describe kudzu and the animals in the poem? Are his descriptions emotional or factual? Or are they both? Why? Explain your answer by citing specific lines from the poem.
4. Contrast the description of kudzu in Dickey's poem with that in "Living with the Kudzu" on pages 18–20. What do these two models have in common?
5. How does Dickey's description and view of kudzu contrast with the description and point of view found in the county agent's information sheet (pages 27–28) and the report of the experiment in "Influence of Time of Herbicide"

(pages 30–35)? What kinds of details does Dickey include that would be inappropriate for the reader of these models? What kinds of details does he omit that they would need?

SUGGESTIONS FOR WRITING

1. Assume you are a county agent who has to prepare a brief description of the problems local farmers are having with snakes because of kudzu. Taking pertinent information from Dickey's poem, write a short factual report for an operator audience.
2. Rewrite this poem as a short article for nonspecialists. The article should be similar to "Living with the Kudzu" (see pages 18–20) in format, but would be likely to be published in a more general interest magazine.

WORD PROCESSORS

Published in Trial: The National Legal News Magazine, *the following article is intended for attorneys who may have little or no knowledge about word processing systems. Although they are experts in their profession, the readers are addressed as consumers, potential buyers of word processing equipment. This article is a good model of how to write for individuals who belong to many different audiences. An attorney himself, the author is writing to "acquaint trial lawyers with some of the technology—specifically office automation—that can make their work a little easier." In shaping his discussion to meet this particular audience's needs, the author explains how a word processor can expedite an attorney's paperwork. The article concludes with sound, realistic advice about selecting an appropriate word processor for the attorney's office.*

Word Processors for Attorneys

Reese I. Joye, Jr.

2001. It is not so far away anymore. What will it be like practicing law then? If technological growth in the next twenty years is anything like that in the last twenty, not only will attorneys have to have their skates on, but the skates also will have to be jet propelled. 1

Trial attorneys must keep up with the times if they are to service their clients as their clients expect and demand. The written word must be produced faster, cheaper, and in greater quantities. Change is inevitable. 2

This article will acquaint trial lawyers with some of the technology—specifically office automation—that can make their work a little easier. 3

ORIGIN OF THE SPECIES

Trial attorneys benefit tremendously from automated office equipment; time is saved not only for attorneys, but also for support personnel. This increases overall productivity, which reduces costs. Office automation takes several forms: telephones, dictation equipment, copiers, typewriters, and calculators. Today, the office is in the era of the word processor and the computer. 4

How did the word processor evolve? 5

Word processing technology is only about 17 years old and is a combination of two older technologies: the typewriter and the computer. Every two years, technological knowledge doubles itself in new and better features. 6

The typewriter was invented in 1874 by Christopher Shoals. Typed correspondence, although produced at a slower rate than handwritten copy, brought prestige to many offices that had used clerks to hand-write documents. Over the years typewriters became more sophisticated than Shoals' clumsy typing machine, and gradually they became a productive, efficient part of the work environment. Today an office is not found without one, and electric, correcting, and memory typewriters are commonplace. 7

The first commercial computer went into operation in 1951; it also has undergone a transformation resulting in a more practical size and weight. Computer technology, and specifically the microprocessor, is now the core of many other products, such as calculators, word processors, communication and video equipment. 8

EFFECTIVE MANAGEMENT TOOL

A word processor, however, is not simply a typewriter with computer pieces. It can be a management tool as well by taking advantage of different software and hardware features. Programs can be adapted easily to serve needs exclusive to attorneys. Some of the things a word processor can do quickly are calendars, case control, time control, client's ledger, general ledger, telephone charges, and billing. 9

Calendars. Every trial attorney is aware of the demand by professional liability insurers for deadline control. This can be accomplished very easily with a word processor. Slips of paper 1½" x 8" are printed with the following categories stretching across the slips' 8" width: 10

1. Deadline Date.
2. Case.
3. Action.
4. Attorney.

These slips are filled in by attorneys and support personnel and collected at the end of each day for each item or deadline needing attention or action. The first thing the word processing operator does the following day is "call up" the calendar on the CRT screen and insert the data from the slips where appropriate. An 8½" x 14" sheet is printed out and distributed each morning to everyone in the office; it contains a two- to three-week schedule of commitments, deadlines, appointments in and out of the office, seminars, personal appointments, holidays, office conferences, trials, etc. Even when someone is sick, he or she calls the word processing operator to get that information on the day's calendar. With this system, everyone in the office knows where everyone else is and what deadlines need to be met. 11

Case control. Does an attorney want to know how many different types of cases came into the office in the last 30 days, or since the first of the year, 12 or last year? How many cases are open or closed at a given time? How does he or she keep up with conflicts of interest within the firm?

13 This is possible with software "sort" capabilities. Although many different variables can be used, there are seven bits of information fed into the word processor for this application:

1. Date opened.
2. File number.
3. Client.
4. Adverse party.
5. Type of case: will, domestic, tort, criminal, etc.
6. Attorney assigned.
7. Status: open or closed.

14 The word processor can sort this information by the clients' or adverse parties' names alphabetically, advise which cases each attorney has assigned and whether they are open or closed, and how many cases the firm has or has had for a particular client. The possibilities are unlimited. The word processor can sub-sort the types of cases into as much detail as desired, for example:

15 "Tort—Automobile Collision—Rear End—Passenger Injury."

16 The word processors can print out a list of the cases according to the length of time they have been open in the office; the word processor also can warn when the statute of limitations is about to run out.

17 *Time control.* Each time an income survey is published, attorneys are admonished that those who keep accurate account of their time, bill more and therefore generate higher incomes. There are hundreds of systems for keeping track of time. However, they all require the following information:

1. Date.
2. Client.
3. Time spent.
4. Services provided.
5. Attorney performing services.

18 If these five items are input into a word processor that has "sort" and "math" capabilities, the result will be bills or statements produced automatically in addition to weekly, monthly, or quarterly printouts of who is producing what for the office. The word processor can sort by client, dates, attorney, and type of services performed.

19 The billing information can be coupled with the *Case control,* which will record how much income each attorney has generated over a given period of time as well as how much income the firm realizes in particular types of cases.

20 If the date of preparation and a statement number is entered when statements are being prepared, accounts receivable can be monitored easily. The word processor can be programmed to diary-aged accounts every 30 days.

21 *Client's ledger.* The next logical step is to input monies received against the billing and also to input costs paid by the client. This, then, results in one-half of the bookkeeping functions being performed by the word processor.

22 *General ledger.* Since keeping records of income and expenses is so easy, there is no reason why the word processor cannot be programmed to provide a general ledger for the law office's bookkeeping. The procedure is virtually the same. Simply couple the "profits" from the *client's ledger* format above and program the office's cost of operation, and now the entire bookkeeping function is available weekly, monthly, and at the end of the year for income tax purposes.

23 *Telephone charges.* In order to get "time and charges" recorded, an operator must assist in making the call. This type of call costs the client at least double and, in some cases, triple the cost of station-to-station, direct-dialed calls. The majority of attorneys delegate the receipt of time and charges from the operator to a telephone receptionist or secretary. It is much easier to write the following down in a log kept by the telephone:

1. Date.
2. City called.
3. Telephone number (and area code).
4. Client's name.
5. Attorney placing call.

24 This information can be entered into the word processor and sorted out at the end of the month by day, client, city, and attorney; or the telephone numbers can be arranged by the word processor in numerical order in a matter of minutes. This will greatly assist the bookkeeper, receptionist, or clerk in matching the charges to each client's ledger sheet. The telephone bill can be input into the word processor with the number, date, city, and charge, and a program can be written to match the charges to the proper client's ledger sheet.

AN INDIVIDUAL CHOICE

25 The needs of a particular law office must be evaluated to determine exactly how a word processor can benefit it. Some present work habits must be considered before the decision is made to purchase word processing equipment, such as:

1. Average length of documents typed in the office.
2. Frequency or repetitivity of certain documents, phrases, names, or words.
3. Complexity of the documents.
4. Number of documents produced each day.
5. Demand for perfection in documents.
6. Typing skill of the secretary.
7. Competition.

26 There are approximately 70 word processors on the market with differing features, capabilities, and prices. Some are pure word processors, some are word processors/quasi-computers, and some are computers with word processing

capabilities, but almost all word processors consist of four basic hardware parts: the cathode ray tube (CRT) screen; one or two disk drives; a printer; and a keyboard.

27 *CRT screen.* A full-page display is the easiest with which to work. Studies have indicated that a full page with black characters on a light background allows the operator to perform on the system with minimum eye fatigue.

28 *Disk drives.* Most systems use two disk drives with a total document capability at any time of 600,000 characters, or 300 pages. Of course, there may be any number of diskettes in the law office library that are being worked with, and the operator may change to the appropriate one when necessary. Floppy disks are the latest and most inexpensive storage media, about 7 cents per page, for word and data processing.

29 *Printer.* Most vendors offer impact printers capable of printing up to 540 words per minute while the operator uses the CRT screen for another function. With an acoustical hood, the printer is quiet and non-distracting.

30 *Keyboard.* The keyboard is designed the same as a typewriter keyboard with the exception of some special-function keys usually located on either side of, or above, the letter keys.

31 There are many different capabilities of the word processors despite their similar hardware design. Features to make note of include:

32 *Software-based.* A software-based word processor can always be upgraded to include the most up-to-date word processing features, eliminating the need to purchase costly hardware.

33 *Simultaneous input/output.* This means that the system can print a document at the same time the operator is typing other documents on the CRT screen.

34 *Hyphenation selection.* A good word processor should have hyphenation selection capabilities. The operator can select between manual hyphenation, automatic word wrap-around, and fully automatic hyphenation according to dictionary spellings.

35 *Automatic centering and underlining.* With the touch of one or several keys, many word processors will automatically center a line of copy on the CRT screen, or automatically underline it as it is being typed.

36 These are basic text-editing features that speed up the preparation of documents. Other, more complex features to make note of are: automatic decimal alignment and ten-key calculator pad; automatic paragraph assembly; program and insert features; text merging; and selection and sort.

THE SELECTION PROCESS

37 These three steps will make the choice easier:

1. Consult the word processing feature analysis charts found in computer and word processing publications. These charts list hardware and software features of the equipment and offer price comparisons.

2. Find out which of those machines are available in your geographical area. If a certain machine is neither sold nor serviced locally, it should not be considered. Quick and reliable service is a must.

3. Obtain a list from those vendors of the law firms in your area that have their equipment. Determine which of the firms are most similar to yours in size and kind of practice. Take time to visit the firms and get a firsthand look at the equipment and operation of it. Avoid being the first on your block with a new toy—it could be a costly mistake.

Once a potential purchaser of word processing equipment is familiar with these features and has done the homework, selection of the equipment is the final, and often the easiest, step. 38

Securing the right word processor and using it properly can be more rewarding to you as a trial attorney than any case you have ever won. 39

QUESTIONS FOR DISCUSSION

1. What types of questions likely to be asked by consumers does Joye's article answer?
2. Where does Joye appeal to attorneys as legal experts and where does he address them as potential consumers of word processing equipment? Are there any places where the two overlap? Identify particular descriptions and examples to support your answer.
3. Is the level of technical information that Joye gives his readers adequate for their needs? Should he have included more information about the way word processors work or are designed? Explain.
4. In what way is the organization of Joye's article suitable for the audience's needs and interests?
5. Joye's sentences are easy to read. Why? Describe his style—e.g., sentence length and structure, the use of the active or passive voice, number of sentences per paragraph.
6. Contrast Joye's description in "Disk Drives" with that in "Formatting Floppies" included later in this chapter, pages 55–57. How do the two discussions differ in audience, purpose, level of technical detail and language.

SUGGESTIONS FOR WRITING

1. The general content of Joye's article is applicable to almost any professional group. Write a discussion (500–600 words) on the advantages of word processors for one of the professional groups listed below. In your paper, select appropriate examples and explanations for an audience that knows very little about word processing, as Joye has done.

a. accountants
b. engineers
c. health-care professionals
d. urban planners
e. agribusiness executives
f. criminal-justice officials
g. any other group

2. Select one or two of the uses of a word processor in the attorney's office that Joye identifies. Rewrite that description for an attorney who is very knowledgeable about word processors—i.e., who has used them in his or her office and who wants to know how other attorneys may have used them, too.

The following memorandum was written for a decision maker who is an engineer. It was written by a student for her business writing class at Purdue University and subsequently won a business writing prize at Purdue's 1982 Literary Awards Contest. The memorandum is an internal proposal to a decision maker suggesting that a detailed plan of action be approved and implemented by a specified date. (See Chapter 8 for a description of proposals.) In writing for her audience, the student anticipated a decision maker's questions and provided persuasive answers.

Memo about Word Processing Needs at a Corporation

Rannie Fox

MEMORANDUM

TO: Les Axelrod
Automation Engineering Manager

FROM: Rannie Fox R.F.
Word Processing Supervisor

SUBJECT: Proposal for Study of Word Processing Needs, Distributors, and Capabilities

DATE: July 20, 1981

In January we rented two DEC 78 stand-alone word processors for use by the Research & Development and Technical Services departments. Since that time, the word processing needs of both of these departments have increased, and two other departments—Marketing and Training—have started using the processors whenever they are available. The Purchasing and Sales departments are also interested in beginning word processor use. 1

Our present equipment cannot meet the needs of all these departments. We have to increase our word processing capacity and also add some capabilities that were not available to us in January. I propose a study to determine the specific needs of the various departments and the word processing equipment that best suits those needs. This memo summarizes anticipated report content and specifies the procedures, schedule, and costs for the project. 2

Report Content

As a result of this study, you will receive a detailed report that includes our overall anticipated word processing needs for the next three years, alternatives for meeting these needs, and my recommendations for rental or purchase of the appropriate equipment for our company. 3

Research Procedures

To gather information for my report, I will interview department managers and leading word processor distributors. First, I will contact all department managers and discuss 4

Les Axelrod
July 20, 1981
Page 2

their present and anticipated word processing needs with them. Specifically, I will want to know:

- approximate word processor time needed (hours per day, present and future);
- special capability requirements, such as columns, list processing, charts, and graphs;
- type of printing needed, such as letter quality and computer type.

Next, I will contact the leading word processor distributors 5
in our area and obtain the following information about their equipment:

- rental cost, with or without purchase option
- purchase costs
- delivery time and arrangements
- capabilities
- training provisions
- service availability and service contract terms

Time Schedule

This study will take four weeks to complete from the time of 6
your approval:

- Week 1 Discussing department needs
- Weeks 2–3 Obtaining information from distributors
- Week 4 Compiling data and making recommendations

This schedule allows time for mail delivery of information 7
and for scheduling mutually agreeable meeting times. I will have the final report ready for your review by August 17.

Les Axelrod
July 20, 1981
Page 3

Approximate Costs

The only costs I anticipate for this study are for telephone charges and postage. These should not exceed $50. 8

Since our backlog has been steadily increasing and we are presently nine days behind schedule, the need for additional equipment is becoming critical. Please let me know as soon as possible if this proposal meets your approval. I will appreciate any suggestions you have for conducting this study. 9

QUESTIONS FOR DISCUSSION

1. Judging from the information in Fox's memo, how much knowledge of word processing does Axelrod have? Should Fox have given him more information?
2. What functions do the first two paragraphs and the last paragraph of this proposal serve for the decision maker who has to keep an organization running smoothly?
3. What types of information relevant to her audience does Fox supply under the four separate headings in her memorandum?
4. Is the order in which Fox presents her information the most useful for the reader's needs? Can you suggest another possible order?
5. Explain how the format of Fox's memorandum makes her work easy to read, reread, and act upon.
6. Several sentences in Fox's memorandum are especially persuasive. Find three examples, and comment on why she included them.

SUGGESTIONS FOR WRITING

1. Assume you are an operator in Fox's department. Write a one-page memorandum to her explaining the problems you have faced because of inadequate word processing equipment or time.
2. Assume you are the head of either the Marketing or Training departments at Fox's company. Write a one-page memorandum to her listing your problems because of inadequate word processing equipment or time.

The following model is a set of instructions taken from the Apple II Installation and Operating Manual. *These instructions are intended for anyone who purchases the Apple II—experts, decision makers, and consumers. However, when readers examine these instructions they will function as operators. Specifically, they will read this model to find out how to install the Super Serial Card (SSC). This card allows users to communicate with printers, terminals, and other computers. The placement of the card in the computer also determines the type of language readers can use to write their programs. It is essential, therefore, that readers know how to install the SSC properly.*

Apple II Super Serial Card Instructions

INSTALLATION PROCEDURE

This section explains how to install the SSC and its internal cable in the 1
Apple II. If the cable clamp is not already assembled, do so now, following
the instructions given in Chapter 1.

> Before connecting or disconnecting anything on the Apple, turn off the power with the switch at the back left corner of the Apple case. THIS IS ABSOLUTELY NECESSARY. If you try to connect or disconnect anything from the inside of your Apple when the power is on, you are likely to damage the circuits.

Do not unplug the Apple, just turn it off. If you unplug the Apple, you will 2
isolate it from earth ground and leave it vulnerable to static discharges.

Remove the Apple cover by pulling up on the two back corners of the 3
cover until the two corner fasteners pop apart. Slide the cover back until it is
free of the case and then lift the cover off.

Look inside the Apple and locate the power supply case—the rectangular 4
metal box along the left inside the Apple II. To avoid damaging the SSC, touch
the power supply case with one hand; this discharges any static charge that
may be on your clothes or body.

Along the back inside edge of the Apple you will see eight long narrow 5
slots called *connector slots.* The connector slots are numbered from Ø at the
left to 7 at the right. The numbers are printed along the back edge behind the
connector slots. For use with Pascal and a modem, install the SSC in slot #2.
For use with BASIC, install the SSC in any slot from #1 through #7.

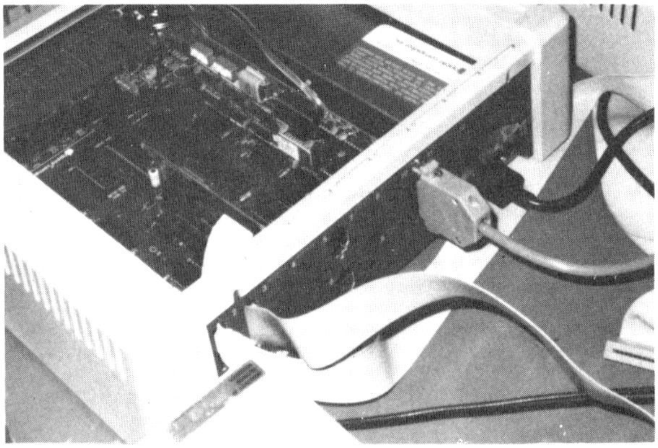

Figure 1. SSC in slot #2 and clamp assembly in notch.

Handle the Super Serial Card as you would handle an expensive phonograph record. Grasp it only by the corners or edges, and do not touch the components or pins, especially the gold fingers on the edge connector.

There are three deep notches along the back of the Apple II case. Temporar- 6
ily set the SSC down near the desired slot. Then take the clamp assembly and slide it down into the notch closest to the slot that the SSC will be in. Tighten the screws until the connector assembly can no longer be moved in the opening.

Grasp the upper corners of the SSC and insert the gold fingers of the edge 7
connector into the slot in the back of the Apple, rear edge first. Gently push the front edge of the card down until it is level and firmly seated. Figure 1 shows how the SSC looks when installed in slot #2.

Note that the outer ends of the screws in the clamp assembly can act as 8
nuts. They are threaded and can receive screws from the printer or terminal connector, to ensure a good connection with the Apple.

Slide the Apple case top plate in place and press down on the rear corners 9
until the corner fasteners pop into place. The Super Serial Card is now installed.

External Cable and Connector

The SSC cable connector you installed in the notch is a standard DB-25 connector 10
with 25 pins. Ten pins of the connector are connected internally to the SSC.

You will need a cable to connect the modem or other device to the SSC 11
connector on the Apple II. Cables with 25-pin connectors on one end are available from your Apple dealer.

The cable must have internal shielding, with the shielding properly termi- 12
nated at both ends, to prevent electromagnetic interference to nearby radios, television sets, and communication equipment. This shielding is necessary for the system to comply with Class B Federal Communications Commission limits

as defined by Subpart J of Part 15 of the FFC rules. Unshielded cables are not recommended.

☞ Make sure that all devices are connected to the same grounded AC power circuit (three-wire wall outlet) as the Apple II. Connecting ungrounded equipment to your Apple II can cause severe electrical damage.

QUESTIONS FOR DISCUSSION

1. In what ways is this model appropriate for several audiences?
2. How much (and what type of) knowledge does the writer assume on the part of the reader? Are any terms or procedures defined for the reader? If so, explain which ones, how they are defined, and why.
3. In the role of operator, a reader has to follow a number of steps. Explain how the writer has identified these steps and made them easy to follow for the operator-reader.
4. Is the author's format appropriate for the reader's purpose? Explain.

SUGGESTIONS FOR WRITING

1. Find a set of instructions written for the operator-reader. Write a one-page analysis explaining how the author designed his or her work expressly for an operator audience.
2. Using the information in these Apple instructions, write a memo to a decision maker proposing that your firm purchase Apple II computers for your staff. Emphasize the particular usefulness of the Apple II hardware for your company.

The following discussion about the construction and use of floppy disks appeared in Word Processing & Information Systems, *a trade journal intended for business managers seeking information about the applications of computers in business and industry. But the reader of this article is not addressed exclusively in the role of decision maker, for he or she can be expected to have considerable experience with text-editing machines. This article shows how technical writing can be directed to individuals who, because of their background and purpose, have the needs of more than one audience—in short they are experts who are also decision makers. The author, an associate editor of the journal, must write appropriately for his audience's background and interest if he expects them to remain satisfied subscribers.*

Exploring Technical Aspects of Formatting Floppies

David A. Rivers

Many text-editing machines on the WP market today utilize floppy disks for information storage purposes. Although users might assume that all floppy disks are much the same except for size (mini vs. standard) and recording capacity (single vs. double sided/density), there is a major difference, invisible to the human eye, that is necessary for machine operation. That is the way disks are coded or formatted. 1

In effect, formatting is a kind of filing system that is written on each disk. It tells the text-editing machine's central processing unit where various sectors of the recording tracks begin and where to read and write information. The computer will often use the first one or two entire tracks as a directory on which to log the locations of material stored on the disk. This space penalty is well compensated for by the high-speed random access capabilities that result from the formatting. 2

Because each manufacturer's equipment requires media formatted to its exact specifications, not all floppy disks can be accepted by all text-editing machines. Furthermore, some text-editing machines can format their own disks; others require disks preformatted by the manufacturer or by a designated supplier. 3

THE MEDIA

To understand how formatting (some people use the term "initializing") works, it is helpful first to visualize the media itself. Standard single-sided, single-density 4

Schematic Representation of a Floppy Disk

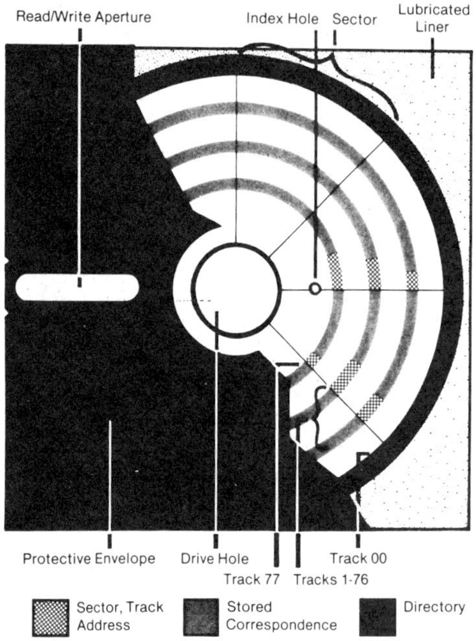

disks are 7.8 inches in diameter. Each is made of a flexible mylar material coated with iron oxide to give it its magnetic recording properties, and each is encased within an 8-inch square protective envelope that is lubricated on its inside surfaces to let the disk spin freely.

Standard floppy disks have 77 concentric tracks upon which digital informa- 5
tion may be recorded. These tracks are divided into pie-shaped wedges called sectors, of which there are two types—hard sectors and soft sectors—depending upon the kind of equipment being utilized.

Hard-sectored disks usually have 32 pin holes punched equidistantly around 6
the 1½-inch center, or "drive" hole, giving them 32 sectors per track upon which to store information. In addition, an index hole is punched between two of the sector holes. Starting from the index hole, the computer counts sector holes to determine where the sector begins on which it has been instructed to read or write information. Because this type of sector indexing requires physical alteration of the media during manufacture, the sectors are fixed and the sectoring format cannot be changed, hence the term "hard-sectored."

More common and more versatile are the soft-sectored floppy disks. These 7
also have an index hole but no sector holes. Again, starting with the index hole, the computer determines sector locations by measuring increments of time against rotation of the disk. There may be any number of soft sectors. IBM-compatible floppy disks, for example, may have eight sectors per track (providing 616 data blocks for information), 15 sectors per track (1155 data blocks), or 26

sectors per track (2002 data blocks). There are other combinations of sectors and tracks as well, depending upon equipment and application.

PREFORMATTING

Soft-sectored floppy disks can often, but not always, be formatted on the text-editing machines that use them. In some cases, the procedure involves calling up a service routine on the terminal or utilizing a special program to format new disks. However, text-editing equipment vendors do not normally advertise this fact and, consequently, many users are not aware that they may be able to save money by purchasing blank disks and formatting them on their own equipment. Instead they must purchase preformatted disks, generally from the equipment manufacturer or from a supplier designated by the manufacturer. 8

The difference in price between formatted and unformatted disks varies greatly according to supplier and quantity purchased. Even without formatting capability, text-editing machine users can often find high quality preformatted disks on the open market at prices significantly lower than those quoted by designated or authorized suppliers. Users should be cautioned, however, that this practice may be frowned upon by the equipment manufacturer and could result in poor user/vendor relations. 9

Legal questions have arisen regarding unlicensed reproduction of some manufacturers' formats. In an attempt to get around the problem, some media manufacturers change a few bits of uncritical information in the format so that it isn't exactly like the original. It will work the same way, in some cases even a little better, but it won't be exactly the same format. 10

Even if there are no specific caveats from the manufacturer regarding use of media, some WP equipment users ask if it is really safe to purchase preformatted floppy disks from suppliers not licensed, designated, or otherwise recommended by the vendors of their equipment. The answer is a qualified "probably." The consensus among experts in the field is that most media manufacturers produce floppy disks of reasonable quality and reliability. 11

QUESTIONS FOR DISCUSSION

1. How does this article fit the purposes of both decision makers and experts? Support your answer with examples from the piece.
2. Judging from the terminology and technical detail, how has the author analyzed his audience's roles and experience? Explain.
3. Why does the author divide the article into "The Media" and "Preformatting"? Is this division helpful for the reader? Why or why not?

4. The author's paragraphs are relatively short. Is that a plus or minus for his reader? Explain. How has the author organized information within and between paragraphs?
5. How is the visual useful to the reader of this article? Would it be inappropriate for the lawyer audience addressed in the Joye article on pages 40–45?

SUGGESTIONS FOR WRITING

1. Write a brief (two or three paragraph) letter from a vendor to a customer discouraging the customer from formatting his or her own disks. Use the information from this article in addition to any other facts you think are relevant for the audience.
2. Find a piece of technical writing directed to a single audience. Adapt the piece to meet the needs of several audiences. Then write a brief analysis identifying the audiences and describing how the piece serves multiple purposes. Bring your example to class with the analysis.

part 2

Basic Strategies

chapter 2

Physical Description

Physical description allows a reader to visualize any object or mechanism, regardless of size—a strip mine, a compressor, a microchip. This chapter is devoted to the technique of physical description, though your study of it will not end here, for description plays an indispensable role in all types of technical writing. Usually, description is not a goal in itself, but is given as part of a larger report. As you will see in later chapters, accurate descriptions are essential to all types of technical writing. In investigative reports specific equipment must be precisely identified and described for readers who may want to duplicate the investigation. In feasibility studies, where particular options are presented, readers will need exact descriptions of these different options—tools and other pieces of equipment or sites. And in proposals, where a specific plan of action is recommended, descriptions of the kinds of equipment and tools used to accomplish the job must be described in detail.

A precise description includes all important physical characteristics of its subject. Which details are important, however, will depend on your audience and its purpose in reading your description. Sometimes it can be helpful to anticipate a reader's questions when you are deciding which details to include. The following questions suggest some of the kinds of information generally sought in technical descriptions:

1. What is it (name, model number, order code)?
2. What is it used for (and by whom)?
3. What does it look like (size, shape, color, weight, density, height, age)?
4. What is it made of?
5. What are its major parts (moving, external/internal) and subparts?
6. How is it assembled or taken apart?
7. How long does it take to do the job (how many cycles does it go through)?
8. Where is it used (under what conditions, in what departments, in what environments)?
9. Are there any special advantages or precautions in using it?

BASIC ELEMENTS OF CONTENT AND STYLE

To be effective, a description needs to be comprehensive, appropriate for its audience, objective, and graphic.

Be comprehensive. Become thoroughly familiar with the object you are describing. If you are unclear about what something looks like, imagine the difficulty readers will have, especially if they have never seen or used the object before. Record relevant details about size, shape, color, finish, location, resistance, capacity, and function. Identify the parts and explain their relationship to one another. Whenever possible, inspect, measure, weigh, test, and analyze. You may even want to disassemble the object and reassemble it. For some mechanisms, it may be wise to operate them yourself or at least to watch them in operation. For a description of a site, you will want to inspect it firsthand.

Determine your audience's needs. What information will your reader need in the description? Will he or she use your description to distinguish one type of object from another? Will your description be read in order to build, operate, repair, or purchase the object? Is your reader primarily interested in what the object looks like from the outside or will he or she need to understand its internal parts as well? If so, you may have to explain the operating principles. Remember that different readers need to know different things. For example, a large radio used in electronics work would be described one way for a technician and another way for someone in a shipping department. Where the technician would need details about antenna length, the ground for the radio, and the capacity of the transmitter and receiver, the delivery person would need to know the number of cartons in a load, the weight of each carton, whether anything was fragile, and the date of delivery.

Be objective. Technical descriptions should be *objective,* not subjective. An objective description gives specific facts about sizes, shapes, materials, and parts—all things that can be seen and measured. The writer's task is to present the object as it is. Subjective details—impressions, feelings, unsubstantiated opinions—can mislead or bias the reader and have no place in technical writing. Impressionistic words such as *big, colorful, delicious, good,* or *handsome* are not specific enough to be useful. By using facts rather than impressions, you help the reader to visualize an object clearly. The following description of the Portable Recompression Chamber (PRC) for U.S. Navy divers is a model of objectivity.

> The PRC is 106 inches long with an inside diameter of $21^{11}/_{16}$ inches, proportionally designed to fit conveniently through an SSN 637 class submarine torpedo latching hatch. It is constructed of aluminum and weighs 350 pounds. It is equipped with a communication system and is capable of being operated in an open-circuit or semi-closed circuit (recirculation) mode. The PRC is an air-only system utilizing standard SCUBA tanks (72s or 90s) as the high pressure compressed air source. A mating adapter system is included with each PRC and makes possible the transfer-under-pressure of an injured diver, while undergoing treatment, to a standard Navy two-lock aluminum RCC. The PRC Treatment Table (a modified Table 1-A) provides

> for timely vents and a method of surfacing and transferring to a larger RCC if the mating adapter cannot be utilized.[1]

This description is absolutely free of any emotional slant. The writer has given the reader nothing but clear, straightforward facts about the PRC: dimensions, materials, capabilities, features.

One way to achieve objectivity in your writing is by using concrete language, words that name objects that can be seen, touched, heard, smelled, or tasted. Notice how many concrete words and quantitative details the writer uses in the PRC example: SSN 637 class submarine torpedo latching hatch, 350 pounds, aluminum, SCUBA tanks (72s or 90s). The more specific your words, the easier it will be for the reader to see the object you are describing. Observe in the following examples how the general, low-content phrases on the left can be replaced by the more precise terms on the right:

dash of salt	level $\frac{1}{16}$ teaspoon of salt
nearby valve	shut-off valve adjacent to rear assembly housing
fast rate	7000 rpm (for a rotary blade) 125 beats per minute (for a tachycardic pulse)
device	rubber-tipped ear plug
signaling mechanism	microprocessor-based vehicle counter
extended period of time	45 days beyond completion date
maladjusted	in the last percentile of the personality adjustment scale

Another pitfall to avoid is empty verbs. Empty verbs are those that do not portray specific actions. The following sentence, for example, contains the empty verb *consider* in a context that requires a description of a precise action:

> The lab technician considered the platelets.

What exactly the technician did is unclear, for there are any number of things that might be done with such a specimen. Notice how the revision below specifies a particular activity.

> The lab technician counted the platelets.

Use graphic language. Not only should your words be objective and precise, they should be graphic as well. Select words that draw pictures. Notice how the following adjectives help you to visualize the objects—*sparkling red fluid, hexagonal steel bars, spring-wound motor.* Comparisons and analogies are espe-

[1] James H. Bloechel, "Portable Recompression Chamber Update," *Faceplate* 12, (Fall 1981): 24–25.

cially helpful in that they give the reader a sense of the complex or the unfamiliar through references to the well-known. For example, a radio tube might be compared to a faucet: current flows from the first and water from the second. In a government pamphlet, *Medicine for the Layman,* the lungs are described through the following analogy: "Imagine the lungs as a train terminal. The cells of the body represent the stations along the route. The heart is the locomotive. It pumps the cargo of oxygen-carrying red blood cells through the blood vessels (arteries, veins, and capillaries)." In the same way, a writer for *Computer* magazine uses analogy to describe an optical scanner as the eyes of a computer.

ORGANIZATION

There are many ways to organize a description. You can describe an object from top to bottom, from front to back, from right to left, from outside to inside, or in other ways. A description of a living organism, such as a crustacean that has commercial value, might be organized according to its size, growing cycle, mating and spawning seasons, distribution, or principal economic uses. Mechanism descriptions will frequently first identify major parts and then examine each in detail.

Before you start writing your description, determine very carefully (1) the functional or structural elements of the object, and (2) the sequence of presenting these elements that will best meet a reader's needs. For example, it would be difficult for a reader to follow a description of a car that began with the roof (the top, outside), then switched to the carburetor (the motor, inside), and concluded with the doors (sides, outside). Instead, it would be more helpful in such a description to group related parts of the car (major units) together.

In most cases, the organization of your description should be determined by the audience's needs. For example, let's assume you wish to describe a crash cart (a portable emergency cart used in hospitals for cardiopulmonary resuscitation) for a readership of nurses. The most logical organization for this audience would be from top to bottom, because the materials on the cart are arranged according to the order in which they are handed to the rescue team by the nurse. Such a description would begin, therefore, at the top of the cart where the cardiac monitor is located and then move to a discussion of the arrangement and contents of the cart's four drawers: drawer one—airway materials (endotracheal tubes, laryngoscope, blades); drawer two—intravenous materials; drawer three—drugs; and drawer four—accessory items. A different emphasis and order would be necessary for individuals whose task it is to build the cart. For this audience it would be most helpful to begin with a description of the frame and then the drawers. Still another organization might be necessary if you were describing it for employees working in cart logistics, who are responsible for stocking and checking carts. In each case, the reader's needs—the way in which he or she will use the cart—determine the most appropriate way to describe it.

Similarly, a description of a desk calculator or an electronic circuit for the operator might begin with the inside and move to the outside, since this individual will be responsible for assembling the internal parts of these machines. For consumers, however, you might organize your description of the calculator from the outside to the inside and stress the attractiveness and ease of reading the instrument.

VISUALS

Visuals hold special importance in description. A photograph or drawing can show clearly and immediately what something looks like in a way that words alone sometimes cannot. In some cases, a visual is the primary means of conveying information and the writer's actual words are secondary. The interior of a machine or an electrical circuit are two examples of subjects that can be portrayed more accurately and succinctly with an illustration than with words. And as anyone who has ever assembled a piece of equipment will attest, written instructions sometimes depend a great deal on accompanying illustrations. Do not include visuals just for decoration. Be sure each visual provides the primary means of information or helps to support or explain the written text.

You can use a variety of visuals to make your description clear and useful. As with your choice of words and the amount of detail you include, select or design visuals with your reader's needs in mind. The following visuals are especially helpful in description:

Photographs give an exact and realistic portrait and are invaluable for an audience who has never before seen the object. A photograph of a technician using a piece of equipment can reveal its size, shape, maneuverability, and important details about the setting in which it may be used. The photograph of the Freedom 200 on page 73 displays the video and keyboard for potential purchasers of the computer.

Line drawings can be more helpful than photographs in that they show as much or as little of the object as a reader needs to see. Line drawings can also present an object from a number of different perspectives (side, front, back). As the drawing of the emergent trap on page 69 reveals, a writer can also use *callouts,* or labels attached to individual parts, to direct the reader's attention to different sections of an object.

Cutaway drawings remove the external covering to reveal the internal components. The cutaway drawing of the shock absorber on page 80 shows what the piece of equipment looks like inside and how the parts are arranged.

Exploded drawings blow an object up and apart so that the reader can see how it is constructed—the details of each part and how all the parts fit together. The figure on page 194 is one example.

Maps show very large areas (a city, a country) or smaller details of topography, population, or roads. The geologist's map on page 87 gives the reader information about land forms for a proposed industrial site. For an introduction to some kinds of maps used in technical studies, see the model "Kinds of Maps and Kinds of Map Data" on pages 100–106.

Tables arrange numerical data into rows and columns. Tables are used in *specifications,* or descriptive, detailed lists of parts, dimensions, and materials that manufacturers prepare for their sales literature. The specification for Shell Rimula Oil (page 400) uses tables effectively.

Graphs represent the relationship among different sets of numbers by means of a curve or line. See how a graph is used on page 4.

Flow charts use a series of block diagrams, pictures, and line drawings connected by arrows to reveal the sequence in which events or stages of an operation occur. Computer specialists also use flow charts to show the steps that must be followed in a program. For an example of a flow chart, see page 162, which depicts the stages involved in making concrete.

Schematics are diagrams showing electrical or mechanical systems or any interior configurations. See page 367, which contains a schematic of a heating/cooling system, and page 462, which presents a schematic representation of the interior of a C130 aircraft.

The following guidelines will help you to use visuals clearly and effectively:

1. Make sure your visuals are clear and complete but as simple as possible. Be careful about including too much detail—focus on what's important to your description.
2. Label the important parts accurately, especially in an exploded or cutaway drawing. The labels should be horizontal whenever possible.
3. Supply accurate and complete specifications in tables and graphs. Keep measurements and terminology consistent both within the visual and between the visual and the text. For instance, do not use metric units for one part of an object and English units (pounds or inches) for another.
4. Be sure that all visuals are referred to in the text.
5. Place the visual as close as possible to its reference in the text. It should not precede this reference.
6. Most visuals should have a title (also called a caption or cut-line). If you are including several visuals, you should number them consecutively as they appear.
7. If your visual is a scale model of the original, provide a scale. If there is a chance that your reader might not be familiar with the object, indicate if the visual is life-size or if it is enlarged or reduced.
8. If a visual comes from another source, acknowledge that source, usually after the caption and visual number.

A CHECKLIST FOR WRITING PHYSICAL DESCRIPTION

1. Determine your audience's needs—what details does it require? How will it use your description?
2. Familiarize yourself thoroughly with the object you are describing—name, model number, dimensions, materials, function, components.
3. Be as objective as possible in your description.
4. Be specific, not impressionistic.
5. Use graphic language that allows the reader to visualize the object.
6. Follow an organization logical to the subject of your description and suitable for the audience's needs.
7. Select visuals appropriate to the purposes of the description and the audience.
8. Introduce, label, and correlate visuals with the text.

This model is a physical description of a static object—a floating trap to catch aquatic insects. It appeared in the Progressive Fish-Culturist, *a government journal read by biologists and individuals in the aquiculture industry. The author's aim is to describe the modifications he made in existing floating traps. He assumes that his readers—experts who here serve as operators—will use his description to build their own trap to collect insects for various research projects. For this reason precision and clarity are essential to his discussion.*

New Floating Trap for Capturing and Preserving Emerging Aquatic Insects

Terence P. Boyle

Emergence traps for aquatic insects have been used since the early 1900's (Needham 1908) for species surveys (Scott and Opdyke 1941), studies of life histories (Miller 1941) and population dynamics (Judd 1974), and as a means of estimating secondary production (Speir and Anderson 1974). Several studies in which the efficiency of different types of emergence traps was compared showed that floating traps caught the greatest number and diversity of species, and that floating tent and box traps had similar catches (Kimerle and Anderson 1967; Guyer and Hutson 1955; Morgan et al. 1963; Mundie 1956). All trap designs tested by these investigators had to be serviced every 1 or 2 days and some required the use of mouth-operated entomological aspirators to collect trapped flying insects. 1

The frame construction of the emergence trap described here is like that of a pyramid trap (Fig. 1). It is made of mitered lumber 2.5 × 5.0 cm, which is both nailed and glued together with waterproof epoxy cement. The 0.5-m^2 area covered by the trap is optimal, as discussed by Morgan et al. (1963). The frame is covered with a sheet of clear polyethylene sheeting 0.1 mm thick (or fine netting if desired) and stapled in place. Plastic or nylon is photolabile and must be replaced at least once every 6 months. The base of the pyramid is pierced with nails projecting downward into Styrofoam floats 5 cm thick, which are also glued into place with silicone rubber cement. 2

The collecting portion of the trap, based on a modified Malaise design (Townes 1962; Marston 1965), is made of a widemouthed funnel with a series of holes 1.5 cm in diameter drilled around the top. The stem of the funnel is shortened and the lid of a half-pint Mason jar (without liner) is attached to the lower part of the funnel with epoxy glue. The funnel is attached to the pyramid frame with wood screws. A flat circular piece of glass is glued to the top rim of the funnel with silicone rubber. The edge of the glass plate is attached 3

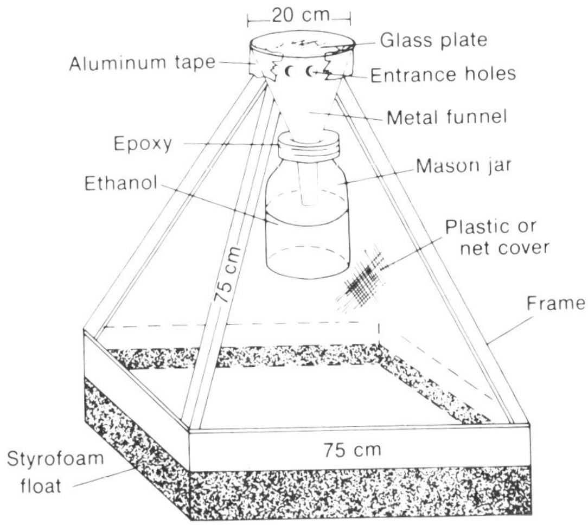

Figure 1. Side view of floating emergence trap for collecting insects.

to the plastic material covering the frame by a strip of adhesive-backed aluminum duct tape 10 cm wide. The trap may be anchored by two screw eyes in opposite corners of the bottom frame.

The collection fluid in the Mason jar is 80% preserving ethanol and a small amount of glycerine. Evaporation of the ethanol will not damage the specimens, since the glycerine will remain in the jar and keep the integument of the insects soft for some time. If netting is used in place of plastic sheeting to cover the trap, the collecting efficiency may be improved by placing a toxicant such as ethyl acetate or dichlorvos in the ethanol. 4

Thirty-six floating pyramid traps with the newly designed collecting device were tested in 12 ponds. The traps were serviced weekly with adequate results. No insects, dead or alive, were found outside the collecting jar. Although the traps have survived high winds, rain, and hail with no damage, their use probably should be restricted to calm waters. The trap is selective only against insects too large to enter the holes around the top of the funnel, such as dragonflies; larger insects may be accommodated by enlarging the holes. 5

Acknowledgments

I thank Norman Marston for bringing the Malaise design to my attention and Phillip Lovely for his suggestions on construction. 6

References

Guyer, G., and R. Hutson. 1955. "A Comparison of Sampling Techniques Utilized in an Ecological Study of Aquatic Insects." *J. Econ. Entomol.* 48:662–665.

Judd, W. W. 1953. "A Study of the Population of Insects Emerging as Adults from the Dundas Maub, Hamilton, Ontario, During 1948." *Am. Midl. Nat.* 49:801–824.

Kimerle, R. A., and N. H. Anderson. 1967. "Evaluation of Aquatic Insect Emergence Traps." *J. Econ. Entomol.* 60:1255–1259.

Marston, N. 1965. "Recent Modifications in the Design of Malaise Insect Traps with a Summary of the Insects Represented in Collections." *J. Kans. Entomol. Soc.* 38:154–162.

Miller, R. B. 1941. "A Contribution to the Ecology of the Chironomidae of Costello Lake, Algonquin Park, Ontario." *Univ. Toronto Stud. Biol.* 49:1–63.

Morgan, N. C., A. B. Waddell, and W. B. Hall. 1963. "A Comparison of the Catches of Emerging Aquatic Insects in Floating Box and Submerging Funnel Traps." *J. Anim. Ecol.* 32:203–219.

Mundie, J. H. 1956. "Emergence Traps for Aquatic Insects." *Mitt. int. Vertheor. angew. Limnol.* 7:1–13.

Needham, J. G. 1908. "Report of the Entomological Field Station Conducted at Old Forge, New York, in the Summer of 1905." *N.Y. State Bull.* 124:167–168.

Scott, W., and D. F. Opdyke. 1941. "The Emergence of Insects from Winona Lake." *Invest. Indiana Lakes Streams* 2:4–14.

Speir, J. A., and N. H. Anderson. 1974. "Use of Emergence Data for Estimating Annual Production of Aquatic Insects." *Limnol. Oceanogr.* 19:154–155.

Townes, H. 1962. "Design for a Malaise Trap." *Proc. Entomol. Soc. Wash.* 64:253–262.

QUESTIONS FOR DISCUSSION

1. Why does the author include references to earlier studies in his description of the trap? How do these references help him and the reader?
2. Which of the questions on page 61 does this description answer for the reader?
3. In what order does the author present details about the trap? Would another organization be more effective?
4. The author's language is highly objective and concrete. Find examples of this objective writing and explain why such objectivity is necessary for a reader of *Progressive Fish-Culturist.*
5. Is the visual included in this article appropriate for the reader's purpose? Why or why not? Would another kind of visual—a photograph, for example—be more useful?
6. Many of the sentences are in the passive voice. Does that make them more difficult to read than if they were written in the active voice? Why might the writer prefer the passive?

SUGGESTIONS FOR WRITING

1. Write a physical description of one of the following objects or one approved by your instructor. Provide an introduction explaining your reason for describing the object, the audience you are writing for, and how exactly your description will assist them.
   a. an air or oil filter for a car
   b. a kitchen exhaust fan
   c. a stethoscope
   d. a double-edged saw
   e. a birdhouse
   f. a storage shed

g. a gauge
h. a spark plug
j. a trowel
i. a carpenter's level

2. Find a photograph of an object that interests you and that you know something about—a camera, a guitar, a bicycle, a VCR—and compose a description that would be useful for a consumer readership.

The following description of the Freedom 200 computer is taken from the "System Components" section of Computer Design. *Readers of this journal are engineers who design digital equipment and systems and decision makers who purchase such equipment. Manufacturers such as Liberty Electronics, the makers of the Freedom 200, supply trade journals with information about their new products to attract potential customers' attention. The description of the Freedom 200 provides objective details about the computer as well as sales information for advertising purposes.*

The Freedom 200

The Freedom 200 offers a full array of features in a single package. Consisting 1
of a 12-in. display and a detachable keyboard, the terminal's points of interest range from a nonglare screen to setup modes and programmable functions keys.

The CRT uses a 24 × 80 display format. The 25th line is user accessible 2
for status within a file or other information. A full tilt (+ 15-degrees, − 5-degrees) and swivel (70-degrees) console houses the etched green phosphor screen.

In native mode, display memory is one page with the second page optional. 3
In emulation mode, memory reaches two pages with additional pages optional. The unit emulates both TeleVideo 950 and Lear Siegler ADM 31 terminals.

Advanced video features include nonembedded character attributes. In vi- 4
sual mode these are normal, reverse, half intensity, underline, blink, and blank. In data entry mode they include protected, alphanumeric, alpha only, numeric only, and extended numeric.

Either jump or smooth scroll is available. Smooth scroll is programmable 5
with rates of 2.5, 5, and 10 lines/s. The user can elect to scroll only parts of the screen by defining a scrolling region.

Four communication modes are provided: block, conversation, monitor, 6
and local. Block can be immediate or deferred and sent from command. Conversation mode transmits both half and full duplex. Monitor mode displays control codes in programming applications, while local mode operates standalone.

The detached, low profile keyboard clusters 106 sculptured keys into func- 7
tional groups. Twenty programmable function keys (using the shift) produce up to 256 Kbytes of code sequences. These keys can also send 20 preprogrammed default sequences. The 11 cursor control keys assume an inverted "T" format. In addition, there are nine editing keys and 12 functional command keys.

Handshaking protocol is x-on/x-off with programmable characters and DTR. 8
Fifteen transmission rate selections range from 50 to 19.2k baud. Parity choices are even, odd, or none, with word structure at 7 or 8 data bits with 1 or 2 stop bits. Interface is via EIA RS-232-C.

The terminal is priced at $745 for quantities of one to nine; $670 for 25 9
to 99 units. Liberty Electronics, 625 Third St., San Francisco, CA 94107.

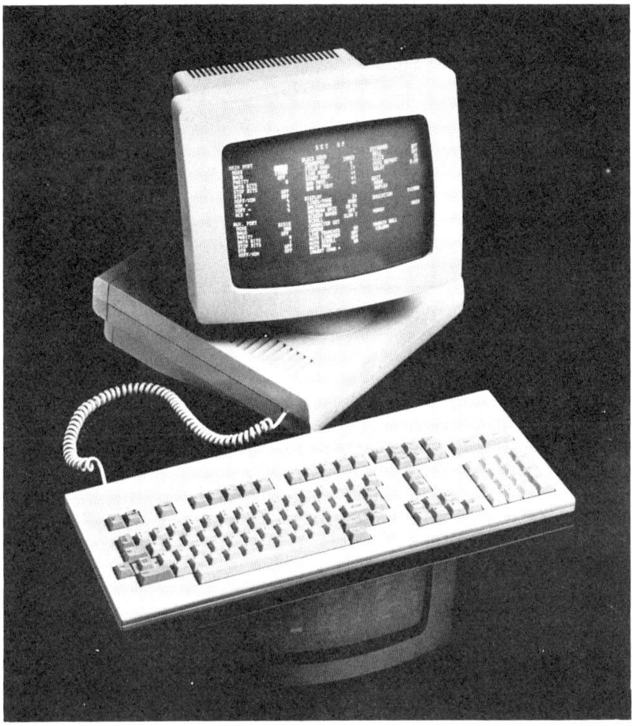

QUESTIONS FOR DISCUSSION

1. How is this description organized? Is the organization consistent with the purpose of the description or would another way be more effective? Explain.
2. Identify the parts of this description that provide purely factual information and those parts that provide an interpretation of these facts to sell the Freedom 200.
3. Which parts of this description would be of most interest to managers? to design engineers? Has the writer met the needs of both audiences?
4. What is the function of the photograph that accompanies this description? Does it simply duplicate information in the written description or does it add information not contained in the text of the description?

SUGGESTIONS FOR WRITING

1. Write a short description (200–250 words) of a piece of equipment used in your field. Give information that would be of interest and use to both experts and decision makers. Include an appropriate visual, or indicate what sort of visual would be helpful.
2. Rewrite this model on the Freedom 200 for a consumer audience with little or no background in working with computers.

The following model is taken from Jack Pine: An American Wood, *a pamphlet published by the U.S. Department of Agriculture Forest Service and available for a nominal fee from the U.S. Government Printing Office. A pamphlet like* Jack Pine *is intended to meet the needs of a wide range of readers—consumers, students preparing papers, county extension service readers, commercial growers, and forestry consultants. Although the authors use a few technical terms* (wings, shelterwood cut, butt-scarring, bole), *their description is easy enough for nonspecialists to understand and appreciate. This model uses description for a variety of purposes; it provides objective, graphic details about location, size, shape, height, weight, color, parts, age, and function.*

Jack Pine . . . an American Wood

Eugene M. Carpenter and Donald A. Perala

DISTRIBUTION

The natural range of jack pine (*Pinus banksiana* Lamb.) extends from the Maritime Provinces west to Alberta and the District of MacKenzie in Canada, and southward to the Lake States of Minnesota, Wisconsin, and Michigan. Local stands are found also in northern Illinois, northern Indiana, northern New York, central Maine, Vermont, and New Hampshire. Perhaps 70 percent of its natural range can be considered commercial range. Jack pine has been planted outside its natural range in windbreaks. Commercial stands have been established on strip mine spoil banks in the east central United States and in the sandhills of Nebraska and central North Dakota. . . . 1

DESCRIPTION AND GROWTH

Mature jack pines normally are 55 to 65 feet tall, and 8 to 10 inches in diameter at breast height. The largest jack pine reported in the United States is 91 feet tall and 25 inches in diameter. Average merchantable yield at age 60 in well-stocked, unmanaged stands, ranges from about 1,600 cubic feet per acre on poor sites to about 2,900 cubic feet on good sites. 2

The species generally develops a widespreading and only moderately deep root system. However, a taproot frequently develops which may penetrate below 9 feet. Trees without distinct taproots usually have lateral roots which grow downward as they approach other trees. The bulk of the root system consists of laterals confined largely to the upper 18 inches of soil, and most of these are found in the top 6 inches. Roots may spread laterally at least 28 feet. 3

Figure 1. Needles and cones of jack pine.

Jack pine needles grow in pairs ¾ to 1½ inches long. They are yellow 4
green, flat or slightly concave on the inner surface, divergent, stout, and are often twisted (fig. 1). They last for 3 years. The bark of jack pine is thin, brown, slightly tinged with red or dark gray, and is irregularly divided into scaly ridges (fig. 2).

Jack pine cones are 1½ to 2 inches long, oblong-conic, light brown, usually 5
pointing forward and often strongly incurved with the scales well developed only on the outer face (fig. 1). Seeds are $\frac{1}{12}$-inch long, triangular, black and roughened, with wings about ⅓-inch long. The cones may remain tightly closed for 25 years or more over most of its range. Along the southern edge of their range jack pines bear cones which open promptly upon ripening, dispersing

Figure 2. Scaly ridged bark of jack pine.

their seeds in autumn. Closed cones are usually opened by a temperature of 120°F induced by fire or by close contact with the ground surface warmed by the sun, although some seeding occurs from cones which do open every year. This accumulation of seed in closed cones accounts for the dense jack pine regeneration that often follows after the parent stand is killed by fire.

Jack pine stands from 10 to 185 years of age produce viable seed, reaching 6
a peak at 40 to 50 years. Where closed cones are common, a good seed supply is always available at any time of the year. Commercial seed averages about 123,000 sound seeds per pound.

Jack pine seed germinates best in partial shade on mineral soil when maxi- 7
mum air temperature reaches about 65°F and moisture is adequate. Shade and reasonably abundant moisture also favor initial establishment, but later height growth is greatest at one-half or more of full sunlight. Summer droughts and high surface soil temperatures frequently kill or injure young jack pine seedlings.

In well-stocked stands, jack pine develops into a rather slender tree with 8
a narrow open crown. As the stand closes and matures, up to two-thirds of the lower crown may die. Self-pruning proceeds moderately well in dense stands, but dead branches may last for many years in partially stocked stands. Open-grown jack pines have poor form and the crowns assume a spreading appearance with live branches persisting almost to the ground.

Jack pine is relatively windfirm although wind breakage may be common, 9
especially in conjunction with glaze or ice storms. Early spring fires can readily kill reproduction as well as young trees up to 4 inches in diameter. Hot surface fires can completely kill pole-sized stands or cause serious butt-scarring. Mature jack pines are relatively fire-resistant. The jack pine budworm may cause damaging defoliation, with areas as large as 50,000 or 60,000 acres being heavily infested. Several sawflies, as well as the pine tussock moth, may also cause serious local problems.

Under management, precommercial and intermediate cuts can approxi- 10
mately double the total yield of jack pine stands. Natural regeneration can be obtained in advance on poor sites by a shelterwood cut 10 years prior to the final harvest cut. All sites may be regenerated after the final harvest cut by disturbing the soil surface or by prescribed burning followed by scattering cone-bearing slash, direct seeding, or planting. Where brush or hardwood competition becomes severe, release by herbicides may be necessary. . . .

CHARACTERISTICS AND PROPERTIES

The sapwood of jack pine is nearly white and generally makes up at least 50 11
percent and often more of the volume of the bole. This is true even in older trees. The color of the heartwood varies from light brown to a distinct orange cast. The wood is moderately light in weight (30 pounds per cubic foot at 12 percent moisture content), moderately low in bending and compressive strengths,

lacking in stiffness, moderately soft and moderately low in shock resistance. It also has moderately small shrinkage. In workability with tools it ranks average. It is more apt to split when nailed than red pine, and is slightly below red pine in nail-holding power. Jack pine has very limited natural durability when exposed to conditions favorable to decay. It is moderately difficult to penetrate with preservative, although incising the wood prior to treatment usually gives satisfactory results. In paint-holding ability jack pine rates below white pine. For most of the above characteristics, jack pine ranks slightly above or comparable to the spruces and true firs.

Jack pine lumber is generally knotty, has a coarser textured, more resinous 12
appearance than white or red pine, and is often mixed with the lower grades of these species for distribution. Jack pine lumber is somewhat prone to warp and check in drying.

PRINCIPAL USES

Jack pine is used principally for pulpwood, and while it is suitable for the 13
manufacture of paper pulp by several processes, most is pulped by the sulfate method. Pitch is a problem with both the sulfite and groundwood methods. The sulfate process yields a very strong pulp useful for wrapping papers, high-grade printing paper, fiberboard, and hardboard. Jack pine pulp is often used in mixture with groundwood and other pulps for a wide range of products.

QUESTIONS FOR DISCUSSION

1. In what order do the authors present information about jack pine? Do the headings suggest a pattern? Can you suggest a better order?
2. What organizational principle guides the authors' selection and presentation of details in the section labeled "Description and Growth"?
3. The authors use precise and graphic language throughout this model. Cite three examples, and explain how they help a reader who may never have seen the tree to identify one.
4. Point out three or four descriptions where you think the authors have done a good job in adapting technical material to match their diverse audience's needs. Indicate why they have chosen the language and types of details they did.
5. Note how the authors' view of jack pine changes in the section labeled "Characteristics and Properties." The appearance of jack pine has been altered for the reader. Why have the authors done this?

SUGGESTIONS FOR WRITING

1. Rewrite one of the sections of this model for one of the following audiences:
   a. botanist
   b. business executive
   c. pulpwood factory worker
   d. homeowner using a product made of jack pine
   e. contractor using jack pine for rough local construction
2. Choose an object or topic and describe it as you would in an informational pamphlet for consumers: e.g., poison ivy, a ceiling fan, an automatic bank teller, a fuse box, a plant, a water pump, a bicycle, an air conditioner. Include suitable visuals.

The description of shock absorbers that follows is taken from a textbook for student mechanics, who will read this selection as operators. The author has correctly assumed that the reader is familiar with basic automotive terms and procedures. The description provides information about the function, appearance, parts, and operation of shock absorbers. After reading this selection, the audience should better understand the importance of properly removing and mounting shock absorbers.

Shock Absorbers

Herbert E. Ellinger

An oscillating automobile is very difficult to control. It is dangerous to drive, because the effective weight on the tires keeps changing. This type of oscillation could allow the tires to bounce clear of the road so all control would be lost. It would also allow the body to roll steer, making directional control difficult. Shock absorbers are installed on a suspension system to rapidly dampen the natural vehicle spring oscillation and thus improve ride, vehicle controllability, and vehicle handling. 1

The spring supports the weight and the shock absorber controls the spring oscillation. Contrary to many people's ideas, the shock absorbers do not support weight. Some shock absorbers have accessories added to them to absorb weight, such as external springs or a compressed air chamber. New standard shock absorbers cannot lift a sagging suspension system. 2

A shock absorber (Figure 1) is basically a cylinder with a piston moving in it. Both sides of the piston are fitted with valves and sized openings or orifices that allow fluid to flow through the piston at controlled rates as the piston moves back and forth in the cylinder. The cylinder is surrounded by a fluid reservoir tube. A valve between the cylinder and reservoir chamber controls the flow of fluid between them. The bottom of the shock absorber housing is mounted on the suspension through rubber bushings. A rod extending from the piston is connected to the vehicle frame through rubber bushings. The piston is forced downward in the cylinder during jounce and upward in the cylinder during rebound. Different valve and orifice designs control resistance to piston movement. 3

The energy absorbed by the shock absorber is converted into heat. The heat warms the fluid and dissipates it through the housing into the surrounding air. When the shock absorber is larger, it can absorb more heat and can, therefore, absorb more energy than small-diameter shock absorbers. 4

Rear axle shock absorbers are connected between the spring seat on the rear axle housing and the frame. The shock absorber may be mounted vertically or it may be canted inward and slightly forward in a "sea leg" type mounting, 5

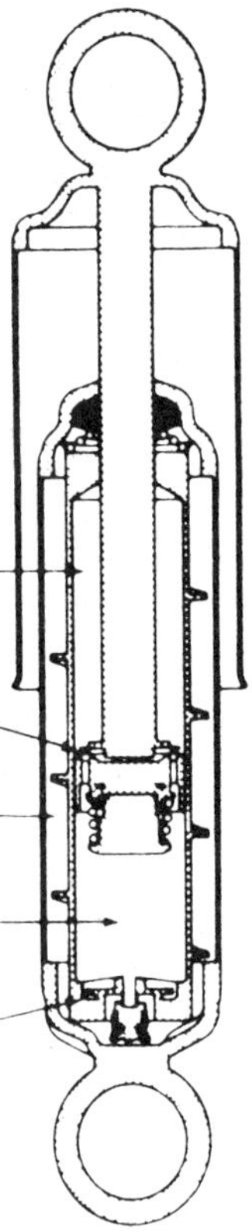

Figure 1. Cross section of a typical shock absorber (Monroe Auto Equipment Company).

to add control to body roll, handling, and roadability. The mounting methods are shown in Figure 2.

Front shock absorbers are mounted inside the spring when coil springs are used. This location provides good mounting attachments, uses otherwise wasted space, and provides protection from damage. The shock absorber is mounted in the same location between the lower control arm and frame when using coil springs above the upper arm and torsion bar springs, as it is when using a coil spring mounted on the lower control arm. 6

Many shock absorbers look identical. Their difference lies in their internal valving, which is designed to match the vehicle suspension dynamics. Shock absorbers of the correct part number must be installed to produce satisfactory ride and handling performance. 7

Shock absorbers are designed to rapidly dampen suspension oscillation. Standard passenger car shock absorbers are usually designed to provide greater control on extension as the vehicle rebounds than it is on compression during jounce. These shock absorbers would be specified as 20–80 (20% control during compression to 80% control during rebound), 30–70, etc., as percentages of the 8

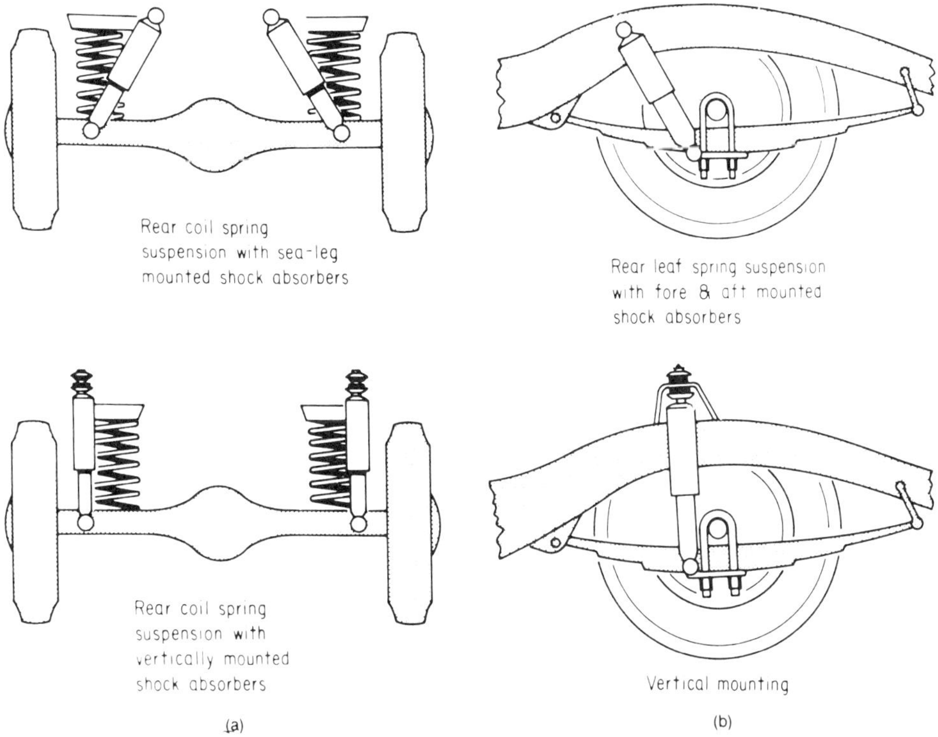

Figure 2. Typical rear shock absorber mountings. (a) Front view, (b) side view (Monroe Auto Equipment Company).

shock absorber's full control capacity. The ratio is entirely dependent upon the vehicle design, ride requirements, handling needs, and driving conditions. In general, racing cars have heavier control than passenger cars, but they also have a harsh ride, reduced driver comfort, and less vehicle durability.

During shock absorber extension, the oil trapped above the piston in the 9
rebound chamber meters through the piston to the lower compression chamber below the piston. The piston rod displaces fluid in the upper rebound chamber and there is no rod in the lower compression chamber. Because of this, during extension some fluid must also flow from the reservoir into the lower compression chamber to make up for the fluid displaced by the piston rod. During rebound, fluid is, therefore, flowing through both piston valves and reservoir valves. During compression, a reverse fluid flow occurs to both chambers.

Fluid control during compression and extension uses metered holes and 10
flutter valves. Flutter valves are discs held with light springs. Oil flow in one direction seats the disc to restrict flow. Oil flow in the reverse direction lifts the flutter valve from its seat and allows free flow. Because these valves are flow valves, rather than pressure valves, shock absorber control forces change as the shock absorber stroke rate is changed. In general, the faster the shock absorber is stroked, the more control force it applies. This allows the vehicle

to have a soft, easy ride and, at the same time, provides heavy control over sudden severe jounce or rebound. The valve design also controls the rate at which control build-up occurs. Rapid build-up gives a solid and somewhat harsh ride while a gradual build-up gives a soft ride. As a result of this valve action, it is not possible to determine shock absorber control or condition by hand operation. It only shows up as the automobile moves over a rough road.

QUESTIONS FOR DISCUSSION

1. What are the functions of paragraphs 1 and 2?
2. What are the differences between the description of a shock absorber in paragraph 3 and the description of the same object in paragraphs 7–10? What is the author's purpose in including these two descriptions? What kinds of information separate them? Why?
3. What kinds of technical terms and procedures does the writer expect the audience to know? Give some examples and explain why this information is necessary to help a reader visualize the mechanism.
4. In addition to a physical description of the shock absorbers themselves, the author includes information about their installation and the kinds of precautions mechanics must take. Why are these details important in this description, and how does the writer introduce and integrate these details into his physical description of the mechanism?
5. Most of the sentences are short and follow a subject-verb-object pattern. The writer also uses coordination, or the joining of two independent units (phrases and clauses), a great deal. Explain how these stylistic elements make his description easy to read and appropriate for the audience's needs.
6. Are the two visuals effectively correlated with the text? Are they suitable for the audience's needs? Would other kinds of visuals have been more helpful? Explain.

SUGGESTIONS FOR WRITING

1. Using the information in this model, write a description of a shock absorber for a consumer who needs to replace that part in a car. Use terms and details that this audience will easily understand and need.
2. Write a description either of one of the following mechanisms in operation or of one approved by your instructor.
   a. sphygmomanometer
   b. dehumidifier
   c. lawn mower
   d. can opener
   e. windshield wiper
   f. light meter
   g. camera shutter
   h. window shade
   i. a part of the body (lungs, elbow, kidney)
   j. stapler

The next model shows how physical description is used in a piece of on-the-job writing. It is a technical report in memo format written by an industrial geologist for his employer, a decision maker at a large railroad. (See Chapter 12 for information about memos.) In his report, the geologist describes the site to assess its potential water supply for a proposed industrial plant. The geologist's employer needs this information to convince a potential client—Jones Consolidated Industries—to locate its plant in this area, which is served by the railroad. Physical description becomes one of the geologist's most important tools.

Geologist's Memo about the Groundwater Potential at an Industrial Site

Richard J. Councill

SEABOARD SYSTEM RAILROAD

SALES & MARKETING DEPARTMENT
500 Water Street
Jacksonville, Florida 32202
(904) 359-3100

TO: J. R. LeGrand
Vice President, Industrial Development

FROM: Richard J. Councill R. J. C.
Chief Geologist

RE: Groundwater Potential
South Wilmington Area
New Hanover County, North Carolina

DATE: June 20, 1982

As you requested, the following report examines the groundwater potential in the Wilmington area to see if it is adequate for the development being considered by Jones Consolidated Industries. I have concluded that the groundwater found in the Wilmington area easily meets Jones's needs for water of constant analysis and temperature. 1

The area can be described as follows. The lower part of New Hanover County below Wilmington is a peninsula formed by the Cape Fear River on the west and the Atlantic Intracoastal Waterway on the east and south. An area consisting of more than 4,000 acres south of Barnards Creek to Telfairs Creek, east of the Cape Fear River and west of U.S. 421, contains two large industrial sites referred to as the "Allied Site" (1,700 acres) and the "Mott (Todd) Creek Site" (1,450 acres) of similar physiographic and geologic setting. Figure 1 shows the location of the properties referred to in this report. 2

Both sites are characterized by rolling, sparsely-wooded fossil sand dunes attaining a maximum elevation of 25-30 feet and lowlands ranging from near 0 feet to 10 feet. Depending upon elevation, the total thickness of the dunes is 45 to 60 feet. The surface and subsurface materials are essentially homogeneous and consist of interbedded strata of 3

J. R. LeGrand
June 20, 1982
Page 2

fine to medium quartz sand, fine to coarse sand containing thin clayey sand and clay seams. Beneath these geologically young materials are intercalated beds of fine to very coarse quartz sand, glauconite, thin discontinuous broken shell beds, and clay-silt beds of the Pee Dee formation.

Both the younger and older units contain appreciable amounts 4
of water available to wells, but the older Pee Dee unit generally contains highly mineralized water, often analyzing upward to 3,000 parts per million chloride and almost equal amounts of other dissolved solids. As a source of industrial water the lower unit must be dismissed. However, the overlying younger materials provide a substantial aquifer for supplying large amounts of groundwater of good quality.

Because the fossil dunes are highly permeable and are 5
covered by little vegetation, practically all water falling as rain in the general area enters the ground and, unless withdrawn through wells, is lost. There is no runoff in the dunes area and evapotranspiration losses are thought to be minor. Therefore, it can be assumed that 85 percent of precipitation in the form of rain enters the ground and is held under water table conditions.

Based on this information and figuring an average of 50 6
inches of precipitation annually, it is safe to say that an acre of the property covered by dune materials receives 1.3 million gallons of water. Approximately 1.1 million gallons of this rainwater enters and is stored in the water table aquifer. Correspondingly, over an area of 1,000 acres, 1,100 million gallons of water enters the aquifer annually. This amount of water is equal to a maximum withdrawal rate of approximately 3 million gallons per day assuming no other water is available to wells. However, the amount of water available per area unit is considerably higher because through pumping a well, the water will flow into the cone of depression formed by withdrawal.

Previous studies in the area indicate the above findings to 7
be true with the aquifer having a transmissibility value of 52,540 (gpd/ft drawdown) so that when wells are drilled to a depth of 50 to 60 feet and spaced 1,000 to 1,500 feet apart

J. R. LeGrand
June 20, 1982
Page 3

each could safely yield with a drawdown of approximately 12 feet in each well with no overlap of cones of depression.

Although an analysis of water from the upper aquifer alone 8
is not available from the site, one from an identical dune
area to the north northwest shows in the table below the
following dissolved solids content, pH value, and
temperature.

	PPM
SiO_2	4.9
Fe_2O_3 and Al_2O_3	0.5
Ca	1.9
Mg	0.5
Na	3.3
HCO_3	6.7
SO_4	2.6
Cl	3.9
NO_3	Trace
Hardness (as $CaCO_3$)	6.9
Total Hardness	23.0
pH	5.90
Temperature	64°F

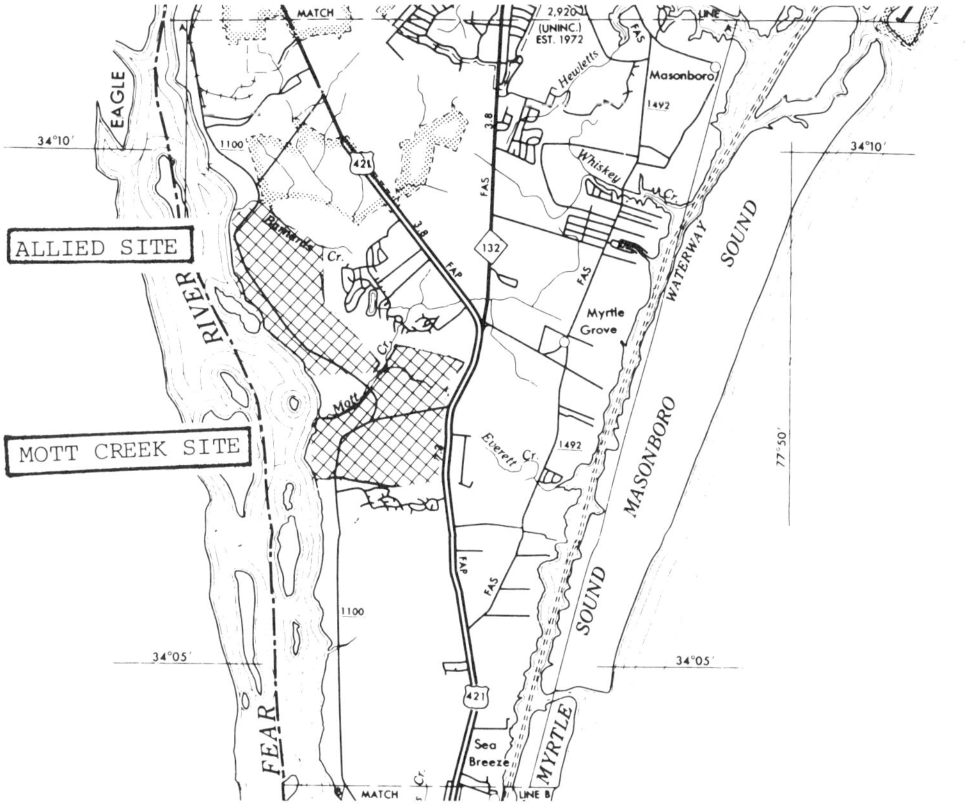

NEW HANOVER COUNTY

NORTH CAROLINA

PREPARED BY THE

NORTH CAROLINA DEPARTMENT OF TRANSPORTATION
DIVISION OF HIGHWAYS—PLANNING AND RESEARCH BRANCH

IN COOPERATION WITH THE

U.S. DEPARTMENT OF TRANSPORTATION
FEDERAL HIGHWAY ADMINISTRATION

SCALE

1 0 1 2 3 4 MILES

0 0.5 1 MILE

SCALE FOR ENLARGEMENTS

QUESTIONS FOR DISCUSSION

1. How much technical knowledge of geology does the writer assume on LeGrand's part? What kinds of language and detail does he use, and how does it relate to his knowledge of the reader?
2. How does the writer organize his description? Can you determine the organizational principle by dividing the memorandum into two parts? How does the writer view the site (e.g., from above, from east to west, another way)?
3. What features of the area are of most interest to the reader and how does the writer restrict his discussion of them?
4. Why doesn't the writer give details (authors' names and titles, for example) of the "previous studies" mentioned in paragraph 7? Is that a good decision?
5. How does the writer introduce and use the two visuals—the map and the table—in his report? Would the information these visuals contain be more difficult to comprehend in words alone? Explain.

SUGGESTIONS FOR WRITING

1. Assume you are LeGrand. Based on the information contained in Councill's description, write a letter to the manager at Jones Consolidated Industries explaining the results of Councill's investigation and urging Jones to develop the Wilmington site.
2. Write a description of an industrial site in your town or a description of a site on campus for a consumer or expert audience.

chapter **3**

Classification and Partition

Two techniques of analysis that can be very effective in technical writing are classification and partition. They are useful for developing discussion about a complex subject in that they divide the subject into logical parts so that each part can be considered separately. Classification sorts a number of units into categories; partition divides a single unit into component parts. With either technique, the writer must use a systematic principle of analysis. That principle is the key to the analysis and must be chosen with both the writer's purpose and the audience's needs in mind.

Here is a short list of the types of questions readers will raise when you use classification or partition.

1. Why do I need to know why something is broken down into categories or parts?
2. Into how many categories or parts is something broken down?
3. In what order are these categories or parts listed? Why?
4. What is the relationship of these categories or parts to each other?
5. Are there any subcategories or subparts? How do they relate to the major categories or parts?

Both classification and partition thus help readers to understand concepts, equipment, and different kinds of organizations.

But how exactly does partition differ from classification? Classification sorts multiple items into groups to show their relationship, most often their similarity. Classification groups corn, squash, potatoes, lettuce, and artichokes into classes of vegetables. Partition, on the other hand, concentrates on a single item to divide it into component parts in order to explain it. Partition peels away the layers of a single artichoke to show, say, its structure. As you read this chapter, here is an easy way to keep the two techniques straight. Only plural nouns (plural in form and meaning—sinks, fish, plastics) can be classified, and only singular nouns (a sink, a fish, a plastic object) can be partitioned.

CLASSIFICATION

Classification creates a workable system to explain diverse materials. It is like replacing a group of random numbers with a neat computer printout divided into organized columns. Or, to use another comparison, using classification is like running an optical scanner to sort mail. As zip-coded letters pass through the scanner, they are routed to appropriately labeled bins, from which they will be shipped to their destination. A letter without a zip code is automatically kicked out of the system. Many kinds of mail in different sizes and shapes pass through the scanner and are sorted according to numerical patterns of zip codes. When you classify, you act like a scanner. You sort, or group, different items (the letters) into categories (the bins) to reveal how the items are alike (have clustered zip codes) and why they belong together (in the same truck for delivery to adjacent areas).

Classification is indispensable to the process of discovering and organizing knowledge. The Dewey decimal system, for example, classifies all books into ten main groups and numerous subclasses. If libraries did not follow some classification system, it would be impossible to find books in them. Imagine the chaos that would result if all the books in an engineering library were simply shelved alphabetically according to title. Other reference works—indexes, abstracts, almanacs, codes—are also classified systematically to help users locate information quickly and accurately.

The scientific method is firmly grounded upon classification. The Linnean taxonomic system classifies animals into phylum, class, order, family, genus, and species, thus showing how animals are related to one another. The Linnean system explains how seemingly diverse animals are related; human beings, the whale, and the duck-billed platypus, for example, are all mammals; snakes and turtles are both reptiles. In the same way, the periodic table of the elements classifies the elements according to their atomic number; engines are classified on the basis of internal or external combustion; and plastics can be divided between those that can be melted and recycled (thermoplastics) and those that cannot (thermosets).

In writing, classification helps a reader to see relationships and to draw conclusions based on those relationships. Consider the bewildering array of financial transactions (accounts payable, accounts receivable) that a large construction company incurs with each new job—items for architectural plans, materials, equipment, energy, transportation, and labor. If these transactions were unclassified, it would be difficult, if not impossible, for a crew chief to write a progress report, for a business manager to pay subcontractors in full, or for an accountant to balance the books at the end of the fiscal year. When organized into categories, the company's expenses are easier to monitor and remit.

Establishing the Basis for Classification

Your classification should be appropriate to your audience's needs and the purpose of your communication. Remember that one group of items may be classified in many different ways depending on your purpose and your reader's needs. Bridges in your town, for example, could be classified in a variety of ways. An engineering firm responsible for their construction or repair would classify them according to structural types—truss, suspension, cantilever, continuous, etc. Truck drivers, however, would find it helpful to have the same bridges classified according to height or the amounts of weight they support. Similarly, for the consumer, thermometers might profitably be classified according to use—cooking (candy and meat being subgroups), assessing illnesses, or measuring outdoor temperatures. Physicists and chemists, however, would find a classification based on the method of operation (changes in a liquid, expansion of a gas, fluctuation of electrical resistance) more suitable.

How to Use Classification

1. *Select one basis of classification.* This will be your basis for analyzing a large and sometimes diverse group of items, the principle by which you will sort individual items into groups. The basis might be chronological, geographic, structural, functional, numerical—to name some of the possibilities. In some instances, a group of items may be classified according to two bases, as the speech disorder model on pages 107–110 is. But make sure readers know why.

2. *Use a principle of classification that is meaningful to your audience and purpose.* Classifying domestic beers according to one individual's preference would create an idiosyncratic system inviting disagreement. An insignificant classification will result from selecting a basis too elementary for, or unimportant to, your audience. For electronics specialists, you would not classify antennas by color, construction materials, or place of manufacture. This audience is more concerned about operating frequencies or wavelengths than in manufacturers' addresses or the color of antennas.

3. *Maintain consistency among categories.* Each category should be similar in type (a category list of engineering firms should not include a technical college). Compare and contrast your categories to see how they are alike or dissimilar. As you start assembling items into different groups, you will need to determine whether you are establishing a major category (coordination) or a subcategory (subordination). For example, a classification which includes France, Nigeria,

Israel, England and Washington, D.C., elevates one member of that group (Washington, D.C., a city) to the level of the other major classes (countries). Similarly, using a brand name to represent an entire group destroys the consistency of your categories. Classifying common cleaning agents into soap, alcohol, iodine, iodophors, and pHisoHex is misleading, for the last entry is a brand name and should be replaced with the more inclusive heading (hexachlorophene) to which it is subordinate.

4. *Make sure categories do not overlap.* In other words, they should be separate as well as equal. Classifying types of household siding as frame, aluminum, stucco, vinyl, and wood would be misleading because of duplication: Wood and frame refer to the same class of items. A zoning report classified into commercial, residential, business, and industrial sectors would also contain misleading duplications. A company included in both a commercial and business zone might be faced with following contradictory ordinances. A better zoning method would narrow the categories and make sure they do not overlap. For example, Residential I (lots under fifty feet of frontage), Residential II (lots over fifty feet of frontage), Commercial, Medical, Light Industry, and Heavy Industry are more logical categories because each is mutually exclusive.

5. *Be inclusive.* Be sure that your classification comprises all essential categories. A classification of the new (artificial) sweeteners that omitted sorbitol or aspartime (Nutra Sweet) would be incomplete. Occasionally you may have a group that cannot be subdivided, such as a genus with only one species. But be sure to include relevant items. For example, a study that classified methods of food preservation by refrigeration, freezing, canning, and fermentation would be incomplete. Dehydration, salting, addition of chemicals, and radiation should be included to make the classification complete and timely. Radiation has been added recently as a method, thus showing that classification systems change with time. It may at times be impractical to list all members, and every category of a class. For large subjects, determine how many items you can realistically offer your reader. For instance, it would be impossible to list every type of steel ever produced, as the authors of *The Making of Steel,* published by the American Iron and Steel Institute, suggest:

> But to think of steel—even in its raw stage—as a single product is most misleading. Actually there are several hundred grades of standard steels and the addition of proprietary grades would bring the total up into the thousands. The grades just mentioned [earlier in the text] are by chemical classifications alone. Shaping these steels and heat treating them by a wide variety of facilities brings the ultimate number of these steels almost to infinity. Quite literally, it is possible to order steel to perform under most conditions known to man. There are special grades of steel to perform at ultra-high temperatures and very low temperatures; steels for strength; steels for electrical conductivity and electrical resistance; steels to resist impact or abrasion or corrosion; steels to take and hold a fine edge; steels to shape other steels.

At times you will have to exclude items no longer in use or techniques too specialized for your audience. Whenever your classification is restricted, tell the reader why and choose an appropriate title: "Selected Characteristics of North Sea Dredging" or "Major Features of Plant Toxicology in Southern Illinois."

6. *Present your categories in a logical order.* Order should be systematic, not random or haphazard. The order of your categories should help a reader to understand and follow them. Sometimes your topic will suggest the order to follow: chronological or spatial, for example. Categories might also be arranged from most to least familiar or from the most frequently used to the least.

PARTITION

Just like classification, partition is a basic technique of analysis, although it operates entirely differently from classification. Unlike classification, which groups many and diverse objects, partition helps readers to see how a single object is broken into its components. We saw (pages 89–90) that a classification system like the Dewey decimal system sorts books into categories. Using partition, an author would divide a single book into individual chapters to help organize information for readers and to make it easier for them to follow and find. Dictionaries, too, would be impossible to use were it not for the principle of partition that divides information in them according to alphabetical letters. In business and industry, a large company is partitioned into many different levels or offices—management, engineering, maintenance, training, sales, public relations—that can be graphically represented on an organizational chart.

To be effective, both classification and partition must follow a principle consistently, but keep in mind that it is the number of objects being studied according to that principle that separates classification from partition. For example, a classification of light bulbs could be made according to filament dimensions (for electronics experts), by wattage or safety factors (for someone in maintenance), or by shape and color (for interior decorators). If you were to partition one incandescent lamp, you would take its component parts and discuss each according to a specific purpose and design. For example, your partition might be organized according to the way each piece is constructed. The cutaway drawing and callouts (labels identifying the parts) in Figure 3–1 graphically illustrate one way of partitioning a lamp.

Partition and classification are frequently used together. Read the following information that appears on a can of paint:

Pigment (by wt)		39%
Titanium dioxide (Type III)	15%	
Silicates	20%	
Silica	4%	
Vehicle (by wt)		61%
Styrene butadiene copolymer resin	12%	
Water	49%	
	100%	100%

Sorting the various ingredients into the large categories of pigment and vehicle employs the classification process. However, partition is used when listing the individual ingredients that compose each separate class—titanium dioxide, silicates, and silica for the pigment, and resin and water for the vehicle. Both techniques would be employed in a feasibility report to a utility company outlining the advantages and disadvantages of purchasing certain types of generators. To evaluate the generators according to their energy efficiency would require classification; but the report writer might also use partition to show the internal mechanism of an individual generator.

Establishing the Basis for Partition

An object can be partitioned in many ways, depending on the principle you use, which in turn depends upon your audience's needs. If you were partitioning the ingredients in the previously mentioned can of paint for a chemist, you

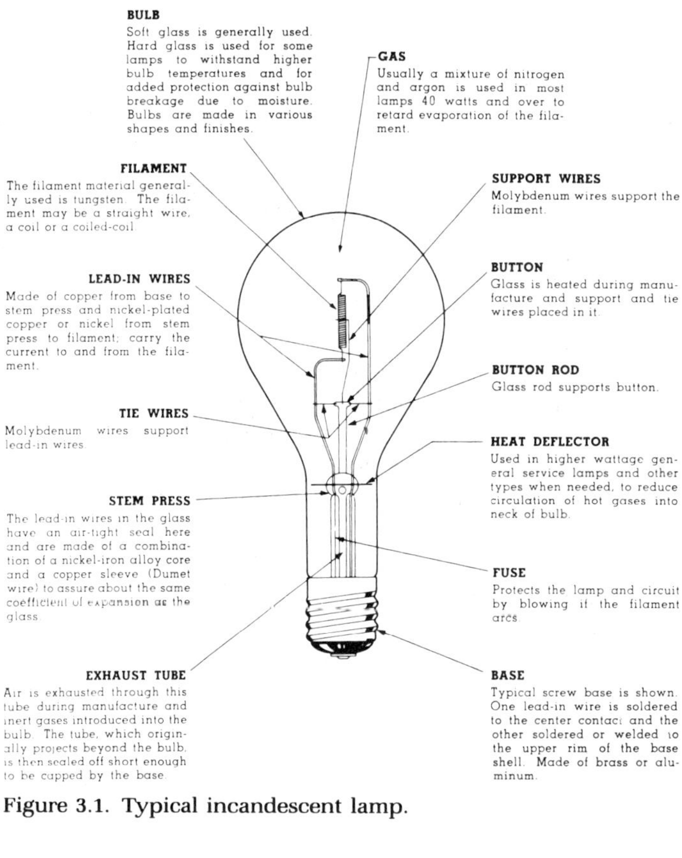

Figure 3.1. Typical incandescent lamp.

would analyze each of them according to a principle more rigorous than their percentage of volume in one can of paint. For example, each component (including every silicate) might be divided and subdivided according to molecular structure, porosity, or elasticity. The various systems of the body might be classified according to physiological function: respiration, circulation, excretion. These systems can in turn be partitioned. The nervous system, for example, might be partitioned in at least three different ways:

1. *Cellular partition*
   a. Neurons—units that generate and conduct nerve impulses, as well as synthesize neurotransmitter substances and hormones.
   b. Glial elements—non-neuronal cells that serve to protect, to modify impulse conduction speed, etc.
2. *Structural partition*
   a. Central nervous system—the nerves within the brain and spinal cord.
   b. Peripheral nervous system—in general, any part of the nervous system not within the brain and spinal cord.
3. *Functional partition*
   a. Voluntary—the part of the nervous system that can be consciously controlled.
   b. Autonomic—the part of the nervous system that controls the various "automatic" functions of the body, such as digestion or sweating. This part is further subdivided into the sympathetic and the parasympathetic.

The entire nervous system is represented by any one of these three ways of partitioning it. Each way incorporates the components of the nervous system but partitions them according to a different organizational principle (by cell, by structure, or by function).

How to Use Partition

1. *Select a single object to partition.* A bicycle, clock, egg, computer, pocket calculator, trombone, or stereo are all examples of single objects. Be careful, though, that you are not misled by a term that looks singular but is really plural—crime, energy, soil. These terms include numerous and complex classes that would be impossible to break down if they were treated as subparts of a single object. If you were asked to write about the products of your company, it would be misleading to bring in details about a subsidiary or parent company.

2. *Divide the components according to a systematic principle that is consistent with the audience and your purpose.* In the example of the incandescent light on page 93, the author is writing to a group of lighting specialists who are interested in the construction of the individual parts of the light. To partition

these parts according to cost or place of manufacture would not help the audience as much as the principle the author chooses. If you were writing about a copying machine for a decision maker, however, you might partition the individual parts according to their replacement value or according to their life span.

3. *Make sure that components do not overlap.* One of the most misleading things for readers is to see a part listed in two different places. Partitioning a lawn mower into the base, the motor, the carburetor, and the handle is misleading because the carburetor is actually a part of the motor. Similarly, listing the cursor as being a part of the keyboard and the video display screen is an example of overlap.

4. *Make sure that you include all components.* If you omit a part, readers will not have a clear and complete picture of the object. A partition of a minicomputer that omitted the disk drive would be incomplete. Note that even though the gas in the incandescent light on page 94 is unseen, the author includes it as an important part of the object. If a part is optional or found on only a certain model or piece of equipment, tell readers so.

5. *Follow a logical order of presentation.* For a piece of equipment you might arrange your discussion from the inside to the outside. A chronological order would be most appropriate for partitioning a plant according to its life cycle.

A CLASSIFICATION AND PARTITION CHECKLIST

1. Use classification for analysis of multiple items; use partition for analysis of a single item.
2. Decide upon a basis for analysis with your purpose and the audience's needs in mind.
3. Be sure that your categories are consistent.
4. Be careful that categories do not overlap.
5. Make certain that categories are inclusive.
6. Decide upon an order that is logical and helpful to the reader.

CLASSIFICATION

This table appeared in a publication entitled Health and Safety Guide for Bulk Petroleum Plants, *published by the National Institute of Occupational Safety and Health (NIOSH). This guide is read by safety engineers and technicians responsible for the transportation, storage, and use of bulk products. As the note at the bottom of the table explains, these products are classified primarily according to their* flash point, *or the lowest temperature at which combustible vapors ignite in the air. The boiling point is not a consideration for items in Classes 1C, 2, and 3. Being mixtures, products in these classes display a wide range of boiling points from very low (naphtha) to very high (camphor oil). Moreover, it is their flash point (not boiling point) that determines their potential danger.*

NIOSH Fact Sheet

NIOSH **Fact Sheet**

CLASSIFICATION OF SOME TYPICAL BULK PRODUCTS

CLASS		FLASH POINT	BOILING POINT
1	FLAMMABLES—Class 1a, 1b, 1c Regular Automotive Gasoline Premium Automotive Commercial Aviation Gasoline Military Aviation Gasoline Jet Fuel JP-4	Lower Than 73 °F	Lower Than 100 °F
	Acetone Benzene Coal Tar Oil Denatured Alcohol Methyl Ethyl Ketone (MEK) Naphtha, VM & P Crude Petroleum Toluene	Lower Than 73 °F	Above 100 °F
	Turpentine Whiskey	Between 73° and 100 °F	
2	FLAMMABLES—Class 2 Camphor Oil "Cellosolve" Solvent Fuel Oil No. 1 Fuel Oil No. 1-D Jet Fuel JP-5 Jet Fuel JP-6 Kerosene Mineral Spirits Naphtha Stoddard Solvent	Between 100 °F and 140 °F	
3	COMBUSTIBLES—Class 3 Asphalt Brake Fluid Castor Oil Coal Tar Pitch Corn Oil Cottonseed Oil Creoste Oil Fuel Oil No. 2 Fuel Oil No. 2-D Fuel Oil No. 4 Fuel Oil No. 5 Fuel Oil No. 6 Gas Oil (65 Deisel Index) Diesel Fuel Transformer Oil Linseed Oil Motor Oil	Above 140 °F	

QUESTIONS FOR DISCUSSION

1. How does the format of this table help readers to understand easily the system of classification?
2. Why is the classification of these diverse products by flash point the most useful way to group them for the intended audience? What are some ways the audience will most likely use this table?
3. How does this system of classification make use of partition?

SUGGESTIONS FOR WRITING

1. Prepare a table dividing one of the following groups into classes. Identify a specific audience for your table and supply an explanatory note similar to the one found in the NIOSH table:
   a. soil types
   b. computer languages
   c. insulation materials
   d. rocks
   e. types of mortgages
   f. horses
   g. foods
   h. cameras
2. Write a memo to an imagined supervisor summarizing the information in the fact sheet.

The model below classifies six principal varieties of maps and briefly describes each of them. It was reprinted from Maps for America *issued by the Department of the Interior to inform nonspecialists (the educated layperson) about the accomplishments of its various agencies (e.g., the U.S. Geological Survey) and the types of information produced about America by these agencies. A three-paragraph introduction precedes the discussion of the various classes of maps. As the second paragraph of the introduction clearly points out, the specific purpose of this discussion is to give readers information about those kinds of maps they are least likely to be familiar with, those used for technical and economic planning.*

Kinds of Maps and Kinds of Map Data

Most map users have one motive in common: They are seeking information. 1
But the information they seek may range from the location of likely fishing spots to the site features for planning a giant hydroelectric plant. The kind of map needed may be a rough sketch showing only a few general features, or it may be an extremely accurate topographic map showing all the terrain features of an area in minute detail, or it may be any variety of map lying between or beyond these examples. Map samples reproduced throughout this book are ordinarily taken from the U.S. Geological Survey 1:24,000-scale 7.5-minute quadrangle series. Any exceptions are noted on that particular map.

Americans are usually familiar with road and street maps (showing how 2
to get from place to place), political maps (showing the boundaries of countries and subdivisions and the locations of communities), and physiographic maps (showing the locations of mountains, valleys, rivers, lakes, and seashores). They are less likely to know of the many map products that are needed for technical and economic planning and countless specialized uses. The primary emphasis in *Maps for America* is on topographic maps and map data of the kind produced by the U.S. Geological Survey; however, other kinds of maps are considered.

In the changing setting of modern times, the demand for conventional 3
line maps continues, but it is accompanied by a burgeoning parallel demand for new cartographic products. These products include not only new types of line maps at a broad range of scales, but also accurate photomaps, remote-sensor maps, and cartographic data presented in digital form. The types of maps that can be made for different uses are practically unlimited. Generally, however, maps can be classified in one or more of the following categories: planimetric, topographic, thematic, digital, line, or photographic.

PLANIMETRIC MAPS

Planimetric maps present the horizontal position of selected features but 4
do not show relief in measurable form. Examples of planimetric maps are base, cadastral, line-route, and outline maps.

Base maps are used to plan or to compile data for the production of special- 5
ized maps.

Cadastral maps show the boundaries of subdivisions of land (usually with 6
bearings and lengths and the areas of individual tracts) for describing and recording ownership. One type of cadastral map is the plat, which often constitutes, or is an essential part of, a legal description of a parcel of land. The Bureau of Land Management is the major single producer of land plats.

Line-route maps are used by utility companies. They show the routes and 7
type of construction of pipelines or wire circuits, plus the locations of facilities such as switchboards, valves, and pumping stations.

Outline maps present only the information needed to provide a basis for 8
the compilation of additional data. Outline maps often show only national and state boundaries and major drainage systems.

TOPOGRAPHIC MAPS

In addition to the features shown on planimetric maps, topographic maps portray 9
the shape and elevation of the terrain, usually by contours, formlines, shading, color gradients, or hachures. Any map portraying relief by one of these conventions can be called a *hypsometric map.* A map on which the elevations are referred to a specific datum is called a *hypsographic map.* Standard topographic maps are in both categories.

In the United States the best known topographic maps are the Geological 10
Survey quadrangle series, which range in scale from 1:20,000 to 1:250,000. The *quadrangle series* are used for various purposes, such as selecting industrial sites, planning highways, routing utility lines, selecting damsites, and locating communication facilities. They are also popular in recreation for hunting, fishing, hiking, and camping.

Other types of topographic maps are engineering, flood-control, landscape, 11
and bathymetric. *Engineering maps* are used for planning and for estimating project costs. *Flood-control and storm evacuation maps* are special-purpose topographic maps used to study areas subject to flooding. *Landscape maps* are used by architects to plan buildings that will conform to the topography of the site. Gardeners use landscape maps to maintain parks, playgrounds, and private estates. *Bathymetric maps* show water depths and underwater topography. Water depth ranges are generally portrayed by various colors or shades. Usually, uniform depth intervals are connected by solid lines called bottom contour lines. The Geological Survey and the National Ocean Survey produce bathymetric maps.

Special Role of Topographic Maps

Because of their vital role in the development of the nation's economy and environment, topographic maps are reviewed in greater depth than other kinds of maps, important though these others may be. Some of the uses of conventional topographic maps are indicated in figure 1. 12

Topographic maps are classified generally by publication scale, and each scale series fulfills a range of map needs. Map scale defines the relationship between the measurements of the features as shown on the map and as they exist on the Earth's surface. Scale is generally stated as a ratio or fraction—1:24,000 or 1/24,000. The numerator, customarily 1, represents map distance, and the denominator, a large number, represents horizontal ground distance. Thus the scale 1:24,000 states that any unit, such as 1 inch or 1 cm on the map, represents 24,000 of the same unit on the ground. Figures 2–5 show the contrast between maps of large, intermediate, and small scale. 13

Large-scale maps, such as the 1:24,000-scale maps, are especially useful for highly developed areas or rural areas where detailed information is needed for engineering planning or similar purposes. 14

Intermediate-scale maps, ranging in scale from 1:50,000 to 1:100,000, cover larger areas and are especially suited for land management and planning. 15

Small-scale maps, such as those made at scales of 1:250,000, 1:500,000, and 1:1,000,000, cover very large areas on a single sheet and are useful for comprehensive views of extensive projects or for regional planning. 16

THEMATIC MAPS

Thematic maps are also called geographic, special-purpose, or distribution maps. They emphasize a single topic, such as geology, climatology, or crop distribution, and the entire map is devoted to presenting this distribution or concentration. Geographers use thematic maps to show the distribution of subjects such as population, languages, crop production, soil, climate, vegetation, land use, and industry. The distributions are shown by several methods, including dot patterns, choropleths, or isopleths. 17

Dots are used to represent quantities such as 1,000 people or 500 acres of corn. The size and value of dots are selected so that the dots coalesce in areas of densest distribution. Sometimes dots of varying sizes are used for different quantities. 18

Choropleth maps are thematic maps in which sections determined by civil boundaries or other arbitrary divisions are colored, shaded, dotted, or hatched to make darker or lighter areas in proportion to the density of a given subject's distribution. Because of this arbitrary delineation of thematic sections, choropleth maps can be somewhat misleading since they show an abrupt change between sections where in fact the change is gradual. 19

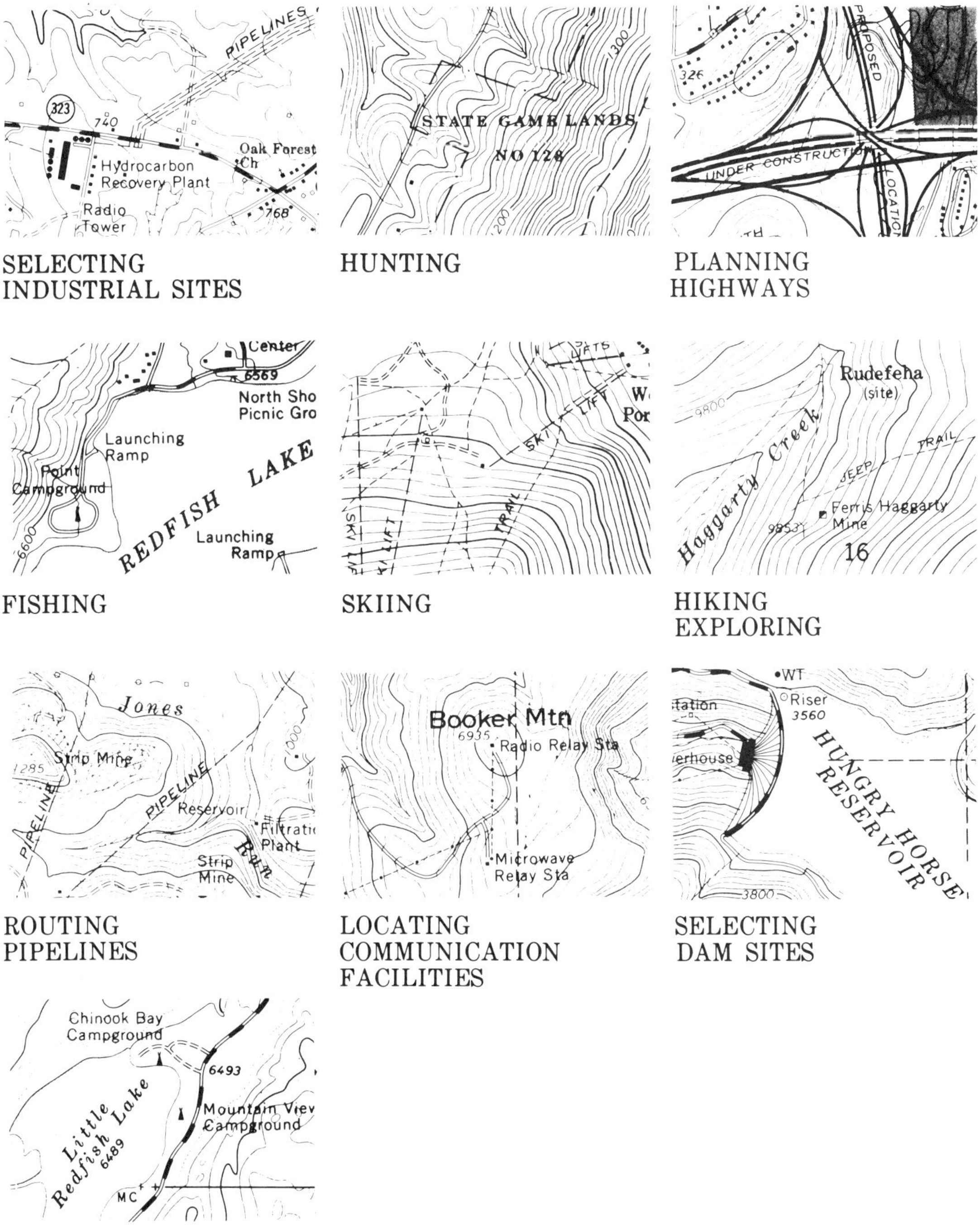

Figure 1. Some uses of conventional topographic maps.

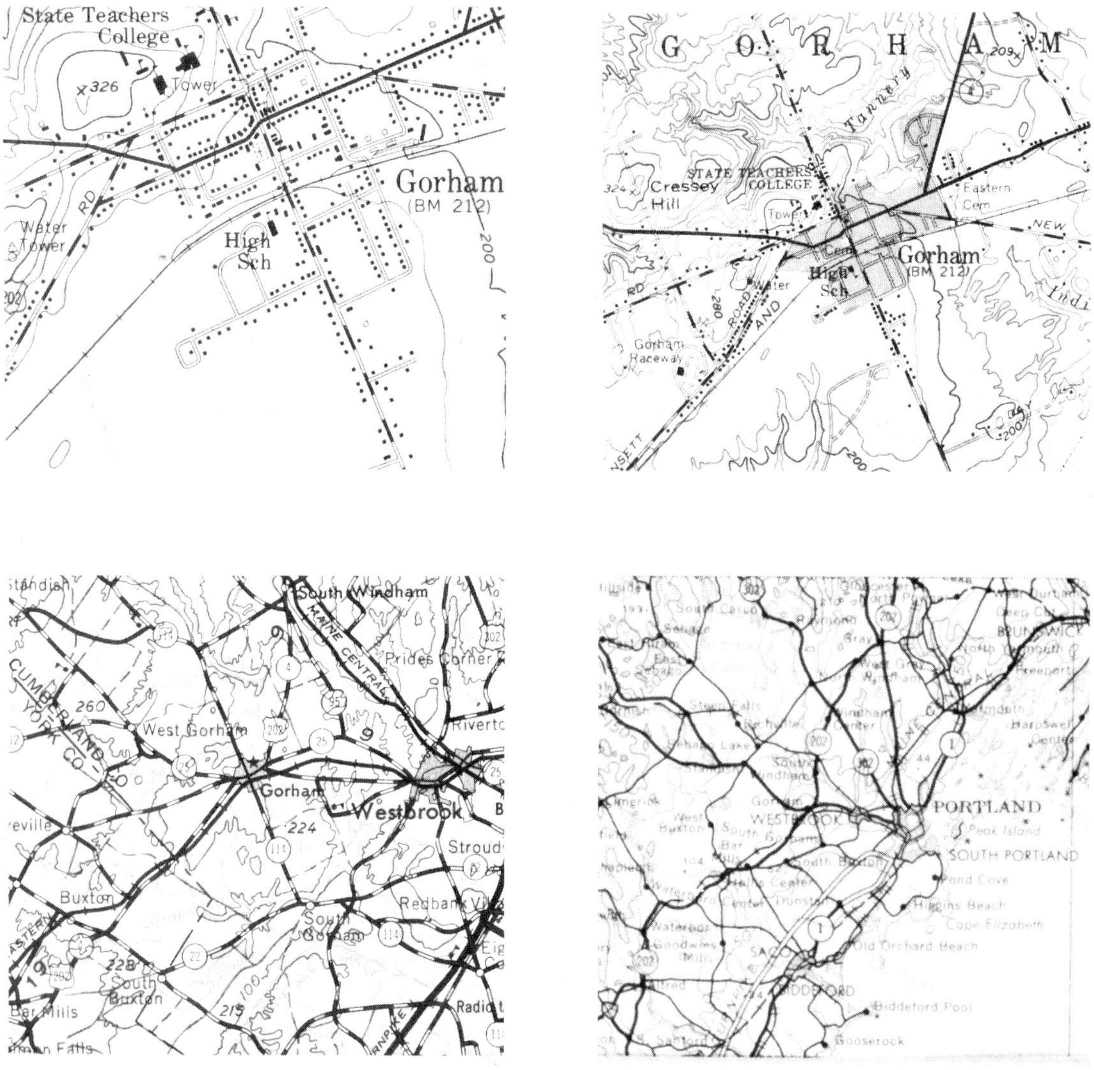

Figure 2. Comparison at various scales of maps covering the Gorham, Maine, area. Top left: 1:27,000 scale, 1 inch = 2,250 feet. Area shown 1 square mile. Bottom left: 1:70,000 scale, 1 inch = about 1 mile. Area shown, 6¾ square miles. Top right: 1:280,000 scale, 1 inch = about 4 miles. Area shown, 108 square miles. Bottom right: 1:1,100,000 scale, 1 inch = about 16 miles. Area shown, about 1,667 square miles.

Isopleth maps are used to show numerical values for continuous distribu- 20
tions (such as rainfall and temperature) rather than discrete variables. Isopleths
are lines connecting places of equal value and have the same inherent virtues
and deficiencies as contour lines. They provide exact information throughout
their entirety but do not provide values in the intervals. Tinting or shading is
sometimes used between isopleths.

Types of thematic maps include geologic, forestry, soil, land use, slope, and historical. *Geologic maps* portray geologic conditions in an area. *Forestry maps* show the size, density, kind, and economic values of trees in a given area. *Soil maps* portray soil conservation parameters. *Land use maps* depict land usage by means of colors, letters, or numbers within small areas. *Slope maps* use colors or shades to represent different degrees of slope graphically and are used in studies related to land use. Slopes can be determined mechanically by the distance between contours on a relief map, or they may be generated by computer printout from digital terrain data. 21

Thematic maps dedicated to the explanation of the past are called *historical maps*. They show such features as battlefields, military routes, and boundary changes. Although all maps are in a sense "historical" because of the time interval between compilation and publication or use, not all historical maps are thematic. Only those intended to illustrate statistics can be classified as thematic. 22

DIGITAL MAPS AND MAP DATA

The development in recent years of powerful data-processing systems has made it possible to store digitized map data in a computer bank and retrieve desired information either in graphic form as a digital map or in numerical form as a body of data. For example, the location and elevation of all bridges in a given area can be obtained from the data bank and automatically plotted on a map or listed in terms of horizontal and vertical coordinates. 23

LINE MAPS

Any map produced from scribed, inked, or pasted-on line copy is considered a line drawing or *line map*. 24

PHOTOMAPS

The photomap is an alternative to the line map—it shows nonselective details requiring photointerpretation by the user. Any aerial photographic image can be considered a photomap. However, although aerial photographs are map substitutes because they show surface features, they may contain serious scale distortions caused by camera tilt or topographic relief. Most photomaps include some cartographic enhancement to help the user—perhaps only marginal information, or an overprinted line drawing, or place names. 25

Image distortions on photographs caused by camera tilt in an aircraft can be removed by a simple rectification process. Distortions caused by relief (the varying heights of ground features) can be removed by an orthophotoscope or 26 other differential-rectification system which produces orthophotographs, correct in scale and relative position.

QUESTIONS FOR DISCUSSION

1. What is the author's principle of classification? Is it the most appropriate one for the reader's purpose? Is this principle followed consistently?
2. The author divides maps into six major classes. Are these classes equal and complete for the reader's purpose? Do any of these classes overlap?
3. How has the author established subclasses that do not overlap and that will not be mistaken for major classes?
4. Are the maps the author chose to include effective and relevant for the reader? Should a representation of a digital map and a photomap also have been included? Why, or why not?
5. The author has a strong sense of audience—both in terms of those people who will use a particular map and those for whom this piece is written. Comment on the ways in which the author has adapted technical procedures and terminology for nonspecialists.

SUGGESTIONS FOR WRITING

1. Other types of audiences could use the information given in this model. Write a paper (400–500 words) in which maps are classified in a way useful for one of the following audiences:
   a. geologists
   b. urban planners
   c. real estate agents
   d. pilots
   e. hikers
   f. firefighters
   g. architects
2. Using the organizational principle by which maps are classified in this model, write a paper on one of the following topics:
   a. bicycles
   b. dams
   c. insurance
   d. windows
   e. energy storage techniques
   f. food groups
   g. types of insulation
   h. catheters
   i. telephones
   j. roads
   k. bandages
   l. recreational areas

Written by researchers in psychology and linguistics, the following brief article classifying speech disorders appeared in a specialized reference work, the McGraw-Hill Encyclopedia of Science and Technology. *According to the Preface to the* Encyclopedia, *articles in it are written "in clear, readable language that is useful to students and nonprofessionals as well as to professionals. The level of writing is, of course, consistent with the degree of complexity inherent in each subject, but is geared to nonspecialists." In less than two encyclopedia pages the authors group and analyze a large number of widely differing speech problems according to principles clearly identified in the first paragraph. Cross references to other related articles in the* Encyclopedia *further help the reader interested in the topic.*

Speech Disorders

Ronald S. Tikofsky
William S-Y Wang

Speech disorders may be classified according to their causes or symptoms. The major causes are organic, imitative environmental, and psychogenic. Organic disorders may result from disease, impairment, or absence of the organs of speech. Imitative disorders occur when the child imitates defective speech. A speech disorder has a psychogenic origin when there is a psychological basis for its presence. A classification includes disorders of articulation, rhythm, voice, and symbolization. 1

Articulation. Disorders of articulation may be so severe that the resultant speech is unintelligible. Specific sounds or groups of sounds may be omitted, added, substituted, or distorted. The following are some examples of this type of disorder: Lalling, involving misarticulated r, l, t, and d sounds, may be caused by poor control of the tongue tip. In lisping, misarticulated sibilant sounds, particularly s and z, are often substituted by the th sound. Delayed speech, involving the absence of many consonants and poor intelligibility, is often caused by slow physical or psychological maturation. Dysarthria, generalized sound substitutions and distortions, is caused by lesions in the peripheral or central nervous system. 2

Rhythm. Disorders of rhythm are characterized by disruptions of the normal rate of speech. Two common disorders of rhythm are stuttering or stammering and cluttering. Stuttering usually begins between 3 and 4 years of age, and occurs most frequently in males. Many theories have been advanced to explain the cause of the disorder but none have been completely accepted. Primary stuttering is the repetition of words, phrases, syllables, or the initial sounds of words, without apparent awareness by the speaker. In secondary stuttering the repetitions become prolonged fixations of the speech musculature, accompa- 3 nied by physical and psychological tension, and often contortions of the face and body. In cluttering, words are slurred and syllables are omitted because of improper phrasing and excessive speed of utterance.

The following examples involve lesion of the nervous system. The cerebral 4
palsied often show disorders of rhythm. One type is spastic speech. This is an inability to make smooth transitions from sound to sound and insufficient breath control to produce polysyllabic words or phrases. Another is athetotic speech, which involves a general jerkiness in speech production which interferes with the normal rate of speech. Similar rhythm disorders accompany Parkinson's disease, multiple sclerosis, and cerebellar tumors. *See* NERVOUS SYSTEM DISORDERS.

Voice. Voice disorders are usually described as defects of pitch, loudness, 5
and voice quality. Improper use of the voice may cause an injury to the vocal folds and intensify an existing voice disorder. Some examples of these disorders are presented below.

Disorders of pitch are characterized as too high, too low, monotonous, 6
and repeated pitch patterns. A high-pitched voice is most often caused by psychological tension. The muscles of the larynx are contracted so that the pitch is raised beyond its normal range. It may be caused also by a small larynx where the vocal folds are short and thin or by the inability to perceive changes in pitch. Monotonous voices and repeated pitch patterns are usually also the result of a failure to perceive the pitch variations.

Voice quality disorders are usually described as hoarseness, nasality, dena- 7
sality, and similar conditions. Hoarseness may be caused by pharyngeal or laryngeal pathologies. Cleft palate speech is a striking example of nasality. It is the result of a failure of the bilateral structures of the palate to unite during fetal life. Nasal speech may also follow a paralysis of the palatal muscles (Fig. 1).

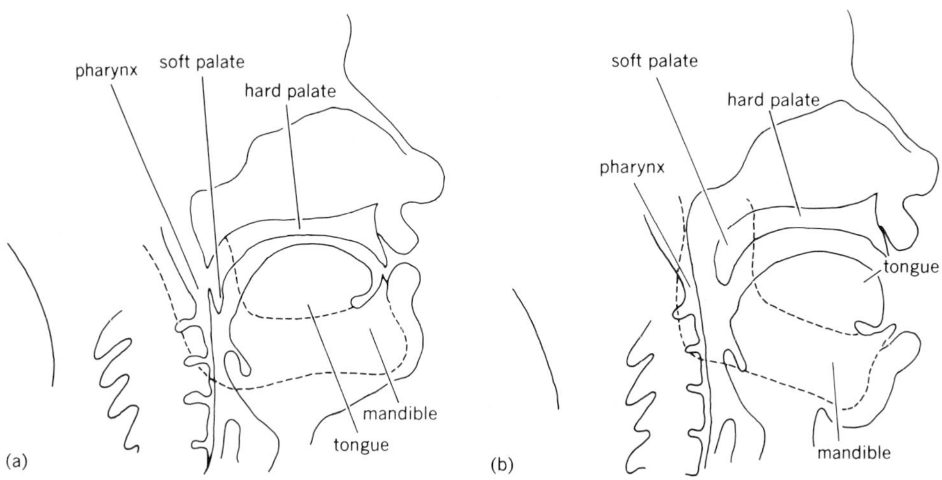

Figure 1. Schematic diagrams showing (a) normal nasopharyngeal closure and (b) failure to achieve closure because of a palatal cleft, paralysis, or injury. (Based on x-rays from the collection of H. H. Bloomer).

The nasal speech results because the palate is unable to aid in achieving nasopharyngeal closure. *See* PALATE.

Aphonia, the absence of voice, does not fall into the above categories. It may result from paralysis, inflammation of the vocal folds, or surgical removal of the larynx. Hysterical aphonia is the result of a marked emotional disturbance. *See* PHOBIC REACTION. 8

Symbolization. Symbolization disorders involve an impairment of language formulation and expression. They may occur without a concomitant impairment of speech production. Common types of these disorders are aphasia and delayed speech. 9

Aphasia is the inability to use or comprehend the symbolic value of language as a result of a brain lesion. The specific types of aphasia usually fall into one of four groups: expressive, receptive, amnesic, or a combination of these three. Expressive aphasia is the inability to express ideas in speech or writing when there is no significant muscular impairment. Receptive aphasia is the inability to comprehend spoken or written symbols when there is no significant impairment of the sensory organs. Amnesic aphasia is the inability to evoke the appropriate names for objects, conditions, or relations. In adults the disorder is usually preceded by a cerebral vascular accident, brain tumor, or injury to the brain. In children these disorders are usually the result of a failure of the development of the brain or of brain damage incurred before, during, or after birth. Many neurologists have postulated centers in the brain which are said to be related to various types of aphasic disorders. Asphasic arrest of speech has been demonstrated by electrical stimulation applied to certain areas of the brain (Fig. 2). 10

Delayed speech may also be considered a symbolization disorder. The symptoms range from complete absence of vocalization to vocalizations which have no communicative value. Speech is considered delayed when it fails to develop by the second year and is judged to be retarded if there are significant deviations from the norm for the age and mental abilities of the child. The causes of delayed speech are varied. Some common causes are deafness, impaired hearing, severe illness during childhood, or emotional disturbances. 11

The patterns of speech development and disorders show great variability among individuals. Research in these areas requires the integrated knowledge of experts in the many fields dealing with language and speech. *See* SPEECH. 12

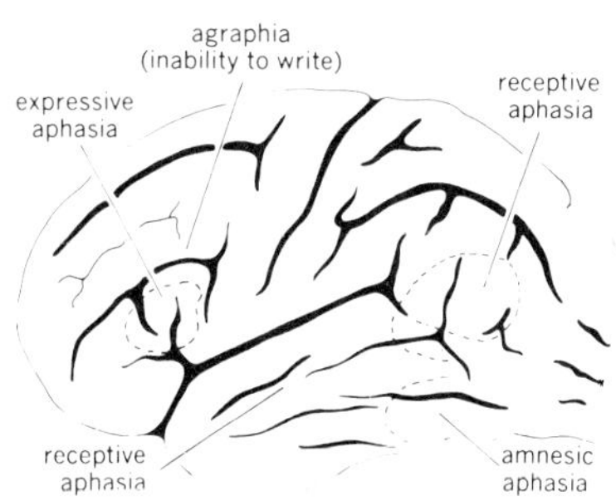

Figure 2. Areas of the cerebral cortex that, when damaged, supposedly result in aphasia. The areas within the dotted lines are areas that, when stimulated electrically, result in aphasic arrest.

QUESTIONS FOR DISCUSSION

1. In this model we see that "speech disorders may be classified according to their causes or symptoms." Do the authors present an accurate yet consistent classification using two principles rather than just one? As part of your answer, describe how both principles are at work in each category.
2. What principle of classification allows the authors to begin with *Articulation* and conclude with *Symbolization?*
3. How do the authors introduce and incorporate subclasses within each of the four main classes of speech disorders?
4. Which parts of this article would be of most interest to specialists? to nonspecialists? How do the authors' style and the two visuals meet the needs of both groups? Refer specifically to sentence length and word choice in your answer.
5. About stuttering, the authors say, "Many theories have been advanced to explain the cause of this disorder but none have been completely accepted." Given the aims of the *Encyclopedia,* would additional information on these theories have improved the authors' classifications for readers?
6. In paragraph seven, readers are told that aphonia does not fall into the previous categories. If it does not, then why do the authors mention it?

SUGGESTIONS FOR WRITING

1. The last paragraph points to the need for "the integrated knowledge of experts in many fields dealing with language and speech." How might speech therapists or psychologists classify speech disorders? Write a classification of the same disorders discussed in this model according to a principle other than causes or symptoms for one of these audiences.
2. Language is a very rich and diverse topic. Select one topic in language (the use of campus slang, pronunciation of a group of related words, jargon in different company sections or units, differences in speech among three members of different age groups) and write a paper on how these differences might be classified.

The following classification of radio frequencies forms a section of From Sundials to Atomic Clocks: Understanding Time and Frequency, *a book issued by the National Bureau of Standards (NBS) for a general audience. The book attempts to offer "a reasonably complete view of the study of time in a limited number of pages." According to the NBS this "descriptive book" is addressed to "the general reader." Elaborating on both the audience and aim of this publication, an NBS official observed in the Foreword that "we often have occasion to explain the NBS time program to interested people who do not have a technical background, and such a book would be an efficient and . . . interesting means of informing them." The authors classify the various frequency bands primarily to show their usefulness in transmitting time signals and to explain the different effects a reader can expect to see in these frequency bands. But although the authors focus on the effects of time transmissions in each category, a much more basic principle of classification underlies their work: the numerical intervals separating one type of frequency from another.*

Radio Frequencies

. . . We've already realized that information about the "correct time" is useless 1
unless it's instantly available. But what, exactly, do we mean by "instantly," and how is it made available to everyone who wants to know what time it is? What can happen to this information on its way to the user, and what can be done to avoid some of the bad things that can and do happen?

CHOOSING A RADIO FREQUENCY

The *frequency* of a radio signal primarily determines its *path*. The signal may 2
bounce back and forth between the ionosphere and the surface of the earth, or creep along the curved surface of the earth, or travel in a straight line—depending on the frequency of the broadcast. We shall discuss the characteristics of different frequencies, beginning with the very low frequencies and working our way up to the higher frequencies.

Very Low Frequencies (VLF)—3–30 kHz

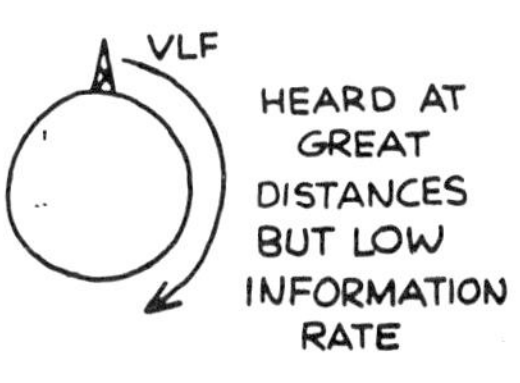

The big advantage of VLF signals is that one relatively low-powered 3
transmitter could provide worldwide coverage. A number of years ago, a VLF signal broadcast in the mountains near Boulder, Colorado, was detected in Australia, even though the broadcast signal strength was less than 100 watts. The VLF signal travels great distances because it bounces back and forth between the earth's sur-

face and the lowest layer of the ionosphere, with very little of its energy being absorbed at each reflection.

Another good thing about VLF signals is that they are not strongly affected 4
by irregularities in the ionosphere, which is not true in the case of shortwave transmissions. This is so because the size of the irregularities in the ionosphere "mirror" is generally small compared to the length of the VLF radio waves. For example, at 20 kHz the radio frequency wavelength is 15 kilometers. The effect is somewhat the same as the unperturbed motion of a large ocean liner through slightly choppy seas.

But VLF also has serious limitations. One of the big problems is that VLF 5
signals cannot carry very much information because the signal frequency is so low. We cannot, for example, broadcast a 100 kHz tone over a broadcast signal operating at 20 kHz. It would be like trying to get mail delivery ten times a day when the mailman comes only once a day. More practically, it means that time information must be broadcast at a very slow rate, and any schemes involving audio frequencies, such as voice announcement of the time, are not practical.

We mentioned that VLF signals are not particularly affected by irregularities 6
in the ionosphere, so the path delay is relatively stable. Another important fact is that the ionospheric reflection height is about the same from one day to the next at the same time of day. Unfortunately though, calculating the path delay at VLF is a complicated and tedious procedure.

One last curious thing about VLF is that the receiver is better off if he 7
listens to signals from a distant station than from nearby stations. The reason is that near the station he gets two signals—one that is reflected from the ionosphere (sky wave) and another that is propagated along the ground. And what he receives is the sum of these two signals. This sum varies in a complicated way as a function of time and distance from the transmitter. So the listener wants to be so far from the transmitter that for all practical purposes the ground wave has died out, and he doesn't have to deal with this complicated interference pattern.

Low Frequencies (LF)—30–300 kHz

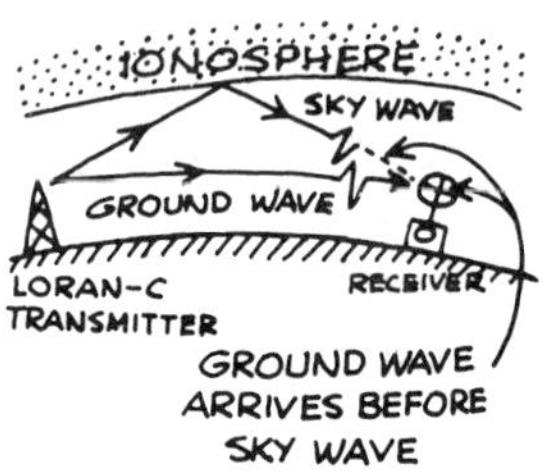

In many respects LF signals have properties similar to VLF. Of 8
course, the fact that the carrier frequencies are higher means that the information-carrying rate of the signal is potentially higher also.

These higher carrier frequencies have allowed the develop- 9
ment of an interesting trick to improve the path stability of signals. The scheme was developed for the Loran-C navigation system at 100 kHz, which is also used extensively to obtain time information. The trick is to send a pulsed signal instead of a continuous signal. A particular burst of signal will reach the observer by two different paths. He will first see

the ground wave signal that travels along the surface of the earth. And a little later he will see the same burst of signal arriving via reflection from the ionosphere.

At 100 kHz, the ground wave arrives about 30 microseconds ahead of the 10
ionospheric wave, and this is usually enough time to measure the ground wave, uncontaminated by the sky wave. The ground wave is quite stable in path delay, and the path-delay prediction is considerably less complicated than when one is working with the sky wave.

Beyond about 1000 kilometers, however, the ground wave becomes so weak 11
that the sky wave predominates; and at that point we are pretty much back to the kind of problems we had with VLF signals.

Medium Frequencies (MF)—300 kHz–3 MHz

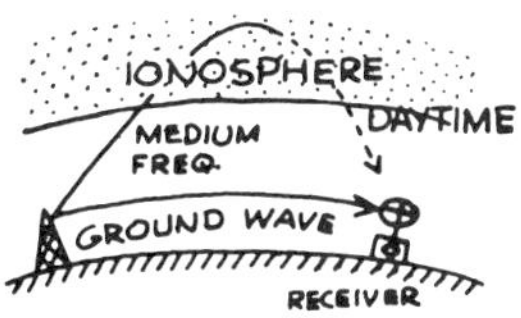

We are most familiar with the medium-frequency band because 12
it contains the AM broadcast stations of this country. During the day the ionospheric or sky wave is heavily absorbed, as it is not reflected back to earth; so for the most part, during the daytime we receive only the ground wave. At night, however, there is no appreciable absorption of the signal, and these signals can be heard at great distances.

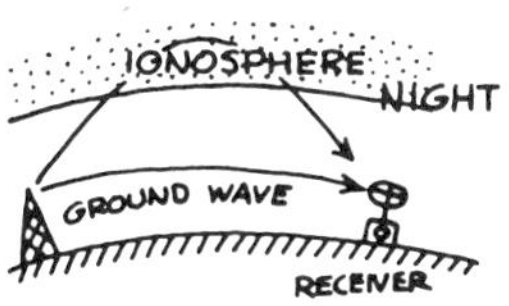

One of the standard time and frequency signals is at 2.5 MHz. 13
During the daytime, when the ground wave is available, the Japanese report obtaining 30-microsecond timing accuracy. At night, when one is receiving the sky wave, a few milliseconds is about the limit. It is fair to say, however, that this band has not received a great deal of attention for time dissemination, and there may be future promise here.

High Frequencies (HF)—3.0–30 MHz

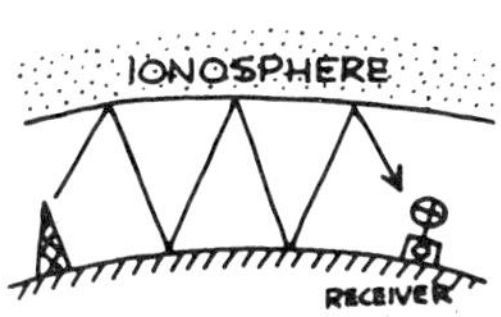

MANY HOPS MAKES DELAY DIFFICULT TO PREDICT. SIGNAL FADES.

The HF band is the one we usually think of when we speak of 14
the shortwave band. Signals in this region are generally not heavily absorbed in reflection from the ionosphere. Absorption becomes even less severe as we move toward the upper end of this band. Thus the signal may be heard at great distances from a transmitter, but it may arrive after many reflections, so accurate delay prediction is difficult.

Another difficulty is that in contrast to VLF waves, HF wave 15
lengths can be of the same order, or smaller, than irregularities in the ionosphere. And since these irregularities are constantly changing their shape and moving around, the signal strength at a particular point will fade in and out in amplitude.

Because of the fading and the continuous change in path delay, accuracy in timing is again restricted to about 1 millisecond, unless one is near enough to the transmitter to receive the ground wave.

Most of the world's well-known standard time and frequency broadcasts 16
are in this band.

Very High Frequencies (VHF)—30–300 MHz

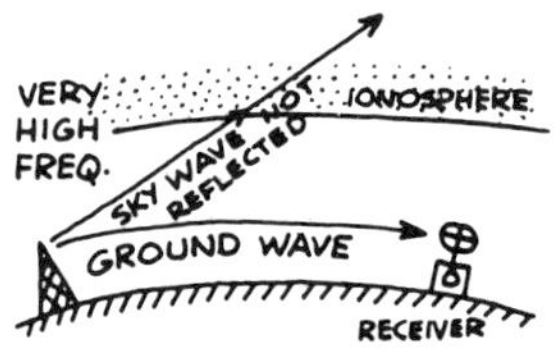

LIMITED TO LINE OF SIGHT.
HAS HIGH ACCURACY.
CAN SEND SHARP PULSES.

From a propagation point of view, one of the most important things 17
that happen in the VHF band is that the signals are often not reflected back to the surface of the earth, but penetrate through the ionosphere and propagate to outer space. This means that we can receive only those stations that are in line of sight, and explains why we do not normally receive distant TV stations, TV signals being in this band. It also means that many different signals can be put on the same channel, and there is little chance of interference as long as the stations are separated by 300 kilometers or more.

From a timing point of view, however, this is bad; for if we want to provide 18
worldwide—or even fairly broad—coverage, many stations are required, and they must all be synchronized. On the other hand, there are advantages to having no ionospheric signals because this means that we can receive a signal uncontaminated by sky wave. We can also expect that once we know the delay for a particular path, it will remain relatively stable from day to day.

A third advantage is that because carrier frequencies are so high, we can 19
send very sharp rise-time pulses, and thus can measure the arrival time of the signals very precisely. Because of the sharp rise time of signals and the path stability, timing accuracies in this region are very good. Microsecond timing is relatively easy; and if some care is taken, even 0.1 microsecond can be achieved.

At the time of this writing the master clock at NBS in Boulder, Colorado, 20
is used as a reference clock for the standard time and frequency broadcasts from WWV near Fort Collins, some 80 kilometers away. A TV signal is used to maintain the link between Boulder and Fort Collins with an accuracy of a small fraction of a microsecond.

Frequencies Above 300 MHz

The main characteristic of these frequencies is that like VHF frequencies the 21
signals penetrate into outer space, so systems are limited to line of sight. There may be problems caused by small irregularities in the path—or "diffraction effects," as they are commonly called—similar to such effects at optical wavelengths. Nevertheless, if a straight shot to the transmitter is available, we can expect good results.

Above 1000 MHz weather may produce problems; this is especially significant in broadcasting time from a satellite, where we wish to minimize both ionospheric and lower atmosphere effects. 22

QUESTIONS FOR DISCUSSION

1. How does the use of classification help the authors to achieve their purpose? Is the use of classification the most appropriate way to present information on radio frequencies?
2. Discuss the effectiveness of the order, exclusiveness, and completeness of the six categories in this model. Cite examples.
3. Do the authors provide subclasses? If so, provide examples; if not, provide an explanation.
4. In their first paragraph, the authors say it is necessary to study the "bad things" that can happen to time information. Do they emphasize the bad more than the good in their classification? What part do the pros and cons play in their classification system?
5. The authors refer to objects and procedures that are more familiar to readers than are radio frequencies (ocean liners, mail delivery, reception of TV signals). Explain how these types of explanations and others help the authors to adapt technical information about time signals for their audience and make it interesting at the same time—the aims of the book in which this model appeared.

SUGGESTIONS FOR WRITING

1. Radio frequencies could be classified in a number of different ways depending upon the needs of the audience. Using the information in this model, write a classification of radio frequencies for one of the following audiences: (a) disc jockeys, (b) ham radio operators, (c) home cable subscribers, (d) junior high school science students.
2. Write a classification paper on one of the following subjects using the same fundamental principle—numerical intervals or sequences—found in the radio frequency model:
   a. storms
   b. growing cycles
   c. sunscreen lotions
   d. swimming or track and field events
   e. airline rates
   f. noise levels in a factory
   g. time zones in the U.S. and/or Canada
   h. layers of the earth's atmosphere
   i. microcomputer memories
   j. local television or radio stations

PARTITION

The following discussion of a bee colony comes from a pamphlet entitled The Honey Bee *published by the Smithsonian Institution, which is famous for, among other things, its collection and preservation of materials in the natural sciences. The purpose of the pamphlet is to give the nonspecialist an informative yet entertaining account of the industriousness and sophistication of this species. The author partitions the colony according to the functions, "the division of labor," performed by each of the three types of residents: the queen, the drones, and the workers.*

The Colony

The honey bee colony is a highly developed community operating much like 1
a superorganism. It has many parts, each with its specialized function. This well-organized life in the beehive is considered to be the highest development of the division of labor found in any animal society lower than man's.

QUEEN

There is only one queen in each colony, although an old queen and her successor 2
may coexist for a brief time (see Figure 1). The sole purpose of the queen, mother of all the colony, is to lay eggs. At the height of her activity, a young queen may lay as many as 2,000 eggs a day, a total sometimes exceeding her own body weight. A queen may continue at this high rate of production for three or four years, when a new queen succeeds her.

When a queen begins to age, workers construct special queen cells for 3
her replacement. These cells are larger than normal and hang vertically from the brood comb. The queen lays fertilized eggs in these cells, or the workers may deposit them there. When the larvae hatch from these eggs, they are fed continuously a special food called "royal jelly" which is secreted by workers. The quantity and quality of this special food develop in the larvae the special characteristics of a queen (sexually complete). Ordinary food (pollen and honey) causes other female larvae to become sexually undeveloped worker bees. In a way, workers are simply undernourished queens.

The future queen and worker bees hatch from their eggs in about three 4
days and remain in the larval stage nine days. During this period both workers and queens are fed over 10,000 times or an average of 1,300 times a day. The

queen, however, receives more and better quality food. At the end of the larval stage, the cells are capped, the larvae spin cocoons and enter the pupal stage. For the well-fed queen, the pupal stage lasts only four days; the underfed future workers take about nine days to develop. The first princess to emerge from her cell makes a piercing noise of challenge to all other contenders in their cells and kills them by a sting.

Between five and ten days after leaving the cell, the virgin queen leaves 5
the hive one afternoon in search of drones on her nuptial flight. Some distance from the hive she will mate, on the wing, with as many as ten drones. The mating act, as is frequently the case in the order *Hymenoptera,* kills the successful male. The queen then returns to the hive and within two days begins to lay eggs. She stores the drone spermatozoa in her special organ, the spermatheca. The male spermatozoa are then used during the queen's life to fertilize the eggs. The queen can either lay fertilized eggs which become workers or unfertilized eggs which become drones. An old queen is displaced within a week after the new queen begins laying. The occasional presence of an old queen in the hive with her wings chewed off suggests that the workers sometimes take a hand in her retirement.

DRONE

The colony playboys are male drones, hatched from unfertilized eggs, whose 6
functions are to mate with new queens and to generate heat in the hive. Depending on the colony, the number of drones varies from a few hundred to a few thousand. Under natural conditions not more than 10 percent of the colony

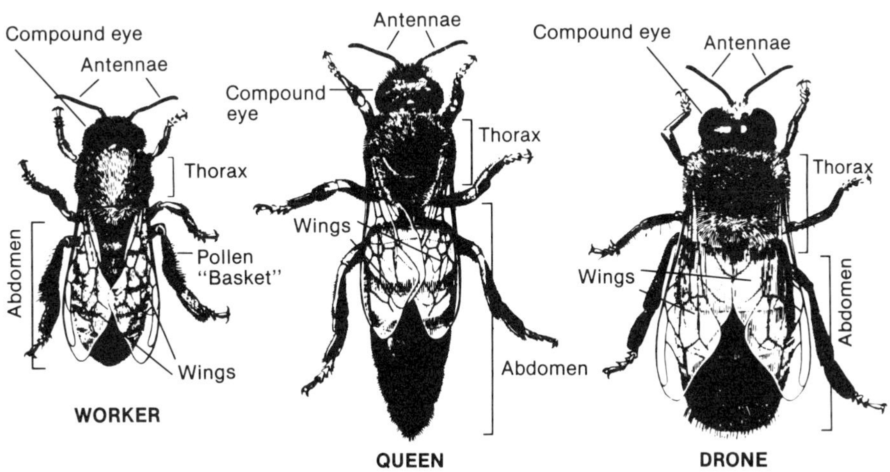

Figure 1. The honey bee. Relative sizes of the worker, queen, and drone. Scale: about 2½ times life size.

will consist of drones, which are easily distinguished in the colony, for they are heavier, fatter, and longer than the workers, but not quite as long as the queen (see Figure 1). Their eyes are huge, and they have large, strong wings. These last assets help the drone find and catch the young queen as he reaches a top speed of perhaps fifteen miles per hour. But the drone is a virtually helpless guest; he has no stinger for self-protection, and he must be fed by the workers. Drones are produced in the spring after supplies of nectar and pollen become available in the fields. As long as the nectar flow continues at a high level, the workers seem to like having some drones for colony "morale." When the nectar flow stops in the fall, however, the drones' days of leisure come to an end. The workers stop feeding them and bite off their wings. Weakened by hunger, the drones are pushed out of the hive to die in the autumn nights.

WORKER

The female workers make up most of the colony (90–95 percent). They perform all hive tasks except egg laying. In an emergency (no queen), a few workers may even produce male eggs. However, the colony would soon die in this situation. 7

Workers have sophisticated organs that equip them to perform the many and often difficult tasks of the colony (see figure 1). Their heads are more triangular than the queen's, and the compound eyes are more widely spaced for navigation. Their tongues are enclosed in a tube designed for sucking honey that will be stored in a honey stomach. Their many body hairs accumulate pollen that is stored in pollen baskets on the workers' hind legs. Their smooth and rounded jaws clean the hive and make wax cells. The jaws are also sharp for the colony's defense. Four sets of glands convert carbohydrates into beeswax. Finally, workers have a poison system (formic acid and alkalai) and a one-shot stinger, which is torn from their body when used, causing them to die. In contrast, the queen's needlelike stinger may be used repeatedly. 8

The first three weeks of the worker's adult life in the hive are spent attending to the queen, acting as a "nurse" to the larvae, cleaning the cells, controlling the temperature and ventillating the hive, plus guarding it against unwanted intruders. Following its apprenticeship, the young worker joins the field force in foraging for food. The worker engages in this activity with such vigor that it is exhausted within three to six weeks and, having expended its usefulness to the colony, is left outside the hive to die. 9

QUESTIONS FOR DISCUSSION

1. What functions does the three-sentence introduction serve for the reader? What would a reader miss if this introduction were deleted and the selection began with the information on the queen?

2. Why does the author list the bees in the order of the queen first, then the drone, and finally the worker? What would be gained or lost by changing this order?
3. The author uses lively, figurative language and comparisons throughout this selection, e.g., the colony playboys are male drones (paragraph 6). Find other examples of the author's lively style and explain how they help the writer to adapt technical information to the intended audience.
4. Discuss the effectiveness of the visual and its caption. How closely is the visual information correlated with the description in the text?
5. The author provides levels of partition—first the colony is divided into its three kinds of residents, and then each resident is described and analyzed. Explain how the author uses the techniques of physical description discussed in Chapter 2 to partition the colony.

SUGGESTION FOR WRITING

1. Following the same principle of organization used in this model, write a paper for a consumer audience on one of the following topics:
   a. a factory
   b. a military unit
   c. a newspaper
   d. a standalone computer
   e. a radio or television station
   f. the human heart or other body organ
   g. a laboratory
   h. an ant hill
   i. a team you played on
   j. any organization to which you belong
2. Write a partition paper on any topic in your major for a specialized audience.

The model below is excerpted from an article in FAA Aviation News *published by the Office of Flight Standards Services of the Federal Aviation Administration "in the interest of aviation safety and to acquaint readers with the policies and programs of the Agency." This model was the second in a three-part series of articles on hot air balloons. The first introduced the topic and the third article discussed "balloon accidents and operating procedures." The author partitions a balloon so that "aeronauts" will understand the components of a balloon and will be able to maintain them. Thus the intended readers are primarily operators.*

Keeping Your Balloon in Shape

Federal Aviation Administration

In the third century B.C. the principle which makes ballooning possible was discovered by the Greek mathematician, Archimedes: *when a gas less dense than air is enclosed in a container the difference between the density of the gas and the air it displaces causes the container to rise.* The modern hot air balloon operates on the same principle, although it uses heated air instead of the lighter gas for a very good reason: economics. 1

To fill a three-passenger balloon with helium, for example, today in the United States would cost about $2500. Hydrogen for the same balloon (dangerously inflammable) would cost about $800. The bill for propane to heat a comparable hot air balloon flight might come to about $10 or $15. This rapid economic descent, with vastly increased safety, is largely responsible for the return to popularity of the ancient sport of ballooning: Although it has no lateral controls, the modern balloon is recognized and certificated as an aircraft and its operator is required to understand its components as well as be responsible for its safe maintenance. Refer to Fig. 1 as you read the following. 2

Starting at the bottom, the first component is the gondola, or basket (so-called usually only if it is made of the traditional lightweight wicker). Aeronauts are such traditionalists that some will trim their aluminum gondolas with wicker for the sake of appearances. Wicker, of course, is a remarkably durable substance for its weight, and has many advantages for ballooning comfort, but it is subject to "embrittling" with age, as well as rot, and it can be damaged by being dragged over rough or stony terrain. Baskets should be checked regularly for signs of damage or loss of resilience, particularly along the floor and at the attachment points for the envelope cables. Aluminum gondolas with plastic liners require less upkeep, but the welds should be looked at after hard use, especially on the bottom. 3

The propane fuel tanks for the burner are usually stored on the floor of the gondola. These could be dislodged by a hard landing, so they need to be looked at regularly after a flight to make certain they are secure and no leakage has developed at any of the fittings, and that the fuel quantity gauge is still 4

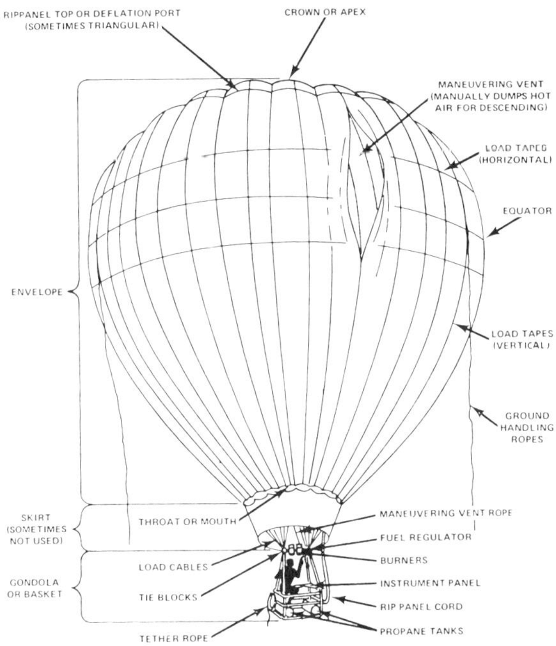

Figure 1. Parts of a typical modern balloon.

operative. The latter is a particularly important source of information to the pilot, and occasionally gets stepped on or kicked out of commission by an excited passenger in the crowded "cockpit."

Additionally the gondola will also contain a drag rope, for landing; a tether 5
rope for holding down the balloon during launch until enough heat has been generated for liftoff; protective headgear, which many flight manuals designate as required equipment because of the possibility of a jarring landing; heavy leather or asbestos gloves for working around hot burners; a sparker, for relighting the burners if they should flame out in flight, plus a cigarette lighter or matches, as backup for the sparker; a portable radio; and some current sectional charts. The latter are useful, not merely for navigation but for identifying radio frequencies along the way. (For flights above 10,000 feet, portable oxygen should be available for all on board.)

The absence of confined storage space in most gondolas makes it essential 6
that necessary equipment be stowed carefully in place after each flight, and checked again before departure.

An abbreviated form of instrument panel is usually located at one upper 7
corner of the gondola. The only required instruments for hot air balloons are a compass, altimeter, rate of climb indicator, fuel quantity gauge and envelope temperature indicator.

Rate of climb or descent is deceptive in a balloon. Because of the gentleness 8
of motion in flight, many aeronauts, especially at altitudes over 1,000 feet, experi-

ence little sense of upward or downward motion until a very high rate of vertical movement has been attained, without the help of an instrument.

Another key instrument is the *pyrometer,* which indicates the temperature at the crown of the balloon envelope (the highest point). Without a pyrometer it is possible to build up more heat in the balloon than it can safely withstand. 9

The old pilot's standby, the magnetic compass, is more useful than some balloonists imagine. The balloon pilot cannot steer by the compass, but it gives him some idea of the direction in which he is traveling, which is not readily apparent under many conditions of weather. 10

The heart of the balloon is the burner, which shoots a flame up toward the open throat of the balloon. It is usually rigged on a brace over the pilot's head, and controlled simply by means of a hand valve. The fuel is liquefied propane, which passes first through preheat coils, where it is vaporized. When the rapidly expanding gas is ignited it roars up toward the envelope, producing enough heat to run a hundred normal household furnaces for an hour. 11

Many balloons have two or more separately functioning burners. Each has a metering valve which gives a fine, low flow of heat for "cruising" at a constant altitude; in addition, each has a blast valve which gives maximum fuel flow only, and is used during flight in short bursts (accompanied by savage roars). 12

Almost all balloons in use in the United States today use airborne heaters fueled with propane, commonly known as liquefied petroleum gas (LPG). At normal atmospheric pressure and temperature, propane is a colorless, *odorless* gas. Propane vapor can be detected (smelled) only by the odor which is artificially added to it by the refiner. Since the type of odor added varies from one refiner to another, it is not possible to identify propane universally in this manner. . . . 13

Propane flows from the tanks to the burners through a combination of flexible hoses, metal lines and metal fittings. Flexible rubber lines are usually used throughout, except in the immediate vicinity of the burners, where the heat might damage the rubber. The metal tubing and fittings are heat resistant but vulnerable to physical damage; careful precautions must be taken to protect them. 14

In the balloon, you have a series of metal fittings connected together into a fairly long assembly with one end rigidly connected to a tank, burner unit, or relatively unmovable part of the rig. If force is accidentally applied to the free end of such a conglomeration (as from strain on a cable, for example, on landing), the fitting closest to the rigid attachment point is likely to break. 15

Furthermore, "quick release" fittings which are sometimes used in hot air balloon fuel systems to facilitate connecting and disconnecting fuel lines, have a history of unreliability. Most of the fittings were not designed to carry liquid or vapor propane, or to operate reliably at the extremely low temperatures that can result from propane vaporizing in the line at the fitting. Such fittings can be easily jarred loose if accidentally brushed by the pilot or a passenger during the flight, or by the jostling that often takes place during a landing. A very serious leak can develop if one of these fittings fails to seal. 16

The LPG standards for propane tanks and fittings are quite high—they 17
must be able to withstand hydrostatic testing up to 1,750 psi. The standard operating pressure for pressure release valves is 375 psi. Fuel systems on sport hot air balloons rarely, if ever, experience pressure greater than 250 psi during flight.

The modern hot air balloon burner is designed to perform three basic 18
functions:

1. Vaporize the liquid propane. 19
2. Mix the propane vapor with air to form a combustible mixture. 20
3. Burn the resulting mixture to form a directional flow of very hot gases 21
which enter the open throat of the envelope.

The burner consists of a cylindrical metal "can," open at both ends, the 22
upper section of which is encircled by a coil of fuel line tubing. The liquid propane is preheated in the coil to vaporization; the vapor emerges from a nozzle at the base of the "can." The nozzle separates the vapor into three or more streams, directed upward, which mix turbulently with the ambient air and are ignited by the pilot light.

The result is a tremendous output of heat—the current trend is toward 23
burners that have a maximum output in the range of eight to ten million BTUs per hour. Earlier hot air sport balloons had no more than half this output. The more powerful burners result in somewhat safer balloons, since the greater rate of heat output available makes the balloon much more responsive and maneuverable.

Fuel management problems occur in hot air balloons; as in other types 24
of aircraft they are usually avoidable if good judgment and procedures are used. It is good practice, for example, to burn off both tanks alternately in the course of a flight. In this manner if one tank proves inoperable, the discovery will come at a time when the course of the flight can be safely altered. It is preferable to have only one tank valve open at a time; this will prevent total exhaustion of fuel, in the event the pilot, enraptured by the joy of floating in the sky, forgets about the passage of time.

Veteran balloonists recommend that when preparing to make a tank switch 25
you add a good burst of heat to the envelope. This assures you that, if you become preoccupied with some problem in the switching, you will not be descending inadvertently into an unfriendly terrain.

Balloon pilots sometimes do get forgetful about which tank valves are open, 26
and many of them are not clearly marked. A good procedure for avoiding confusion is to move the handle back half a turn or so after opening it. This allows the pilot to determine quickly if a tank valve is open by simply "twiddling" it. If it turns in both directions, it is open; if it turns in only one direction, it is closed.

Finally, we have the component known as the envelope, the hot air con- 27
tainer which is not exactly a balloon in that it is open at the throat at all times. ("Balloon" is a term which refers to the complete apparatus.) The envelope is usually made of high-strength, heat resistant, rip-stop nylon, and it is fastened

to the gondola by means of heat resistant steel cables. The envelope is coated internally with a plastic material which helps contain heat. One manufacturer recently tested the heat resistance of his nylon envelope by heating the air in it to 250 degrees F. for a period of 250 hours without any degradation of the fabric.

The individual strips of the envelope fabric are called "gores." Sewn into 28
the seams of the gores are horizontal and vertical "load tapes" which help the balloon maintain its shape when inflated. Load tapes are usually made of heavy nylon webbing, similar to the material used in seat belts. Some balloons also have a nylon "skirt" or "collar" attached to the open throat of the envelope, to shield the burner from wind gusts or shear.

The envelope has two opening ports—one at the crown, or very top, and 29
one high on the side. The crown contains a deflation port, a large circular piece of fabric which can be separated from the envelope by means of a ripcord. It is used on landing to achieve rapid deflation and prevent the balloon from buffeting on the ground. The side port is a smaller vent, also activated by a ripcord, which allows the pilot to dump hot air and decrease altitude in flight. When the cord is released the vent is closed immediately by pressure from inside the envelope. Both of these venting sections should be checked regularly by the pilot, as they are vital to his control of vertical movement of the aircraft.

QUESTIONS FOR DISCUSSION

1. Why is this model on hot air balloons an example of partition rather than classification?
2. What order does the author follow in partitioning the balloon? How does the order used here differ from that employed for the model on the bee colony?
3. How much does the author assume the reader knows about hot air balloons? How does this assumption affect content and style? Cite textual examples in your response.
4. What part of the balloon receives the most attention? Given the reader's purpose and background, is this emphasis justified? Explain your answer by citing examples from the article.
5. The author uses a number of transitional phrases to help readers follow the partition of the balloon. Identify some of these transitions and explain their specific usefulness in the article.
6. How does the visual help to emphasize the relationship of the parts of the balloon? Are all the parts of the balloon displayed in this visual? Would another kind of visual have been more helpful? Explain.

SUGGESTIONS FOR WRITING

1. Using the partitioning of the hot air balloon as your model, write a partition paper on one of the topics below. Make sure that you identify a specific audience for your paper and clearly specify what its purpose might be in reading your paper.
   a. a fireplace
   b. a sailboat
   c. a desk lamp
   d. a swivel chair
   e. a grain elevator
   f. an office building
   g. a lighthouse
   h. a thermometer
   j. a food blender
   i. any piece of equipment you may or will use on the job
2. Using the same topic that you selected for the first writing suggestion, write a paper that uses classification as its organizational scheme. Once again, specify your audience and its purpose.

chapter **4**

Definition

Trouble can arise when a writer assumes that every reader will understand the meaning of a technical term. Too often nonspecialists, or even specialists in another field, are baffled when they see a technical word they do not know; at other times a reader may assume that a specialized term has a meaning different from the one the writer intends. Given the rapid rate at which technology changes, the use of definition is a necessity in technical writing. Definition plays a major role in keeping the lines of communication open between a writer and a reader. It makes communication possible by removing barriers, so that you and your reader can agree on the meaning of key words.

Specifically, definition clarifies the meaning of words and explains difficult concepts. In explaining and clarifying, definition and description often overlap. A reader needs to know what a concrete object looks like in order to understand what it means. An effective definition answers the following kinds of questions for readers:

1. What does this term mean?
2. Where is the term used?
3. How does the term apply to me and my work?
4. What special meanings does the term have?
5. How does the writer's use of the term differ from other previous uses of the term?
6. How does the term differ from similar terms (pardon, parole, commutation; hurricane, tornado, cyclone).
7. When was the term first used? How has it changed meaning over the years? Has its meaning expanded or narrowed?

Your audience and purpose will always affect your use of a definition. It is absolutely imperative to define a term whenever you think your reader will not know its meaning. In general, be careful always to define scientific or technical terms readers are not likely to know—e.g., *cultivars, immunoglobulins, pyrolysis, wrap around mortgage*—and common words that have specialized meanings in your writing—e.g., *crab* (moving hoist on a travelling crane), *stack* (linear list in computer science), *yellowcake* (material resulting from the first processing of uranium ore).

Definitions can be included in the text, in footnotes, or in a glossary. Sometimes they may even comprise an entire report.

This chapter will show you how to write the kinds of definitions that are useful in technical writing. These include:

- informal definitions
- formal definitions
- expanded definitions
- stipulative definitions

INFORMAL DEFINITIONS

An *informal definition* is a brief explanation of an unfamiliar term, usually a synonym or short phrase, included in the text of your report. An informal definition is useful when your audience does not need a great deal of information about the term itself but must grasp its meaning quickly. Such definitions are placed in parentheses or set off by commas or dashes immediately after the unfamiliar word.

You can define a word informally in a variety of ways. One way is to give a more commonly known synonym after the technical one:

> Effective March 16, the U.S. Department of Agriculture officials will require psittaciformes—parrots and related birds—shipped from California to be identified with leg bands.

> The petrous (rock-hard) quality of the material made it valuable for outside use.

You also can point out a defining characteristic that is relevant for readers:

> Lithium, the lightest metal, has many uses in bioengineering, especially in the development of pacemakers.

Finally, you can describe an object's appearance or identify its function:

> The V stem (the tube that connects the exchanger to the storage tank) needs to be lubricated at least once each week.

You will use informal definitions more often when writing to consumers than to experts. But whenever they are appropriate, informal definitions are a handy way to give the reader a small amount of information most conveniently.

FORMAL DEFINITIONS

A formal definition is more precise, complete, and logically structured than an informal one. Whereas an informal definition can be given in a word or phrase, a formal definition is always constructed as a sentence with three separate and necessary parts. The first part, called the *species,* is the term you will define. The second part is the *genus,* or the class in which that term belongs. The third part consists of the *differentiae,* or the distinguishing features that separate the term from all other members of the same class. Think of a formal definition as an equation in sentence form:

SPECIES (the term)	=	GENUS (the class)	+	DIFFERENTIAE (distinguishing features)
A Husky	is	a dog		weighing about 80 lbs. with a heavy, furry coat usually white or gray that was originally bred in Siberia to haul sleds.

To write a formal definition, first place the term into its correct class. A Husky is a dog. Then identify the distinguishing characteristics of the Husky that separate it from every other breed of dog. If the features you assign to the Husky can apply to other dogs, your definition is not restricted or accurate. To identify, as well as single out, one member of a class, consider such features as color, size, function, location, cost, and composition. In the example of the Husky, its size, its type and color of fur, its origins, and its function distinguish the Husky from all other kinds of dogs. After all, collies have long fur and can weigh as much as Huskies, but collies do not come from Siberia, nor are they used principally to pull sleds. In the following examples study the distinguishing features in each formal definition to see how the writer has identified and isolated each term.

SPECIES	=	GENUS	+	DIFFERENTIAE
Artspeak	is	a programming language		used in college courses for nonmath or science students which would enable them to do computer projects after learning a small amount of programming language syntax.[1]
Black lung	is	a respiratory ailment		that is caused by inhalation of coal dust and is a contributing

[1] *Encyclopedia of Computer Science and Engineering* (New York: Van Nostrand, 1983).

			cause in the deaths of many underground miners.
Resilience	is	a property of a material	that enables it to resume its original shape or position after being bent, stretched, or compressed.[2]

Tips for Writing Formal Definitions

The effectiveness of a formal definition depends upon its clarity and precision. Therefore, it is important to avoid making the following common errors:

1. *Avoid repeating key words.* Repeating in the differentiae the word you want to define gives readers a circular definition that leads nowhere.

> An attenuator is a device that attenuates the signal passing through it.

Find another verb which will show what the instrument does.

> An attenuator is a device that reduces the frequency of the signal passing through it.

There are times when you may have to repeat the less technical word in a phrase you are defining. For example, in defining "strobe light," you will most likely use the word "light"; your reader will obviously know what a light is.

2. *Avoid using language that is too technical for your reader.* Defining one technical word with another may baffle even an audience of experts. The oft-quoted definition of "network" from Dr. Johnson, the author of a famous eighteenth-century dictionary, illustrates the problem of using unnecessarily complicated language:

> A network is any thing reticulated or decussated at equal distances, with interstices between the intersections.

Contrast this with the more helpful definition of "network" from the *American Heritage Dictionary,* which uses more familiar terms:

> An openwork fabric or other structure in which rope, thread, wires, or other materials cross at regular intervals.

3. *Be careful the differentiae are not too broad.* Your distinguishing features must be sufficiently narrow so they cannot apply to any other object or process.

> Tennis is a game in which players use rackets to hit a ball back and forth on a court.

[2] *American Heritage Dictionary.*

Aside from omitting one of the most important parts of the game (the net), this definition might also include jai alai, racketball, or squash—all games in which players use rackets and battle on courts.

4. *Be careful the differentiae are not too narrow.* When your differentiae are too narrow, you may exclude members of the class you want to define.

> A chalkboard is a writing surface measuring 3′ by 12′ made out of black slate and found on the front wall of a classroom.

This definition excludes other functions, sizes, colors, compositions, and locations of chalkboards, e.g., a green hand-held chalkboard.

5. *Never start a definition with "is where," "is when," or "is a kind of."* Beginning with "is where" or "is when" violates the formal logic of the definition.

> A hurricane is when turbulent storms cause severe damage.

This sentence does not tell readers what a hurricane is.

> A compass is a kind of navigational instrument.

This definition does not separate it from other types of navigational instruments—radar, sextants, or sonar.

EXPANDED DEFINITIONS

Sometimes readers need more information about a term than a mere synonym or even a sentence alone can supply. In those cases, you need to expand, or amplify, your definition. Expanded definitions can be a few paragraphs long or they may even comprise an entire report. Expanded definitions explain abstract terms (*cogeneration* or *supply-side economics,* for example) that are too complex to be adequately treated in a few words or sentences. This kind of definition gives you room to discuss such things as the origin, significance, and application of the term as well as to provide examples of its use for the reader. Expanded definitions are also necessary when a term (*data processing*) acquires new meanings; in order to explain the new you have to contrast it with the old. Words like *compliance* or *construction management* rapidly change meanings in business and industry; even experts and decision makers need to be told exactly how these words are being used.

Ten Ways to Expand a Definition

Following are ten different ways in which you can write an expanded definition.

1. *Etymology.* Since many scientific and technical terms are derived from Greek or Latin words, one useful way to expand a definition might be with a discussion of etymology (or word history).

Rhinitis—a word derived from two Greek words *rhin* (nose) and *itis* (inflammation)—is an inflammation of the membranous lining of the nose.

If you are defining a word that is an acronym (a word formed from the initial letters of the words comprising it), tell readers what the individual letters stand for:

A laser is a device for producing an intense beam of light. The word is an acronym for *L*ight *A*mplification by *S*timulated *E*mission of *R*adiation.

2. *Origin and Application.* Sometimes it is necessary to supply background information about (1) the discovery of an object or process, (2) its early uses and any major problems, (3) changes in construction, design, or application, and (4) how these changes may affect current and future use. In the following example the Bureau of Mines begins a definition of columbium by surveying the applications of this important mineral:

Columbium (niobium) serves worldwide mostly as an alloying element in steels and in superalloys, and so is in greatest demand in the industrialized countries. Discovery of the strengthening effect of columbium in structural carbon steels has led to a widespread and growing use for columbium in high-strength, low-alloy (HSLA) steels. Two of columbium's more significant applications are in pipeline steels for transporting oil and gas and in structural steels. The use of columbium in the superalloys in jet engines has continued to be prominent; demand has fluctuated along with the fortunes of the aerospace industry.[3]

3. *Exclusion.* Sometimes you can define something by specifying what it is *not.*

Software is the nonhardware components of a computer.

A special order is one that has not been routed through the Purchasing Department.

When you define by exclusion be sure that your audience understands the object or process being excluded. In the examples just cited, the reader would have to know the meaning of hardware or be familiar with the ordering procedures in the company.

4. *Example.* One of the best ways to define an object or a process is to provide an example of it. The following definition of abrasive materials includes many examples of the materials used to preserve historic properties.

Abrasive cleaning methods include all techniques that physically abrade the building surface to remove soils, discolorations or coatings. Such techniques involve the use of certain *materials* which impact or abrade the surface under pressure, or abrasive *tools and equipment.* Sand, because it is readily available, is probably the most commonly used type of grit material. However, any of the following materials may be substituted for sand, and all can be classified as abrasive substances: ground slag or volcanic ash, crushed (pulverized) walnut or almond shells, rice husks, ground corncobs, ground coconut shells, crushed eggshells, silica flour, synthetic particles, glass beads and microballoons. Even *water* under pressure can be an abrasive substance. Tools and equipment that are abrasive to historic building

[3] Larry D. Cunningham, *Columbium* (Washington, D.C.: Dept. of the Interior, Bureau of Mines, 1979).

materials include wire brushes, rotary wheels, power sanding disks, and belt sanders.[4]

5. *Comparison and Contrast.* Comparisons establish likenesses; contrasts emphasize differences. Used together or separately, these techniques assist you in clarifying a term for the reader. The following paragraphs from a news release issued by the Bic Corporation use comparison and contrast to define a new product:

> The roller pen is a writing instrument which combines in one unit many features found separately in older type products, without their individual drawbacks. For example, the roller pen writes smoothly, with the hard point of a ball pen, but does not use full-color, high viscosity ball pen ink. Instead it writes with bright, vivid colors. In this respect it is similar to a fine-line marker pen, but does not have the drawback of a fibrous marker point which may quickly lose its fine writing qualities and wear into a brush. In its place, the roller pen has a durable, hard point which maintains its writing qualities during the entire life span of the pen. It is also capable of making carbon copies, which most fine-line marker pens cannot do.
>
> The roller pen writes with very fluid ink and has the free flowing qualities of fountain pen ink. However, unlike a fountain pen, the roller pen has a large ink supply and never needs to be refilled, cleaned or fussed with in any way. Thus, it is maintenance-free, with all the convenience features necessary in a modern writing instrument.

6. *Cause and Effect.* This technique tells the reader why something happens and what follows from its occurrence. In defining lightning, magnetism, a policy, or a disease, you might begin with the causes and then describe the effects. For example, you might list the conditions that produced a disease and then explain the symptoms as well as precautions. The following definition of a patent, modified from a government pamphlet, explains what patents are (the cause) and then informs the reader about their advantages (the effects):

> A patent is issued to prevent someone else from taking credit for or profiting from the inventor's creation. A patent is a grant, or printed document, issued by the U.S. Patent and Trademark Office, conferring on the inventor the right to exclude others for 17 years from making, using, or selling the invention. After obtaining a patent, the inventor is the sole owner of the original process, machine, or composition of material or plant and can retain or sell the patent to it.

7. *Method of Operation.* An explanation of how something works, is made, or occurs is useful when you have to define a procedure, a manufacturing or construction process, or a natural occurrence. In the following example, the writer expands the definition of an echo sounder for U.S. Navy personnel by explaining how a conversion of energy takes place:

> Most soundings are made by means of an echo sounder. In this instrument a pulse of electrical energy is converted periodically to sound energy and transmitted down-

[4] *Preservation Briefs 6,* (Washington, D.C.: Technical Preservation Service Div., Heritage Conservation and Recreation Service, U.S. Dept. of the Interior, 1979).

ward by a transducer. When the energy strikes the bottom (or any other object having acoustic properties different from those of water), a portion is reflected back to the transducer as an echo. This energy is reconverted to electrical energy for presentation. Because the speed of sound in water is nearly constant, the amount of time which elapses between the transmission of a pulse and the reception of its echo is a measure of the distance traveled, or in this case, depth.[5]

8. *Description.* Physical description (discussed in detail in Chapter 2) can be indispensible to an expanded definition of an object. Often the reader's first question when confronted with a new term for an object is "What does it look like?" The description of the loquat tree that follows comes from a report assessing the ability of new crops to adapt to environmental stress.

> The loquat tree is a small evergreen, reaching 25 feet in height, with a symmetrical, compact, round crown. The leaves are large, alternate, 8 to 12 inches long, relatively narrow and without indentation, leathery and dark green when mature. The flowers are small, ⅜ to ¾ inch across, white, fragrant, and borne in clusters at the end of branches. Blooming occurs from October through February. The fruit is a small pome growing in clusters of 4 to 10. It is oblong-pyriform to spherical in shape, 1 to 3 inches long; when ripe, the skin is yellow or orange and the flesh is firm, juicy, white to orange. The flavor is tart to sweet and there are 1 to 6 seeds in a central core (as in the apple).[6]

9. *Classification.* A good way of explaining a complex term or concept is to break it into logical units and consider each one individually. (See Chapter 3 for a detailed discussion of classification.) In the following definition, fire doors are classified into six distinct types.

> FIRE DOORS—Doors intended to prevent the passage of fire from room to room. They are classified and "labeled" by Underwriters Laboratories according to their time-rated degree of fire resistance. Class A doors are rated for three hours protection and are used in four-hour *Division Walls.* Class B doors are rated one or one and one-half hours and are usually used in two-hour division walls, stairwells, and elevator shafts. Class C doors are rated three-quarter hours and usually used for doors opening into corridors and through one-hour fire separation walls. Class D doors are rated one and one-half hours and used for exterior doors subject to exposure from an outside fire. Class E and F doors are rated three-quarter hours and are used for Class D doors but in areas having less exposure.[7]

10. *Partition.* A complex mechanism or process can be easier to understand if it is explained in terms of parts or stages. (Partition is discussed more fully in Chapter 3.) In the following example, a manual on radar navigation starts with a helpful breakdown of the functional requirements of the system:

[5] Nathaniel Bowditch, *American Practical Navigator* (Washington, D.C.: U.S. Defense Dept., Mapping Agency, Hydrographic Center, 1977), 151.

[6] National Science Foundation, *Feasibility of Introducing Food Crops Better Adapted to Environmental Stress* (NSF-RA 780289), Vol. I, 120.

[7] Hugh Brooks, *Illustrated Encyclopedic Dictionary of Building and Construction Terms* (Englewood Cliffs, N.J.: Prentice-Hall, 37).

> The functional breakdown of a basic pulse-modulated radar system usually includes six major components. . . . The functions of the components may be summarized as follows:
>
> The *power supply* furnishes all AC and DC voltages necessary for the operation of the system components.
>
> The *modulator* produces the synchronizing signals that trigger the transmitter the required number of times per second. It also triggers the indicator sweep and coordinates the other associated circuits.
>
> The *transmitter* generates the radio-frequency energy in the form of short powerful pulses.
>
> The *antenna system* takes the radio-frequency energy from the transmitter, radiates it in a highly directional beam, receives any returning echoes, and passes these echoes to the receiver.
>
> The *receiver* amplifies the weak radio-frequency pulses (echoes) returned by a target and reproduces them as video pulses passed to the indicator.
>
> The *indicator* produces a visual indication of the echo pulses in a manner that furnishes the desired information.[8]

STIPULATIVE DEFINITIONS

Stipulative definition as a method of expanded definition deserves separate discussion. One definition of *to stipulate* is "to make an agreement with an individual to do or not to do something." When you write a stipulative definition, you make an agreement with your reader to use a word or term in a special way by restricting or broadening the meaning of that term in some sense—time, space, application, cost, or personnel. When you define "schools in Hartford" to mean "only junior high schools," you restrict the ordinary meaning of *schools*. But when you define "schools in Hartford" so as to include "bimonthly informal workshops sponsored by the local Junior Achievement Association," you broaden the meaning most readers have of the word *schools*.

Stipulative definitions are usually given at the beginning of a report to alert the reader from the start to how you are using the word. They are always given on the first use of the term.

The following stipulative definition of *food industry* is taken from a recent government report:

> The combination of activities, institutions, technologies, and policies that collectively involve the production of basic raw food materials, the processing, preservation, and distribution of those food materials to the points where they are purchased for human consumption.
>
> This definition includes land and marine based raw material production systems, both natural and man-made systems utilizing traditional and nontraditional technologies. While our concern in this report is limited to the food industry within the continental U.S., there are, of necessity, some international connecting points that are critical to the analysis.

[8] Ernest B. Brown, *Radar Navigation Manual* (Washington, D.C.: Defense Mapping Agency, 1979).

This definition enlarges the meaning of the term to include many different phases of the food industry so that a reader will not assume it to be restricted to restaurants or grocery stores. As the report points out, "The food industry is actually a constellation of subindustries."

A stipulative definition can restrict as well as broaden. One governmental report offered the following stipulative definition of *rail rapid transit:* "an electrified railway system operating in urban areas on exclusive rights of way. Rail rapid transit is considered here to exclude commuter railroad systems and light rail systems, although the technology of train control is similar for all three." Because the report concerns train control, or the process by which the movement of rail rapid transit vehicles is regulated for safety and efficiency, the reader needs to be informed that not every type of rail system will be studied.

VISUALS

A definition can be made clearer and more detailed with visuals. A picture (such as that of the TENS unit on page 146 or of terra cotta on page 151) or diagram can be a helpful adjunct to a description of an object or its method of operation. A chart can help clarify the classes of an object or make a comparison or contrast more specific. When you use visuals, follow the guidelines listed in Chapter 1 (page 16) and Chapter 2 (pages 65–66).

A CHECKLIST FOR WRITING DEFINITIONS

1. Define all terms the reader may not know.
2. Give an *informal definition*—with a word or phrase—when a brief and immediate explanation is needed.
3. For a *formal definition*—one that is complete and precise—write a statement defining a species in terms of genus and differentiae.
4. For terms too complex to be defined in a few sentences, give an *expanded definition,* relying on such strategies as etymology, origin and application, exclusion, example, comparison and contrast, cause and effect, method of operation, description, classification, and partition.
5. When you wish to use a term in a special way, give a *stipulative definition,* specifying exactly how a word is being used.
6. Use visuals when they are helpful in defining a term.

Each of the following two models defines pH—potential of Hydrogen or measure of the acidity or alkalinity of a solution or substance—but for two widely different audiences. The first model is taken from a guide to indoor garden lighting written for consumers—specifically, for home gardeners who have purchased a Sylvania Gro-Lux lamp. This audience will be reading to obtain practical information about using their new lamp with soil that has the proper pH value. This audience does not want or need to know a lot of technical details about pH but only how the concept applies to home gardening activities. The Sylvania definition, therefore, is short and written in easy-to-understand language.

The second model defining pH comes from a specialized reference work, The Condensed Chemical Dictionary, *clearly intended for an audience of experts—chemists. The needs of this group of readers are far different from those of the consumer readers of the Sylvania definition. The audience for the* Dictionary *already knows what pH is; here they are looking for thorough and detailed information in order to perform a variety of technical activities where pH plays a major role. To meet this audience's needs, the* Dictionary *uses technical terms and formulas to explain the chemical derivation, values, and tests of pH.*

The Importance of pH Measurements

Sylvania Corporation

The pH measurement is a convenient expression of the acidity or alkalinity 1
of the soil. In simple terms the pH measurements indicate whether the soil is sour (acid) or sweet (alkaline).

The pH scale is logarithmic and runs from 0 to 14 with 7 being neutral. 2
Values below 7 are acid and those above indicate alkalinity. The fact that the scale is logarithmic means that a pH of 6 is ten times more acidic than a pH of 7. Thus, it is important to have close control.

Most plants prefer a slightly acid soil which falls in the pH range of 6.2 3
to 6.8 with the optimum pH of 6.5. Generally, ground limestone is used to correct acid soil and aluminum sulphate is added to correct alkaline soil conditions. It is generally a good practice to add limestone when mixing the soil. Add a 4-inch pot of ground limestone for every 3 quarts of soil. Common limestone is made up of calcium carbonate which neutralizes acid soil conditions just as sodium bicarbonate is taken for acid indigestion. The calcium is also helpful as it influences absorption of plant nutrients, and promotes early root formation and growth. Dolomitic limestone may also be used, in which case one is adding magnesium carbonate with the calcium carbonate. Magnesium is an important secondary element for normal plant growth.

The only sure way to ascertain whether a soil has the proper pH and 4
nutrient level is to conduct a soil analysis. This analysis can be obtained, free of charge, from all State Agricultural experiment stations in the country.

A Chemist's Definition of pH

Gessner G. Hawley

pH

pH is a value taken to represent the acidity or alkalinity of an aqueous 1
solution; it is defined as the logarithm of the reciprocal of the hydrogen-ion concentration of a solution:

$$pH = \ln \frac{1}{[H^+]}$$

Pure water is the standard used in arriving at this value. Under ordinary condi- 2
tions water molecules dissociate into the ions H^+ and OH^-, with recombination at such a rate that with very pure water at 22°C there is a concentration of oppositely charged ions of 1/10,000,000, or 10^{-7}, mole per liter. This is commonly expressed by saying that pure water has a pH of 7, which means that its concentration of hydrogen ions is expressed by the exponent 7, without its minus sign.

When acids or hydroxyl-containing bases are in water solution they ionize 3
more or less completely, furnishing varying concentrations of H^+ and OH^- ions, respectively, to the solution. Strong acids and bases ionize much more completely than weak acids and bases; thus strong acids give solutions of pH 1 to 3, while solutions of weak acids have a pH of about 6. Strong bases give solutions of pH 12 or 13, while weak bases give solutions of pH about 8. As the pH scale is logarithmic, the intervals are exponential, and thus represent far greater differences in concentration than the values themselves seem to indicate. (See table).

In acid-base titrations, changes in pH can be detected by indicators (q.v.) 4
such as methyl orange, phenolphthalein, etc. Litmus paper can also be used as a rough indication of acidity or alkalinity. In carrying out titrations the point at which the solution is found to be neutral (pH 7) is not always the correct end point.

pH control is of critical importance in a large number of industrial opera- 5
tions such as water purification, chrome tanning process for leather, preservation of food products, electroplating baths, dyeing, agriculture, and numerous other instances. See also acid; base.

	pH Value	Ratio of H^+ or OH^- Concentration to That of Pure Water at 22°C
Acid side (excess of H^+ ions)	1	1,000,000
	2	100,000
	3	10,000
	4	1,000
	5	100
	6	10
Neutrality	7	1
Alkaline side (excess of OH^- ions)	8	10
	9	100
	10	1,000
	11	10,000
	12	100,000
	13	1,000,000

Liquid	pH Value
Pure water	7
Sea water	8.5–10
Electroplating bath	6.5–5
.01 *N* HCl	2
.01 *N* NaOH	12
.1 *N* acetic acid	3
.1 *N* NH_4OH	11
Gastric juices	2
Urine	5–7
Blood	7.3–7.5
Milk	6.5–7
Soil (optimum for crops)	6–7

QUESTIONS FOR DISCUSSION

1. What details about pH are included in both the Sylvania and *Dictionary* definitions? Compare the presentations of these details in the two definitions in terms of diction and sentence style.
2. What details about pH are found in just the *Dictionary* definition? How are the details given (comment on style, word choice, symbols, the use of visuals)? Why might experts need these additional details and consumers not? Give practical examples.
3. How does the definition of *logarithm* differ in the two models? Why?

4. What kinds of defining techniques are illustrated in the following sentences from the Sylvania model:
   a. In simple terms the pH measurements indicate whether the soil is sour (acid) or sweet (alkaline).
   b. The fact that the scale is logarithmic means that a pH of 6 is 10 times more acidic than a pH of 7.
   c. Common limestone is made up of calcium carbonate, which neutralizes acid soil just as sodium bicarbonate is taken for acid indigestion.
   d. Magnesium is an important secondary element for normal plant growth.
5. Identify four of the strategies that the writer of the *Dictionary* model used to expand the definition of pH. How effectively are these strategies used?

SUGGESTIONS FOR WRITING

1. There are a number of technical words that are not defined in the *Dictionary* definition of pH because the audience knows their meaning. Provide a sentence definition of four of the following terms, making them easily understandable for consumers.
   a. aqueous solution
   b. ion
   c. dissociate
   d. oppositely charged
   e. mole per liter
   f. electroplating bath
   g. methyl orange
   h. titrations
2. Write two expanded definitions of 250–300 words each (one for experts; one for consumers) of one of the following terms:
   a. protein
   b. a microcomputer
   c. friction
   d. affirmative action
   e. a catheter
   f. an electronic game of your choice
   g. vitamin C
   h. a carburetor
   i. epoxy
   j. the metric system
3. Following the Sylvania model, write an expanded definition for consumers of one of the following subjects, which they might have to know about in doing some job in their homes.
   a. ac/dc current
   b. viscosity
   c. corrosion
   d. fragility
   e. toxicity
   f. grade (for poultry, eggs, etc.)
   g. dew point
   h. boiling point

Some terms hold the same meaning from one year to the next: pH, for example. But not every term or word you will have to define will have a constant, single meaning. In the following expanded definition of computer-related crime we can see the kinds of problems a writer must face when defining a word or phrase with multiple and changing meanings.

The definition of computer-related crime comes from a manual prepared for criminal justice officials by the Law Enforcement Assistance Administration of the U.S. Department of Justice. Computer-related crime presents new challenges for the criminal justice community. Unlike other crimes, computer-related offenses are not restricted by geographic constraints; thanks to telephones with attached computer terminals, a crime can be committed in one part of the world from any other part. Moreover, such crimes must be seen in light of a new time scale, for they can occur in less than three milliseconds. These technological realities are an important part of the definition. This model provides insight into the types of questions a writer must ask, and successfully answer, when defining a term that has undergone many shifts in meaning.

A Definition of Computer-Related Crime

Computers have been involved in most types of crime, including fraud, theft, larceny, embezzlement, bribery, burglary, sabotage, espionage, conspiracy, extortion, and kidnapping. Criminal justice agencies having limited experience with computer-related crime have generally thought of it as crime that occurs inside computers. This narrow definition has recently broadened as the proliferation of computers into most societal functions proceeds at an increasing pace. The public media have added to the confusion through sensationalized, distorted, often incorrect reporting by journalists and their sources who do not sufficiently understand the technology. 1

Computer-related crime is not well understood in the criminal justice and business communities, and no consensus on its definition exists. One definition is that it is a form of white-collar crime committed inside a computer system; another definition is that it is the use of a computer as the instrument of a business crime. An application of the former definition is making unauthorized changes to a computer program to transfer funds from inactive accounts into a favored account and then legitimately withdrawing the funds. 2

The definition of computer-related crime should be based on the problem that needs to be solved. The problem addressed is twofold: how to reduce the incidence of any type of crime in which a knowledge of computer technology is needed to understand the intentional acts that result in losses and how to successfully prosecute the perpetrators. Whereas this is now predominantly a 3 white-collar crime, the criminal justice and business communities must be prepared to deal with any illegal acts based on an understanding of computer technology. A study of 669 reported cases of computer abuse over the past 8 years has revealed that computers are involved in an increasing number of crimes of all types except murder and person-to-person street crimes. The proliferation and use of personal small computers make even the latter crimes subject at least to the indirect involvement of computer technology.

The broadest definition of computer crime is called for. The term "crime" 4
is used here (as is usual) as a convenience to mean "alleged crime" because no harmful or antisocial act is a crime until a court declares it so by convicting a party for violating a law. Three terms have been used to describe the subject: "computer abuse," "computer crime," and "computer-related crime." Computer abuse identifies the broad range of intentional acts from which something can be learned to make organizations that use computers more secure. Computer abuse is any intentional act involving a computer where one or more perpetrators made or could have made gain and one or more victims suffered or could have suffered a loss.

Computer crime is a common term used to identify illegal computer abuse; 5
however, it implies direct involvement of computers in committing a crime. Therefore, we adopt the term computer-related crime that conveys the broader meaning of *any illegal act for which knowledge of computer technology is essential for successful prosecution.* This definition is based on the scope and nature of the particular problem being addressed. The crimes and alleged crimes may involve computers not only actively but also passively when usable evidence of the acts reside in computer stored form. The victims and potential victims include all organizations and persons who use or are affected by computer and data communication systems. People about whom data are stored and processed in computers also are included.

Computer-related crime goes far beyond business, white-collar, or economic 6
crime. It could include violent crime that destroys computers or their content or jeopardizes human life and well-being because they are dependent on the correct functioning of computers controlling sensitive processes.

Computers play four roles in crime: 7

- Object—Cases include destruction of computers or of data or programs contained in them or supportive facilities and resources, such as air conditioning equipment and electrical power, that allow them to function.
- Subject—A computer can be the site or environment of a crime or the source of or reason for unique forms and kinds of assets.
- Instrument—Some types and methods of crime are complex enough to require the use of a computer as a tool or instrument. A computer can be used actively such as in automatically scanning telephone codes to make unauthorized use of a telephone system. It could also be used passively to simulate a general ledger in the planning and control of a continuing financial embezzlement.

- Symbol—A computer can be used as a symbol for intimidation or deception. This could involve the false advertising of nonexistent services, such as in dating bureaus.

All known and reported cases of computer-related crime involve one or more of these four roles. 8

The dimensions of the definition of computer-related crime become a problem in some cases. If a computer is stolen in a simple theft where based on all circumstances it could have been a washing machine or milling machine and made no difference, then a knowledge of computer technology is not necessary, and it would not be a computer-related crime. However, if knowledge of computer technology is necessary to determine the value of the article taken, the nature of possible damage done in the taking, or the intended use by the thief, then it would be a computer-related crime. To illustrate, if an individual makes a telephone call to a bank funds transfer department and fraudulently requests a transfer of $10 million to his account in a bank in Zurich, two possibilities occur. If the clerk who received the call was deceived and keyed the transfer into a computer terminal, the funds transfer would not be a computer-related crime. No fraudulent act was related directly to a computer, and no special knowledge of computer technology would be required. However, if the clerk was in collusion with the caller, the fraudulent act would include the entry of data at the terminal and would be a computer-related crime. Knowledge of computer technology would be necessary to understand the terminal usage and protocol.

These examples indicate the possibilities of rational conclusions in defining 9
computer-related crime. However, more practical considerations should not make such explicit and absolute decisions necessary. If any information in this manual is useful for dealing with any kind of crime, its use should be encouraged. Finally, as the criminal justice community has begun to understand from experience the involvement of computers across all types of crime, the definition has broadened. It is expected that the results of efforts funded by LEAA will establish a consensus on a broad definition of computer-related crime to cover the current and anticipated experience.

QUESTIONS FOR DISCUSSION

1. What is the writer's purpose in giving readers the "broadest definition of computer crime" possible? Why do readers need such a definition and what will they do with the information this definition provides?
2. What are the "problems" mentioned at the beginning of paragraph 3? How is a knowledge of these problems important for both the writer and the reader?

3. Why did the writer choose definition as the most suitable rhetorical technique to accomplish the purpose of this section of the report? What other strategies might have been used?
4. Find examples in this model of the strategies of expanded definition, which are listed below. Explain how the writer uses each strategy and how they are used to complement each other.
   a. exclusion
   b. classification
   c. concrete examples
   d. comparison and contrast
   e. formal definition
5. How does the author define "computer abuse"? What role does this term play in defining computer-related crime? Could the author have omitted any references to computer abuse?
6. What exactly does the last paragraph contribute to the overall definition of computer-related crime?

SUGGESTIONS FOR WRITING

1. The writer uses a number of terms without defining them. Write an expanded definition of one of the following terms for a consumer audience. Use information from this model on computer-related crime whenever necessary.
   a. white-collar crime
   b. extortion
   c. telephone scanning
   d. terminal usage and protocol
2. Write an expanded definition for a general audience of one of the following terms that has multiple meanings or whose meaning has changed over the years or both. Include all of the term's meanings in your definition.
   a. calculator (noun)
   b. rollover (verb)
   c. rhetoric (noun)
   d. chip (noun)
   e. terminal (noun)
   f. laser (noun)
   g. fluorocarbons (noun)
   h. telecommunications (noun)

The next model, written by a student for an assignment in a technical writing class, will show you how definition can play a major role in a piece of on-the-job writing. The student, who is majoring in nursing, wants to work in a pain clinic after graduation and has prepared a letter similar to the one her employer may expect her to write. Her letter tells about the TENS unit, a device used by some outpatients of the clinic for pain therapy. Her letter uses many of the strategies of expanded definition discussed in the introduction to this chapter. It also illustrates the kind of "patient teaching" that occurs when an expert (the nurse) must define a technical term for the consumer (the patient).

A Letter to Patients of a Pain Clinic

Carla Waddell

UNIVERSITY PAIN CLINIC

MEMORIAL UNIVERSITY
Shenstone, PA 18343

Dear Outpatient:

Your physician has recommended that you use a TENS unit to help relieve your chronic pain. The purpose of this letter is to provide some background information about the TENS unit you will receive on your next visit to the clinic. 1

TENS stands for Transcutaneous Electrical Nerve Stimulation. Since transcutaneous means "through the skin," you can understand how TENS works. Simply put, a low voltage electrical current is sent through your skin along nerve pathways in the pained area to decrease pain. The intensity of the electrical impulse can be regulated by the user to insure adequate pain control. 2

TENS is safe, easy to use, and has no dangerous side effects. While TENS cannot cure your problem, it will ease movement and thus make life more pleasant for you. 3

What the TENS Unit Looks Like

A picture of the TENS unit is included with this letter. You might want to look at that picture now before you read the following description. 4

The TENS unit is portable, lightweight, and compact. It is small enough to fit into your pocket or purse. Some patients clip the TENS device to their belts, wear it on their shoulders under a coat or dress, or clip it to the inside of a bra or a t-shirt. 5

TENS devices have three major parts. They are (1) the TENS unit itself, (2) the electrodes, and (3) the energy source. 6

TENS unit: The unit, a small rectangular box made of metal or plastic, is approximately three to five inches long, and one inch wide. There are several dials on the unit, the on-off dial and those that regulate the amount of electrical impulse the unit generates. Several small 7

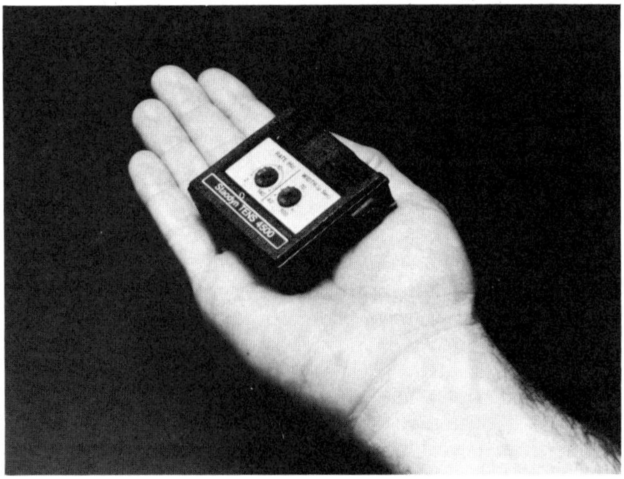

terminals on the unit serve as attachment sites for the electrodes.

Electrodes: These are thin wires running from TENS and taped to your skin. The electrodes conduct electrical impulses from the unit to the site of pain. Placement of the electrodes is of primary importance in determining the amount of pain relief you will receive from TENS therapy. For example, the best placement for relief of pain that radiates from the spine is along the spine itself. To treat localized pain, you would place the electrodes in crisscross fashion directly over the painful area. 8

Energy source: Batteries, located inside the unit, provide the energy for the production of electrical impulses. You should always check the manufacturer's directions included with your unit to know how often to replace batteries for maximum effectiveness. 9

Why the TENS Works

The idea of using electricity to decrease pain is not new. 10
The ancient Romans once filled tubs with water and placed electric eels in the tubs. Pained areas of a patient's body were submerged in the water and the electrical impulse produced by the eels (similar to that which a TENS unit generates) would provide relief from pain. In the 18th

century, the Italian scientist Galvani showed that nerves could be affected by the application of a direct electrical current. Today many athletes use TENS to relieve pain.

There are two major theories of pain control that help to explain why TENS works. They are Gate Control and Endorphin Theory. Some researchers believe both theories play a part in TENS effectiveness. 11

Gate Control: This theory suggests that TENS impulses block pain messages sent to the brain. The electrical current from a TENS unit "closes the gate" so that pain signals cannot reach the brain. The brain no longer interprets the pain as pain but instead perceives it as a pleasant, tingling sensation. 12

Endorphin Theory: Endorphins are natural pain killers found within the human body. Some researchers call them "morphinelike." It is believed that TENS impulses increase the production and release of endorphins to relieve pain without the aid of artificial drugs. 13

What Advantages TENS Has over Narcotics

The pain relief provided by TENS has several advantages over the use of narcotics. Most important, TENS units are not addictive. Many patients who take narcotic drugs become confused, disoriented, and sleepy; TENS does not have this effect. In addition, narcotics often slow down all bodily functions. Use of the TENS unit will not do this. In fact, TENS may speed up some bodily functions, especially digestion. 14

What Precautions to Take

The major disadvantage in using TENS is that it may cause irritation to the skin at the electrode placement sites. You should never use TENS over broken skin. 15

The electrical impulse TENS generates can overstimulate various organs of the body; therefore, be cautious in your placement of the electrodes. Never place the unit over the front of your neck. This may stimulate the windpipe to contract and cause severe breathing problems. Also, TENS should not be used by patients in the early stages of pregnancy. The unit may stimulate the uterus to contract and produce a miscarriage. However, the TENS unit can be safely used during labor. 16

TENS should not be used by patients with heart problems. The electricity generated by TENS may cause pacemakers to malfunction and can aggravate heart conditions. 17

We hope this letter has answered some of your questions 18
about the TENS device and clarified your doctor's reasons for recommending it. When you return to the clinic for your next appointment, you will be given your own TENS unit and we will help you determine the correct placement of the electrodes and the amount of electrical impulse you need for optimum pain relief.

Please call me if you have any questions about TENS. 19

Sincerely yours,

Carla Waddell

Carla Waddell, R.N.
Head Nurse

QUESTIONS FOR DISCUSSION

1. Why would an outpatient read this letter? How does the definition meet the likely needs and purposes of its audience? Point to specific parts of the letter in your answer.
2. Why does Waddell give a description of the unit before discussing why it works? Explain the significance of these two sections for readers and identify various strategies of expanded definitions used in each one.
3. Why is the visual—a photograph of the TENS unit—essential to the definition? How is it appropriate for the audience?
4. How does Waddell effectively define the two theories of TENS relief for her audience? Does the reader need this information?
5. Why does Waddell end with a discussion of the advantages and precautions in using the TENS device? What strategies of expanded definition does she employ in these final sections?

SUGGESTIONS FOR WRITING

1. Write an expanded definition in letter or memo format on one of the following subjects or one approved by your instructor.
   a. a watt meter (assume you work for a power company and have to explain this machine in a letter sent to customers)
   b. computer graphics (you can represent a company trying to convince a decision maker in a letter to buy your product)
   c. robots (write a memo to a decision maker in your own company explaining and defining the use of robots in one part of your firm's operation and conclude with how useful they might be in your own division)
   d. insulating material (write to an operator who must install one of these types—fiber, foil, or polystyrene)
   e. floor polisher (write to a group of operators who must use and maintain this piece of equipment)

The following model on terra cotta (a building material made from fired clay) is excerpted from a Preservation Brief published by the Technical Preservation Services Division of the Department of the Interior. The purpose of these briefs, or technical reports, is to provide federal agencies and state and local governments with "information concerning professional methods and techniques for preserving, improving, restoring, and maintaining historic properties." The audience for this report includes architectural historians, urban planners, government officials, and specially trained individuals employed in city, state, and federal engineering and maintenance departments. These people represent experts, decision makers, and operators. The author, an architectural historian, has developed the report almost exclusively through the types and techniques of definition discussed in the introduction to this chapter. As you read, observe the different levels of definition the author uses—from brief informal definitions for some terms to expanded definitions for others.

The Preservation of Historic Glazed Architectural Terra Cotta

de Teel Patterson Tiller

Glazed architectural terra cotta was significant in the development of important architectural idioms in this country—specifically, the "Chicago School," the High Rise and the Historic or Beaux Arts styles. In fact, glazed architectural terra cotta is one of the most prevalent masonry building materials found in the urban environment today (Fig. 1). Popular between the late 19th century and the 1930s, glazed architectural terra cotta offered a modular, varied and relatively inexpensive approach to wall and floor construction. It was particularly adaptable to vigorous and rich ornamental detailing. However, with changing vogues in materials and architectural styles and rising production costs, glazed architectural terra cotta fell into disfavor and disuse by the mid-20th century. 1

Today, information on the maintenance, rehabilitation and replacement of glazed architectural terra cotta is limited, as are sources of new glazed architectural terra cotta. This report, then, will discuss some of the major deterioration problems that commonly occur in historic glazed architectural terra cotta, methods of determining the extent of that deterioration and recommendations for the maintenance, repair and replacement of the deteriorated historic material. 2

WHAT IS TERRA COTTA?

Generically, the broadest definition of terra cotta refers to a high grade of weathered or aged clay which, when mixed with sand or with pulverized fired clay, 3

Figure 1. Terra-Cotta Detailing. Adaptable to every nuance of color, texture, and modeling, glazed architectural terra cotta was ideally suited to satisfy the tastes of an eclectic age. Its popularity was, however, short lived; it endured only 30 or 40 years after its introduction as a building material late in the 19th century. (*Larry Payne, Houston, Texas*)

can be molded and fired at high temperatures to a hardness and compactness not obtainable with brick. Simply put, terra cotta is an enriched molded clay brick or block. The word *terra cotta* is derived from the Latin word *terra cocta*—literally, "cooked earth." Terra-cotta clays vary widely in color according to geography and types, ranging from red and brown to white.

Terra cotta was usually hollow cast in blocks that were open to the back, like boxes, with internal compartment-like stiffeners called webbing (Fig. 2). Webbing substantially strengthened the load-bearing capacity of the hollow terracotta block without greatly increasing its weight.

Terra-cotta blocks were often finished with a glaze; that is, a slip glaze 4
(clay wash) or an aqueous solution of metal salts was brushed or sprayed on the air-dried block before firing. Glazing changed the color, imitated different finishes, and produced a relatively impervious surface on the weather face of the final product. The glaze on the terra-cotta unit possessed excellent weathering properties when properly maintained. It had rich color and provided a hard surface that was not easily chipped off. Glazing offered unlimited and fade-resistant colors to the designer. Even today, few building materials can match the glazes on terra cotta for the range and, most importantly, the durability of colors.

Figure 2. Webbing. Webbing, or the hollow internal compartment construction of glazed architectural terra-cotta blocks, made them inexpensive to produce, easy to handle and light in construction; these were significant factors in the popularity of the material in the first decades of this century.

TYPES OF TERRA COTTA

Historically there are four types or categories of terra cotta that have enjoyed 5
wide use in the history of the American building arts: 1) brownstone, 2) fireproof construction, 3) ceramic veneer, and 4) glazed architectural.

Brownstone terra cotta is the variety of this masonry material used earliest in 6
American buildings (mid- to late 19th century). The brownstone type is a dark red or brown block either glazed (usually a slip glaze) or unglazed. It was hollow cast and was generally used in conjunction with other masonry in imitation of sandstone, brick or real brownstone. It is often found in the architecture of Richard Upjohn, James Renwick, and H. H. Richardson, and is associated with the Gothic and Romanesque Revival movements through such ornamental detailing as moldings, finials and capitals.

Fireproof construction terra cotta was extensively developed as a direct result 7
of the growth of the High Rise building in America. Inexpensive, lightweight and fireproof, these rough-finished hollow building blocks were ideally suited to span the I-beam members in floor, wall and ceiling construction (Fig. 3). Certain varieties are still in production today, although fireproof construction terra cotta is no longer widely employed in the building industry.

Ceramic veneer was developed during the 1930s and is still used extensively 8
in building construction today. Unlike traditional architectural terra cotta, ceramic veneer is not hollow cast, but is as its name implies; a veneer of glazed ceramic tile is ribbed on the back in much the same fashion as bathroom tile.

Ceramic veneer is frequently attached to a grid of metal ties that has been 9
anchored to the building.

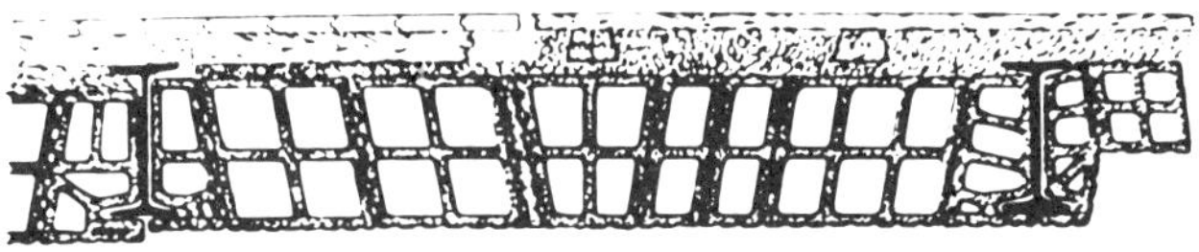

Section Through Typical Arch

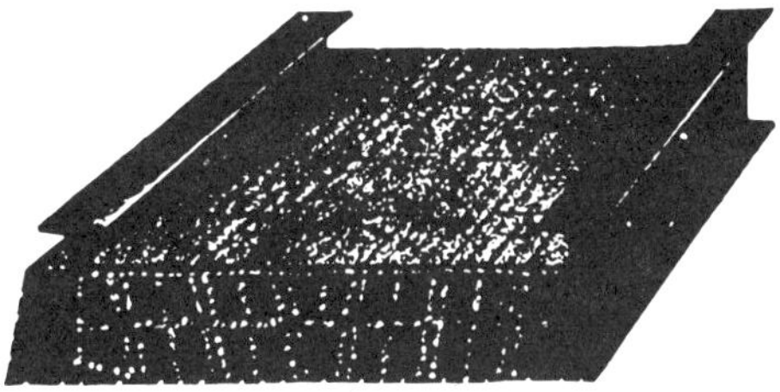

Perspective of Typical Arch

Figure 3. Fireproof construction terra cotta. Perspective and section through fireproof construction terra cotta and I-beam detailing in industrial floor construction. (Detail, *"Sweets" Industrial Catalogue of Building Construction,* 1906)

Glazed architectural terra cotta was the most complex development of terra 10
cotta as a masonry building material in this country. The hollow units were hand cast in molds or carved in clay and heavily glazed (often in imitation of stone) and fired. Sometimes called "architectural ceramics," glazed architectural terra cotta was developed and refined throughout the first third of the 20th century and has been closely associated with the architecture of Cass Gilbert, Louis Sullivan, and Daniel H. Burnham, among others. Significant examples in this country include the Woolworth Building (1913) in New York City and the Wrigley Building (1921) in Chicago.

Late 19th and early 20th century advertising promoted the durable, impervi- 11
ous and adaptable nature of glazed architectural terra cotta. It provided for crisp, vigorous modeling of architectural details as the molds were cast directly from clay prototypes without loss of refinement. Glazed architectural terra cotta could accommodate subtle nuances of modeling, texture and color. Compared to stone, it was easier to handle, quickly set and more affordable to use. Thought to be fireproof and waterproof, it was readily adaptable to structures of almost any height. The cost of molding the clay and glazing and firing the blocks, when compared to carving stone, represented a considerable saving, especially when casts were used in a modular fashion—that is, repeated over and over again. Maintenance of the fired and glazed surface was easy; it never needed paint and periodic washings restored its original appearance.

With the passage of time, many of the phenomenal claims of the early 12
proponents of glazed architectural terra cotta have proven true. There are many examples throughout this country that attest to the durability and permanence of this material. Yet present-day deterioration of other significant glazed architectural terra-cotta resources ultimately belie those claims. Why? Historically, the lack of foresight or understanding about the nature and limitations of the material has, in many instances, allowed serious deterioration problems to occur that are only now becoming apparent.

CHARACTERISTICS OF GLAZED ARCHITECTURAL TERRA COTTA AS A BUILDING MATERIAL

Glazed architectural terra cotta has many material properties similar to brick or stone. It also has many material properties radically different from those traditional masonry materials. It is those differences that must be considered for a better understanding of some of the material characteristics of glazed architectural terra cotta when it is used as a building material. 13

Difficult to Identify

Glazed architectural terra cotta probably comprises one of the largest if not the largest constituent material in some of our urban environments today. However, the infinite varieties of glazing have hidden this fact from the casual observer. One of the attractive features of glazed architectural terra cotta in its time was that it could be finished (glazed) in exact imitation of stone. In fact, many building owners and architects alike are often surprised to discover that what they presumed to be a granite or limestone building is glazed architectural terra cotta instead. 14

Two Separate Systems

Historically, glazed architectural terra cotta has been used in association with two specific and very different types of building systems: as part of a traditional *load-bearing* masonry wall in buildings of modest height, and as a *cladding material* in High Rise construction. As cladding, glazed architectural terra cotta often utilized an extensive metal anchoring system to attach it or to "hang it" onto a wall framing system or superstructure (Fig. 4). In the first instance the anchoring was limited; in the second, the anchoring was often extensive and complex. Likewise, in the first instance, deterioration has generally been limited. However, where glazed architectural terra cotta was used as cladding, particularly in high rise construction, present-day deterioration and failure are often severe. 15

Complexity of Deterioration

Deterioration is, by nature of the design, infinitely complex—particularly when glazed architectural terra cotta has been used as a cladding material. Deterioration creates a "domino"-like breakdown of the whole system: glazed units, mortar, metal anchors, and masonry backfill. In no other masonry system is material failure potentially so complicated. 16

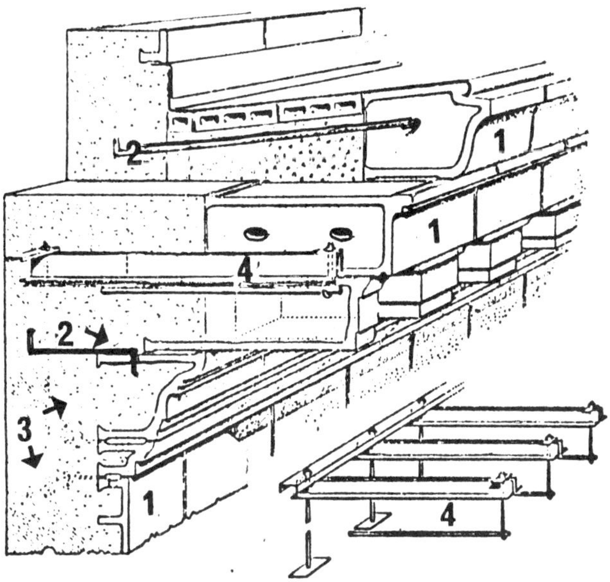

Figure 4. Typical Construction Detail of Glazed Architectural Terra-Cotta Ornament. Construction detailing was often complex. The terra-cotta units (1) which were laid in mortar were fitted with holes or slots to receive the metal anchors (2) (often called "Z" straps or "light iron") which were often fitted directly to the building frame. Masonry backfill (3) (either brick or poured cement) was laid between the terra-cotta units, with the building frame encasing the metal anchor. Overhanging or protruding elements were further secured by metal dowels or outriggers (4). (Detail, *Architectural Terra Cotta,* Charles E. White, Jr., 1920)

Poor Original Design

The root of deterioration in glazed architectural terra-cotta systems often lies in a misapplication of the material. Historically, glazed architectural terra cotta was viewed as a highly waterproof system needing neither flashing, weep holes, nor drips. This supposition, however, has proved to be untrue, as serious water-related failure was evident early in the life of many glazed architectural terra-cotta clad or detailed buildings. . . . 17

SUMMARY

Today, many of this country's buildings are constructed of glazed architectural terra cotta. However, many of these are in a state of serious deterioration and decay. Glazed architectural terra cotta was, in many ways, the "wonder" material of the American building industry in the late 19th century and during the first decades of the 20th century. New technology and methods of rehabilitation now hold promise for the restoration and rehabilitation of these invaluable and significant resources. Restoration/rehabilitation work on glazed architectural terra cotta is demanding and will not tolerate half-way measures. Today's preservation work should equal the spirit, attention to detail, pride in workmanship and care that characterized the craftsmanship associated with this widely used, historic masonry material. 18

QUESTIONS FOR DISCUSSION

1. How does the explanation of terra cotta given in the first paragraph differ from the definition supplied under "What is Terra Cotta?"? Why didn't the writer begin with "What is Terra Cotta?"? In your answer consider the informational needs of the reader at these two stages of the report.
2. Why does the writer define the types and characteristics of terra cotta? How do the definitions in the earlier sections of the report assist the reader in the later parts of the report?
3. Where does the writer use informal or sentence definitions? Cite two examples of each type of definition.
4. The report uses many of the strategies for expanding a definition discussed in the introduction to this chapter. Find at least one example of each of these strategies.
5. Evaluate the use of visuals in this report. How do they contribute to the definition of terra cotta? Where are they placed and how are they introduced?
6. How does the writer make his definition clear and relevant to experts (architects, engineers), decision makers (city commissioners, bureau chiefs, mayors), and even operators (individuals doing the restoring)? Can you suggest how the report might be different in terms of content and style were it intended for experts only?

SUGGESTIONS FOR WRITING

1. Write a report modeled after the terra-cotta brief in which you use most of the strategies for expanding a definition discussed in the introduction to this chapter. In the margins of your report, identify these different strategies with red ink. Identify the audience(s) you want to reach in the first paragraph and state a clear aim for their reading your work.
2. Using information in this report, write a description of terra cotta. Assume that your audience is made up of consumers interested in terra cotta for home-improvement information. Suggest any appropriate visuals.

chapter 5

Process Explanation

A process explanation leads the readers through a series of steps, or stages, to show how something is produced or why something occurs. The readers' goal is to understand, not perform, the process. Process explanation relies heavily on physical description, which is discussed in Chapter 2. When the readers must perform the actual steps of a particular process, you have to write a set of instructions to guide them. It is important to recognize the difference between these two kinds of writing. This chapter looks at the rhetorical technique of process explanation. Chapter 6 shows how the techniques of process explanation are applied in writing instructions.

A process explanation is a narrative that tells the reader what will take place during all stages of the process. The narrative illustrates what happens during each stage and must accurately follow the order in which the stages take place. It may also identify any problems or special conditions, and in some cases may point out any advantages or benefits of the process. After completing such an explanation, the reader should be able to understand how individual steps are interrelated, like links in a chain. To make a process explanation clear and helpful, you may use many of the rhetorical techniques you have already studied—physical description and definition in particular.

Process explanations will vary in content according to the process being described. In general, however, they provide answers to the following sorts of questions:

1. How is something made, manufactured, or produced? (cottage cheese, glass, leather, macramé, paper, perfume, steel)
2. How does something work? (a heat pump, the kidneys, a laser, an oscilloscope, a satellite for a cable television station, a robot, a zipper)
3. Why or how does something happen or occur? (why dough rises, how blood is oxygenated, why an oil spill occurred, how a hiring or training policy is implemented, how a hurricane develops, how cloning results)

4. Are any operators involved in the process? (how an appraisal is made, how a piano is tuned, how an airplane lands, how a safety inspection is made)

As the examples following these questions reveal, process explanations help readers to understand natural phenomena, mechanical operating principles, and business policies and procedures. In order to explain some processes fully, the writer may sometimes have to discuss the role of an operator. For example, an explanation of assembly line production would require a description of the duties of various employees working on that line. For other kinds of processes, it would be pointless or impossible to include an operator. In an explanation of how solar energy cools a house, there is little need to comment on the role of the contractor who supplies the solar collector. Likewise, individuals play no role in the creation of such acts of nature as gust fronts, electromagnetic frequencies, or volcanoes.

Process explanations are an important and common on-the-job writing assignment. It is often necessary for engineers to explain the appropriate mechanical or electrical processes that go with various types of construction work. Health-care professionals are also called upon to use process explanations, whether about the course a disease might run, the performance of medical equipment, or the various steps a surgical process requires. Someone preparing a proposal must explain how the work is to be accomplished and how it will be monitored. Sales personnel sometimes have to explain a company's manufacturing techniques. Finally, investigative reports (discussed in Chapter 7) frequently include process explanations in sections describing the methods researchers used to gather and evaluate data.

CONTENT AND ORGANIZATION

A process explanation follows a conventional organizational plan, with sections serving as an introduction, body, and conclusion. Each one of these three main sections can, of course, be further subdivided as the following discussion will show.

Introduction

All but the shortest explanations require an introduction to help the audience get its bearings. Depending on the complexity of the process and the reader's technical background, your introduction can range from a single sentence or paragraph to a number of pages. An introduction for a process explanation may do all or some of the following:

- state the aim of the explanation
- define the process
- elaborate on the reason for knowing about the process

- describe the point of view from which the process will be explained (expert to expert; expert to manager; expert to consumer)
- explain the principles behind the process
- mention developments that may occur in time because of the process
- specify any operators in the process
- describe any special equipment necessary to the process
- identify any special conditions or training for the process
- preview the major steps of the process

This last point is an important part of some introductions and so merits further comment. Previewing the major steps alerts readers to the number and order of steps they will have to follow in detail. Such forecasting makes the reader's job easier and also reveals the clear organization of your work. List the major steps in a numbered column such as this preview of the process involved in making potato chips:

1. Peeling the potatoes
2. Slicing potatoes into chips
3. Washing the potato chips
4. Frying the potato chips
5. Salting the potato chips
6. OPTIONAL STEP: Adding any special flavors to the potato chips
7. Bagging the potato chips

If the steps are relatively few, you can list them using bullets, as the following example on the way a transducer works in an echo sounder:

- Converts electrical energy to sound energy
- Transmits sound into the water
- Receives returned echoes
- Converts returned sound energy into electrical energy

Make sure that all of the major steps you are previewing are grammatically parallel (each step should begin with the *same* grammatical form). For example, the following steps on the operation of the transducer are not grammatically parallel because each one begins with a different grammatical form of a verb or noun:

- To convert electrical energy to sound energy
- Transmittal of sound into the water
- Receiving returned echoes
- Converts returned sound energy into electrical energy

Body

The body is the most important part of a process explanation. In this section, you show the reader the process step by step. You can use individual lists, paragraphs, flow charts, or any combination of these ways to provide and explain the steps. The amount of information you include for each step, of course, will be determined by the reader's needs and technical background. An explanation of a water filtration plant will differ in the amount of technical detail it contains for city officials than for city engineers, who may have to give a recommendation to the officials based on the process explanation. For city officials you will need to supply definitions and background information. You will also have to omit technical data on water flow dynamics and chemistry.

Following are some guidelines for writing an effective body for a process explanation.

1. Determine the exact chronological order the steps take.
2. Be sure to include all the steps in the process. Omitting a step can distort the process and mislead the reader. Alert the reader to any optional steps such as the one for flavoring potato chips on page 159.
3. Use numbers, headings, subheadings, and transitions to make clear when one step ends and another begins. (See pages 161–162 for more information on style in a process explanation.)
4. Describe what happens in each step. Remember that each step is a process in itself. Tell readers the purpose of the step, describe necessary equipment, alert them to possible dangers, and, where necessary, stress the benefits the step offers, especially when readers are potential customers of your process. A competitor's process may lack a step you include. Conversely, you may have improved upon a competitor's process by eliminating a costly step.
5. Point out any substeps, taking special care to place them at the proper stage of the process. Be careful that you do not mistake a substep for a major step in the process. Also, beware of misplacing a substep by inserting it before or after the major step of which it is a part.

Conclusion

Some process explanations, especially shorter ones, end with the last step. However, many process explanations include a more formal conclusion. Depending on the complexity of the process and the reader's specific needs, a conclusion can give the following sort of information.

- a summary of the major steps of the process
- an assessment of the results of the process—citing advantages and disadvantages
- a comparison of the process with other similar processes
- an indication of how the process relates to other work or processes

STYLE

Style—the words, sentences, paragraphs, and tone you use—is a major element in process explanations. As in other kinds of technical writing, your language should be clear, straightforward, and easy to read. But it must also be appropriate to express the kinds of information a process explanation gives.

A process is a sequence of events that occur in time and space and have distinct causes and effects. To help the reader see and remember these relationships within the process, include helpful transitional words and phrases. These words and phrases, which function like mortar or connecting steel rods in a building, join the different parts of the process together. They are especially helpful in guiding readers from one part or stage of a process to another. Table 5–1 lists some connective words and phrases you will find useful in expressing relationships.

TABLE 5–1. CONNECTIVE WORDS AND PHRASES FOR TIME, SPACE, AND CAUSE

Time	*Space*	*Cause*
after	above	accordingly
afterward	adjacent to	also
again	behind	although
as long as	below	as a result
as soon as	beside	because
at the same time	close to	consequently
before	east, west, north, south of	for this reason
first, second, third	far from	generally
finally	front	hence
immediately	here	if . . . then
in the meantime	higher than	in effect
later	inside	in spite of
meanwhile	lower than	likewise
next	nearby	moreover
now	on the other side of	nevertheless
occasionally	opposite	on the other hand
often	outside	possibly
simultaneously	over	probably
since	parallel to	similarly
soon	rear	therefore
subsequently	side	thus
until	throughout	usually
when	under	
while	within	

Sentences in a process explanation may be in the active or passive voice, but chances are you will use the passive voice more often. Since the emphasis, in most instances, will be on scientific principles or the mechanics of a piece of equipment rather than on a human agent, the passive is more suitable than the active voice. Also use the present rather than the past tense for verbs (*drives, alternates, interlocks, vaporizes*) to stress the timelessness of the process. After all, you are explaining a process performed yesterday, today, or next year.

VISUALS

Visuals are an important part of a process explanation. By graphically depicting the sequence and relationship of events, they help to clarify your explanation of the process for the reader. Make visuals appropriate for your reader's needs. You would not want to include chemical formulas, mathematical equations, or complex schematic diagrams in a process explanation for the general reader. For this audience a simple flow chart might be appropriate.

A variety of visuals can clarify a process explanation. When you explain how something is made or how it works, a *cutaway drawing* is especially useful. It is also helpful to show a visual of a machine in operation. Among the most important types of visuals used in process explanations are *block diagrams,* or *flow charts.* A flow chart is a group of labeled boxes connected by arrows to show readers how a process progresses from start to finish. Flow charts can be organized to be read from top to bottom or from left to right, as the example of the process of removing and transporting seaweed after an oil spill has occurred shows:

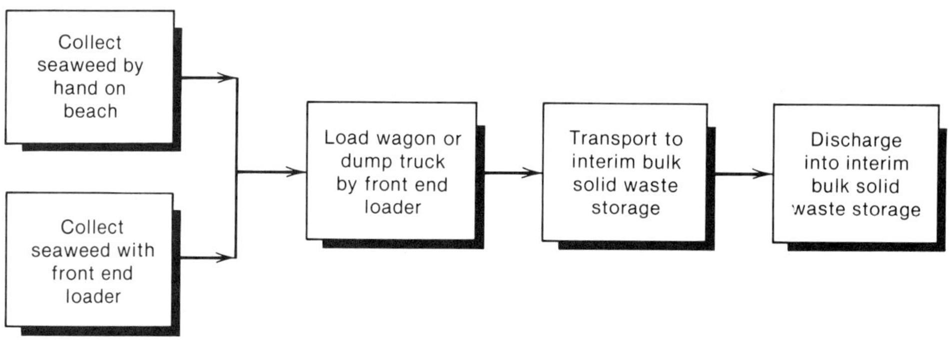

Some flow charts include pictorial representations to illustrate how a process is carried out. Figure 5–1, for example, includes pictures of coal, equipment, and dewatering and power plants to show the reader how coal moves through a slurry pipeline system. The reader can connect different stages of the process with the pictorial representations used in this flow chart. Other flow charts

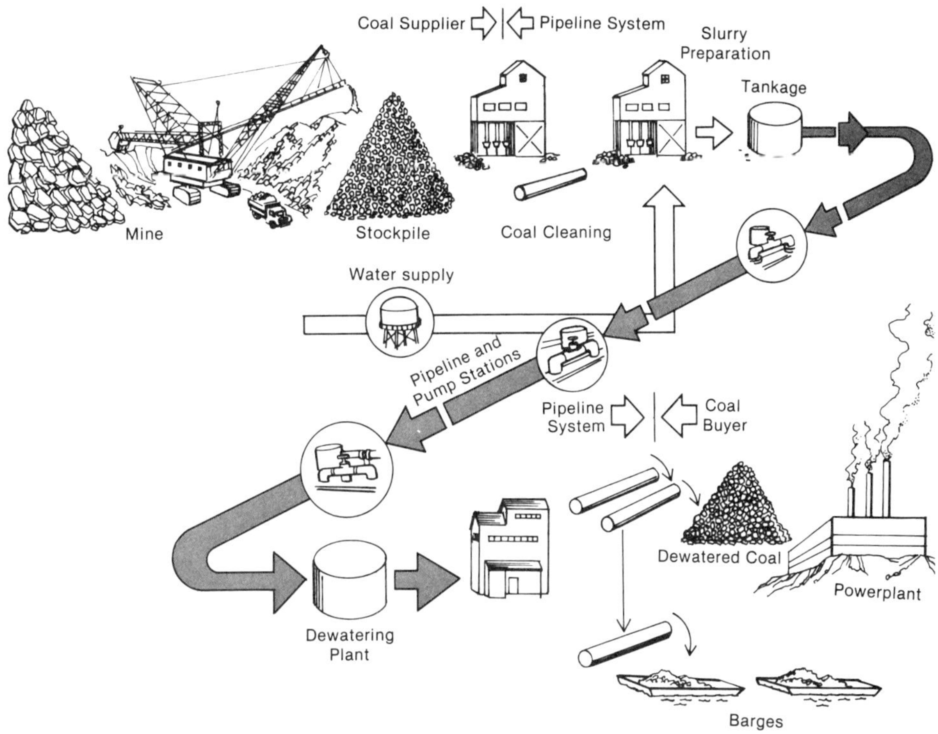

Figure 5.1. Flow chart of a coal slurry pipeline system.

can be more detailed and technical. Flow charts used by computer scientists are complex blueprints for preparing a program. Civil engineers use a critical-path diagram, another type of flow chart, and accountants and engineers employ project-task flow charts to assess an organization's progress on a project.

A PROCESS-EXPLANATION CHECKLIST

1. Divide a process explanation into at least three parts: introduction, body, and conclusion. In some short processes, however, there is no conclusion.
2. In the introduction define the process, identify its function and any operators, describe any equipment, and preview the principal steps.
3. List and discuss all the steps in order in the body; use lists, paragraphs, and relevant headings and subheadings.
4. In the conclusion summarize and evaluate the process, perhaps comparing it with other processes.
5. Link the various steps of the process with appropriate transitions.
6. Use the present tense, and the passive voice when necessary.
7. Select visuals that will help the reader see the various parts of the process—cutaway drawings and flow charts are especially useful.

The explanation of the brewing process below was used in a promotional campaign by the U.S. Brewers Assocation. It was intended for a general (consumer) audience that was unfamiliar with the way beer and ale are made. This audience did not need detailed, technical commentary on the process. Instead, their purpose in reading this model was to understand and enjoy, not perform, the steps in the brewing process. The flow-chart format with eye-catching visuals and succinct explanations is easy to follow and yet offers an accurate explanation of the steps in the process.

The Brewing Process

U.S. Brewers Association

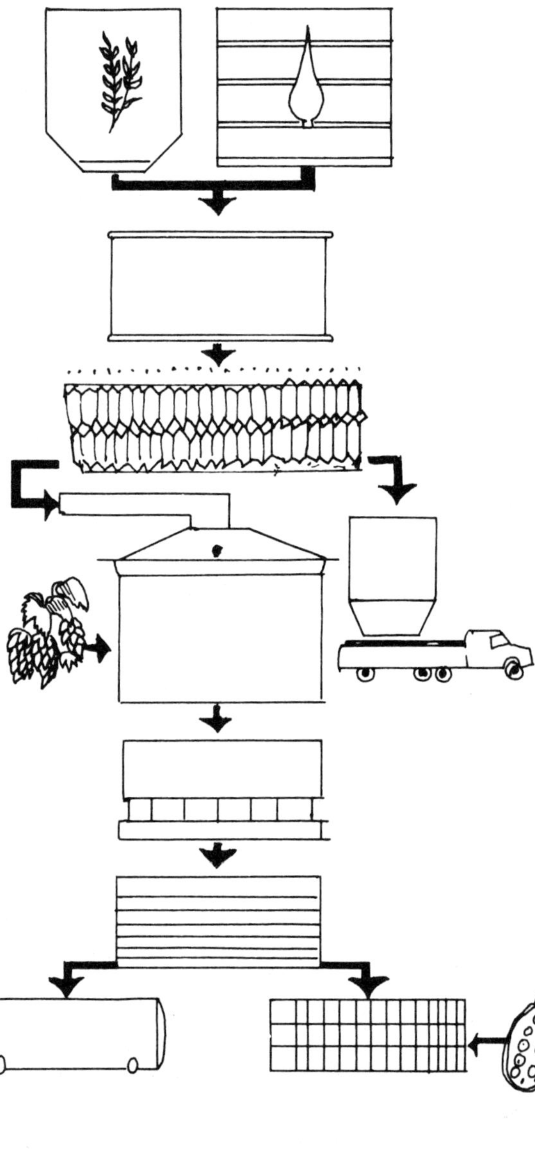

1. The brewing process begins at the malt and cereal storage bins, where malt, made from quality barley, is precisely measured.

2. In the mash tub, the malt is mixed with the heated, sterilized, and filtered water. This is stirred until the malt proteins are broken down.

3. A cereal mash of other grains is boiled in the cooker and then is added to the malt mash. Cooking changes the malt and grain starches into a fermentable sugar.

4. After this mixture is filtered through the lauter tub, spent grains are removed and are shipped out as livestock feed. The remaining liquid is called "wort." The wort flows into the brew kettles, where it is boiled with hops to give the beer or ale its pleasant, tangy flavor. The hops are then strained off.

5. Distinctively flavored, the wort now is pumped to a wort-collecting tank.

6. The wort then flows through coolers, where the temperature drops from about 212°F. to about 50°F.

7. Pumped into the open starter tank, the wort receives a portion of carefully cultured yeast and fermentation begins.

8. Now the wort and yeast go into fermenting tanks, where they remain about a week. Beer is made by a "bottom-fermentation" process, ale usually by "top-fermentation."

9. Fermentation over, the beer is then pumped into the aging tanks and remains in them for many weeks to age properly.

10. After carbonation and a final filtering, beer or ale to be served on draught goes to the racking room to be put into barrels that have been scrupulously cleaned. Beer or ale to be packaged goes to the bottling or canning plant where it is pasteurized after automatic machinery has filled and closed the surgically sterile containers.

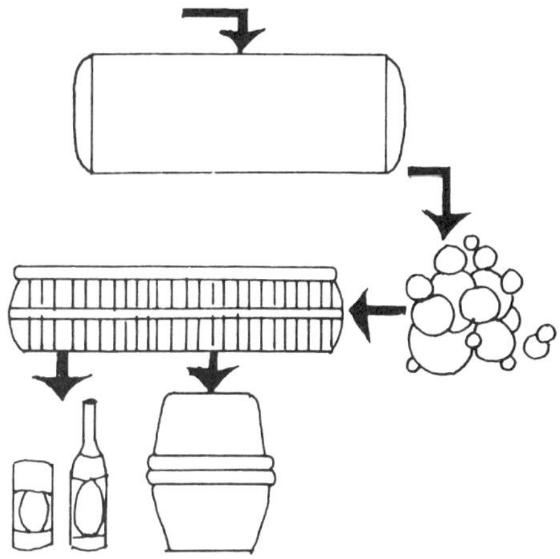

QUESTIONS FOR DISCUSSION

1. How are the individual steps of the brewing procedure distinguished for the reader? Comment on both visual and verbal (e.g., transitional) techniques.
2. Which of the steps of the process contain substeps? How are these substeps identified and why are they important for a reader unfamiliar with the process?
3. Describe the use of visuals in this model. What other kinds of visuals might have been appropriate?
4. How might this process explanation have been different if it were written for operators?
5. Contrast the explanation of the brewing process found in this model with the explanation contained in the ad for the PQ Corporation on page 409. Explain how the audience and purpose of each model guide the writer in the selection and presentation of details.

SUGGESTIONS FOR WRITING

1. Write an introduction for this model in which you explain its purpose, organization, and relevance to the reader. As part of your introduction, alert the reader to an especially important or difficult step.
2. Using "The Brewing Process" as a model, prepare a simple explanation (with appropriate visuals) of one of the following processes:

   a. transmitting a telephone call

   b. developing black-and-white film

c. pasteurizing milk
d. calling up a program or deleting a section of a text in a computer
e. making steel
f. purifying water
g. growing and harvesting a crop
h. bringing a criminal case to trial
i. making paper

The following model is an abridgement of a pamphlet entitled "Medicines and You" issued by the National Institute of Health. Explaining a biological process, the pamphlet is aimed at consumers who want to learn more about how the human body responds to drugs. After reading it, consumers will know, for example, why their doctors sometimes prescribe higher doses of a medication for the first day and smaller ones for subsequent days or why different dosages are prescribed for different age groups. To explain these and other processes for the intended audience, the author uses examples, definitions, comparisons and contrasts, and easy-to-read prose.

How Your Body Deals with Drugs and Other Chemicals

Maya Pines

Whether you swallow a pill, eat an orange, or drink water that is contaminated with insecticides, your body reacts in the same general way: it tries to break down the ingested matter, use parts of it, and excrete the rest. Your body does not know the difference between synthetic and natural compounds, nor between useful and harmful ones. 1

Suppose you have taken a pill. After you have swallowed it and the pill has dissolved, the solution must pass across various body membranes before reaching its target. The first barrier is the lining that separates the contents of your intestinal tract from your bloodstream. This lining is composed of cells whose membranes are made of lipids (fats), proteins, and other chemicals. Molecules that dissolve easily in fat will diffuse freely through the fat-like portions of these membranes, enter some of your blood capillaries, and go into your bloodstream. Other molecules, which dissolve in water, will have more difficulty passing through tiny pores or channels in these membranes, and may eventually be excreted without reaching your bloodstream. 2

The earliest form of the antibiotic tetracycline was not very lipid-soluble, for instance. Only a fraction could be absorbed into the bloodstream; the remainder stayed in the gastrointestinal tract, where it changed the bacterial flora and produced many digestive problems. More recent chemical forms of this group of antibiotics, such as doxycycline, are more lipid-soluble and produce fewer intestinal side effects. 3

Once a medicine starts circulating in the blood, it is distributed to all parts of your body. Simultaneously with this process of absorption, however, your body begins a process of elimination: it tries to get rid of the drug. Your kidneys filter out water-soluble molecules of the drug from the bloodstream and excrete them, unchanged, in your urine. As your blood flows through your liver, lipid- 4 soluble elements are drawn out by the liver's cells. Then your liver—your body's major chemical processing plant—gets to work on these substances and transforms them into more water-soluble compounds, which can be excreted in your urine.

How long, then, does each drug remain active in the body? This depends on the number of steps required for the drug to be eliminated as well as on your biochemical individuality. The time it takes for half of a given amount of medicine to be eliminated is called its half-life. The half-life of some drugs is only a few minutes, while that of others may be several weeks. The same drug can also have different half-lives in different persons. Such differences become particularly important when you need to have a certain amount of drug in your body constantly over a period of time. 5

When you start taking a medicine regularly, its concentration in your body increases with each of the first few doses: some of the first dose is still active in your body when you take the second dose, and so on. But eventually the concentration of the drug reaches a so-called "steady state"—a state of maximum efficacy for that particular dosage—during which as much of the drug is absorbed as is eliminated every 24 hours. It does not matter whether the drug is eliminated through your liver, kidneys, bile or sweat, as long as it leaves your body or is converted to an inactive form. Generally, four to five half-lives must elapse before a drug reaches this steady state. Therefore, if a drug has a half-life of only a few hours and you take it several times a day, it reaches a steady state within a single day. If the drug has a very long half-life, however, the process takes much longer. This is why it takes about a week for a drug such as digoxin (which is derived from digitalis and has a half-life of 1 to 5 days) to produce its maximum effect in increasing the heart's pumping ability. In order to reach a steady state more rapidly, when dealing with drugs with a long half-life, doctors sometimes prescribe a larger initial or "loading" dose. 6

For water-soluble drugs that are excreted in the urine unchanged, such as penicillin, certain heart drugs and lithium, the half-life depends mostly on the state of your kidneys. Kidney function varies greatly with age, deteriorating rapidly as one grows older. By the age of 80, even a healthy person may have only half the kidney function he had at 40. As a result, older people are usually more sensitive to certain drugs, which can accumulate in their bodies, and doctors may need to prescribe smaller doses for them than for younger people. Kidney function can be measured quite easily through "creatinine clearance" tests; based on the results, the dosage of certain drugs may need to be changed. 7

Most chemicals that enter your body are lipid-soluble, however, and must be transformed into more water-soluble substances before they can be excreted. While some of these biotransformations take place in your intestines, lungs or skin, by far the largest number occur in your liver. And here your biochemical individuality plays a major role: the liver's ability to metabolize or transform chemicals varies extremely widely, even more widely than kidney function. 8

Unfortunately, no simple method of measuring your ability to transform 9 specific chemicals yet exists. None of the commonly used tests of liver function provide any clue to it, though some promising experimental tests have been developed in research centers.

Because of the large individual variations in people's ability to metabolize 10
drugs, giving patients an "average" dose of a lipid-soluble medicine implies that it must be inappropriate for some of them. If your metabolism of this particular medicine is very rapid, the average dose may be insufficient to treat you, and if your metabolism is very slow, the medicine may accumulate to toxic levels in your blood.

QUESTIONS FOR DISCUSSION

1. Why do readers need the information contained in paragraph 1 before reading the rest of the explanation? Does this paragraph function as the introduction or is it one of the steps of the process? Explain.
2. How does the author organize her discussion of the process? Into how many major steps or stages is the process divided? Identify key sentences that announce these different stages.
3. How much and what type of technical detail, including information about theory or principles, does the author include about the process for her audience? Is this too much? too little? just enough? Explain.
4. The author is careful to define terms her reader may not know. What terms does she choose to define and which ones does she not? Why is it important to know what the defined terms mean? What techniques of definition does she employ?
5. Evaluate the author's style, including verbs, adverbs of time and cause and effect, transitions, sentence length. Also, explain why the second person "you" alternates with the third person "it" and "they." What reasons might explain this interchange?
6. Evaluate the appropriateness of the conclusion, given the audience and purpose of this process explanation.

SUGGESTIONS FOR WRITING

1. An explanation of the same processes would differ depending on the audience being addressed. Rewrite this process explanation for one of the following audiences: nurses, law enforcement officers who arrest drunken drivers, dietitians, managers concerned about employees' drug or alcohol problems, physiologists, or chemists.

2. Write a process explanation appropriate for consumers on one of the following topics or on another topic approved by your instructor.
   a. how the eyes see
   b. how the ears hear
   c. how the nose smells
   d. how the taste buds work
   e. how nicotine affects the lungs
   f. how static electricity is conducted through the body
   g. how a metal detector works
   h. how a filling prevents a cavity from getting larger
   i. how pollution affects the air we breathe
   j. how a food additive or dye changes the color of an object

The next model illustrates a conventional way of organizing a process explanation. The individual steps of the electrostatic copying process are first listed (here with bullets) and then each step is, in turn, taken up for discussion. Such an explanation helps the reader to see how a machine (a photocopier) operates. This model comes from Word Processing & Information Systems, *a trade magazine printing news and articles on computer applications in business communications, especially on automated dictation and transcription systems. Readers of this magazine are business managers with a knowledge of, and a background in, computer hardware and software, computer training programs, and information management systems.*

How the Electrostatic Copying Process Works

Copiers seem to be such simple machines. You place the material to be copied 1 on a glass surface, select the number of copies desired, and push a button marked print. Seconds later, copies of the document are released one after the other.

Actually, most copiers in use today utilize a fairly complex electrostatic 2 copying process.

The electrostatic copying process generally involves the following steps: 3

- Reflecting the image onto a photoconducting surface.
- Charging the photoconducting surface, thereby rendering it light sensitive so that it can form a latent image of the material being copied.
- Adhering ink particles to the appropriate charge areas of the photoconducting surface and thus creating a visible image.
- Transferring the image to a sheet of paper.
- Making the image permanent.
- Cleaning the photoconductor.

In order to copy a document, an optical system and a light source are 4 needed to see what is being copied. The light that illuminates the document is reflected through a system of lenses and mirrors or, in the case of a fiber optics system, through fine glass or plastic rods that direct the light to the photoconducting surface. (See Figure 1.)

The new intelligent image printers also use the same electrophotographic 5 process to produce a printed page. The only difference is that the light source exposing the photoconductor is a low-powered laser beam instead of the reflected light from a document.

Electrostatic machines employ some type of chemical compound, such 6 as silenium, zinc oxide, or cadmium sulfide on this photoconducting surface,

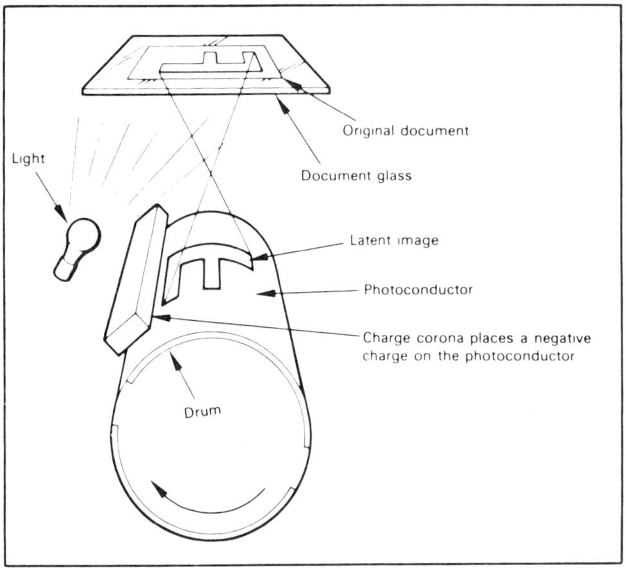

Figure 1. Charging the photoconductor and illuminating the original document causes a latent image to be formed on the photoconductor, which, in this illustration, is a drum.

which may be the paper or the machine's drum or belt. When the chemical compound is subjected to an electrostatic charge, the coating becomes light sensitive.

Illuminating a document causes the white areas to reflect light and reduce 7 the electrostatic charge on the corresponding areas of the photoconductor. The printed areas of the document reflect little or no light and cause the corresponding areas on the photoconductor to remain electrostatically charged. The result is a latent image of the original document electrostatically stored on the photoconducting surface.

To create a visible image on the photoconductor (this can be paper, a 8 drum, or a belt), liquid or dry toner is applied to its surface. The ink adheres to those areas that are electrostatically charged. Putting the ink directly on coated paper is known as a direct electrostatic process. All that remains to be done at this point is to make the image permanent. (See Figure 2.)

Transferring the visible image from a drum or belt to a plain bond sheet 9 of paper is an indirect process. The developed image is transferred onto the paper when the paper comes in contact with the photoconductor. A charge corona places an electrostatic charge on the paper just before it contacts the photoconductor. This charge attracts the toner away from the photoconductor and holds it on the paper as the transferred image.

The paper then moves away from the drum and passes between rollers 10 that may be heated to melt the toner particles so that they adhere to the paper. Cold pressure and evaporation (using a fan) are alternative methods of bonding the ink to the paper.

The copy is delivered and the excess ink removed from the photoconductor 11

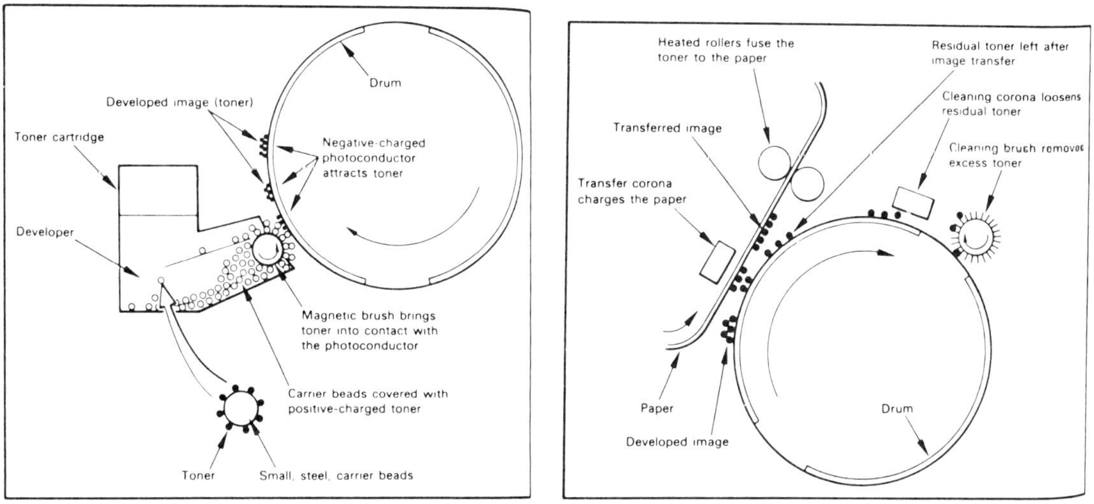

Figure 2. *Left:* illustration shows how carrier beads covered with positive-charged toner are attracted to negative-charged photoconductor areas. At *right,* the image is transferred onto charged paper (unless the paper itself serves as the photoconductor and already contains the image) and made permanent through a fusing process. Any residual toner remaining on the photoconductor after the image transfer is loosened and removed by a cleaning brush and then returned to the toner cartridge for reuse.

before the next copy is made. This is accomplished by passing the rotating photoconductor under a cleaning corona, which removes any remaining electrostatic charge. As the toner loosens, a cleaning brush removes the excess toner from the photoconductor.

The removed toner is then returned to the developer for reuse and the 12
machine is ready to repeat the copying process the next time someone walks up and pushes the print button.

Intelligent copiers use much the same process. However, instead of requir- 13
ing hard copy input, they accept digital information. At the time the text is processed, it is formatted (one page at a time) and stored in the copier's memory (page buffer). When a page is ready for printing, the processor takes the data from the page buffer and translates digital bit information into dot patterns required for the individual characters depending upon the type style, size, and line spacing used. These patterns are then used to modulate a laser beam to form the character on the photoconductor.

The low-powered laser beam scans across the length of the photoconductor. 14
Each scan represents one column of the dot pattern. Scanning is repeated, creating additional columns of the dot pattern, until the entire matrix of information (one page) becomes a latent image on the photoconductor. The latent image is then developed, transferred to paper, and made permanent.

The copy is then ready for distribution. 15

QUESTIONS FOR DISCUSSION

1. Explain how the author has adapted the process explanation successfully for the intended audience. As part of your answer discuss the terms and level of technical information the author uses.
2. Separate within the article each of the procedural steps summarized in paragraph 3. How does the reader know when one step is over and another begins?
3. Where does the writer discuss equipment used in the process? Why is the information important to the reader?
4. Why, in the body of this process explanation, does the author differentiate between a direct and indirect process?
5. What type of conclusion does the author provide? Why are intelligent copiers discussed?
6. Are the three illustrations helpful in explaining the electrostatic copying process? Does the caption to each illustration assist the reader significantly? Does it matter that the captions are a repetition of material from the body of the article? Is that an advantage or a disadvantage?

SUGGESTIONS FOR WRITING

1. Imitating the summary opening technique of this model, explain how one of the following works:
   a. a light bulb
   b. a transformer for an electric train set
   c. a flush toilet
   d. an automatic garage door
   e. a drill press
   f. a milking machine
   g. an escalator
   h. an IVAC machine or IPPB machine
   i. power disk brakes
   j. a transistor radio
   k. a car battery
2. Rewrite the electrostatic copier model for a consumer audience. Specify what, if any, visuals you would use.

This last and longest model of a process explanation is a section of a field operations and enforcement manual for Environmental Protection Agency (EPA) inspectors. The manual describes the major targets of air pollution control at a number of industrial sites (e.g., petroleum refineries, foundries, paint and varnish manufacturers, and cement plants). Because of the complexity of these industrial operations, "training and familiarization on the part of the enforcement officer is required before he or she can assume complete responsibility for the inspection of these plants." The manual provides the necessary background information and explains the relevant inspection procedures in order to "prepare enforcement officers" to enter and investigate "facilities that may be unfamiliar to them." These officers are viewed principally as experts. The model illustrates how process explanation is employed for on-the-job writing assignments, especially for procedure manuals that are major examples of process explanations.

EPA Air Pollution Inspection Procedures for Cement Plants

Melvin I. Weisburd

A. DESCRIPTION OF SOURCE

Portland cement takes its name from a fine building stone it resembles when made into concrete, which is quarried on the isle of Portland, England. Portland cement is manufactured in 180 plants in the United States. The raw materials—limestone, cement rock, clay and iron ore—are either blended dry or mixed with water to form a slurry and then burned into cement clinkers. These two processes account for all of the 500,000,000 barrels of Portland cement produced in the United States.[1] 1

Due to the increasing use of Portland cement, the anticipated annual growth rate of this industry is 5% per year with a trend toward larger plants. 2

Cement plants are located throughout the United States with the greatest production occurring in eastern Pennsylvania, Maryland, Alabama, Missouri, central Texas, California, and the Pacific Northwest.[2] 3

B. PROCESS DESCRIPTION

Basic operations employed in cement plants are crushing and grinding, blending, materials handling, clinker production and cooling, and fuel preparation for 4

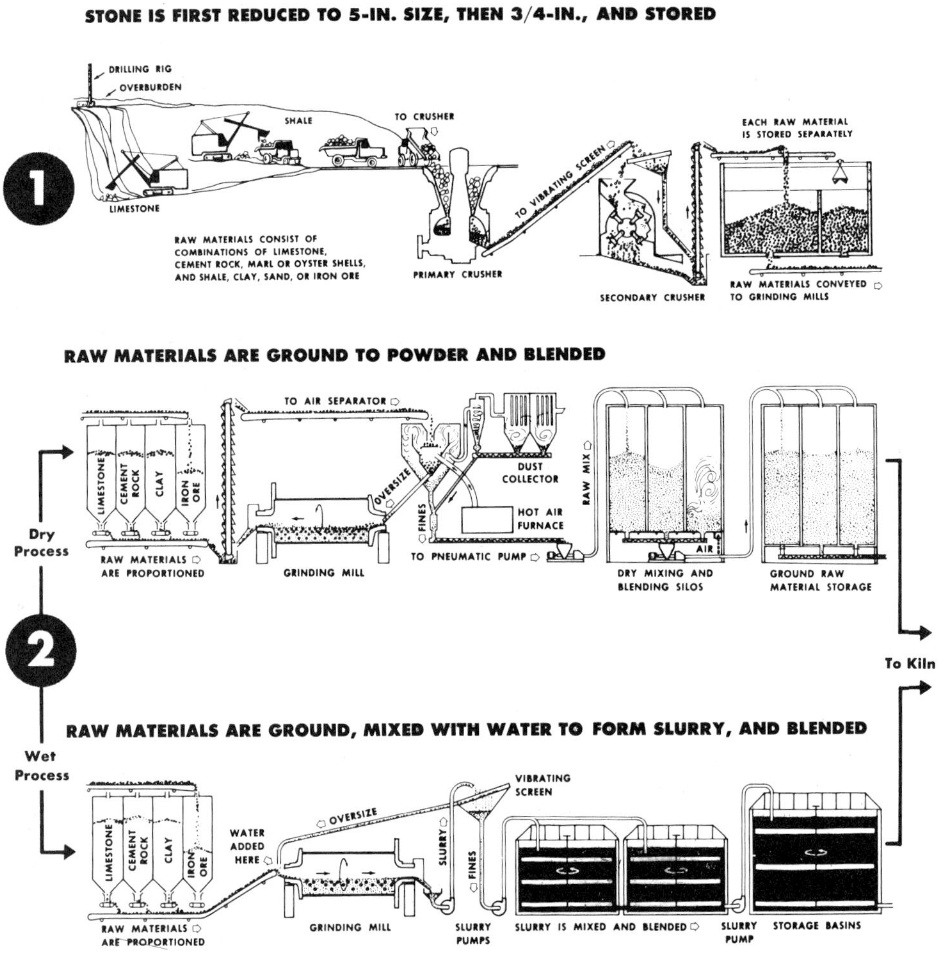

Figure 1. Steps in the manufacture of portland cement.

coal-fired plants. Figure 1 depicts the sequence of operations and major differences between the wet and dry processes.[3]

The process begins with quarrying of the raw materials, transporting by truck, rail, or barge to primary crushers for initial size reduction, and transporting by conveyor to secondary crushers for further size reduction and storage (Figure 1). In the *wet process* the raw materials are proportioned according to the desired cement specifications and mechanically conveyed to a grinding mill where water is added. The resulting slurry is pumped to a vibrating screen, which passes the wet mixture containing the specified size fines to the blending tanks and diverts the oversized material back to the grinder. The slurry is mixed and blended and is then pumped into a final storage vessel before introduction into the kiln.

BURNING CHANGES RAW MIX CHEMICALLY INTO CEMENT CLINKER

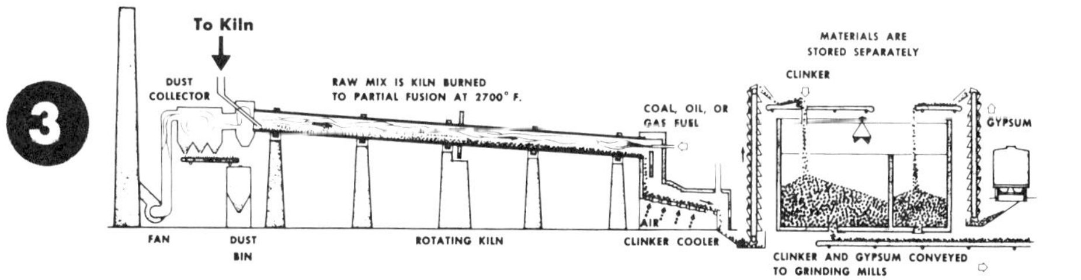

CLINKER WITH GYPSUM ADDED IS GROUND INTO PORTLAND CEMENT AND SHIPPED

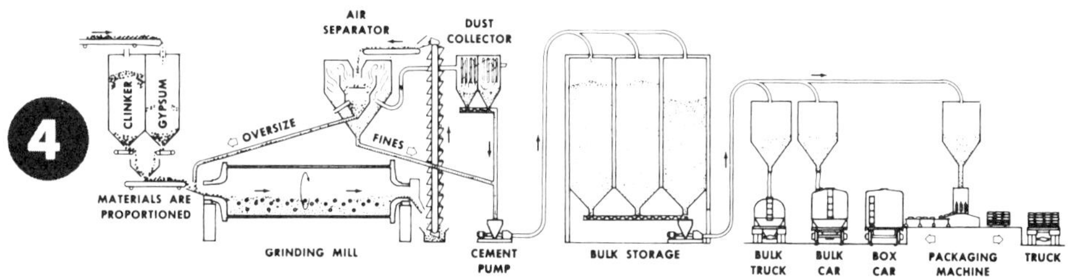

In the *dry process* the proportioned raw materials are run through an air separator (where drying takes place by heated air from a furnace), which allows the proper size fines to be pneumatically pumped to the dry mixing and blending silos and recycles the oversized material back to the grinder. The blended mix is then pneumatically conveyed to the ground raw material storage silos from which the material is introduced into the kiln. In both processes, the slurry or dry mix is fed into the elevated end of the kiln. Firing takes place at the opposite end of the kiln, giving a heat flow counter to the material flow. The fuel can be powdered coal, oil, or gas. The raw mix is burned to partial fusion (to produce a cement clinker) at 2700°F, air cooled, proportioned with gypsum, ground in the same manner as the dry mix process and stored in bulk silos for ultimate bagging or bulk shipment. 6

The heart of the cement plant is the cylindrical rotary ceramic-lined kiln, which is also the largest single source of dust emission in the entire operation. Kilns range in size from 6 feet in diameter and 60 feet in length to 25 feet in diameter and 760 feet in length.[2] The kilns revolve on huge roller bearings and slope from the feed end down to the discharge and firing end at the rate of from ½″ to ¾″ per lineal foot of kiln. 7

C. EMISSIONS AND CONTROLS

Every operation in the cement production process, by its nature, produces dust. 8
Dustfall measurements made in the Lehigh Valley in Pennsylvania, near plants

using modern air pollution control equipment, have recorded a dustfall rate of 35 tons/sq. mi./month. This area has a high concentration of plants producing 31,000,000 bbl/year of cement. Even with collection efficiencies of 90%, as much as 10 tons/day of dust can become airborne from a 4000 bbl/day plant because of the heavy grain loading of the kiln gas.[2]

Test data[4] of dust emissions from cement kilns are as follows: 9

- Wet process—dust from kilns is 1 to 33% of finished cement (4 to 124 lbs/bbl).
  Arithmetic average of dust loadings in exhaust gases is 10.1% of the finished product.
- Dry process—dust from kilns is 1 to 25% of finished cement (4 to 94 lbs/bbl).
  Arithmetic average of dust loadings in exhaust gases is 11.3% of the finished product.

Technological changes for processing equipment will probably be concen- 10
trated in recycling collected kiln dust, which may affect firing rates and the shape of the kiln. Increased collection efficiency of air pollution control equipment to 99.5% + is essential to reduce kiln-dust emissions. Table 1 presents the representative dust loading of kiln gases from the wet and dry process in kiln dust in exhaust gases.[2]

Most of the sulfur dioxide formed in cement kilns as a result of the combus- 11
tion of fuel is recovered as it combines with the alkalies. Tests of a coal-fired kiln burning 2.8% sulfur coal showed a range of 6 to 39 ppm SO_2.[4] There is no test data available on NO_X production, but its formation is a distinct possibility with kiln temperatures in excess of 2600°F. Hydrogen sulfide and polysulfides may also be produced during drying of the slurry or in the dry process when marl, sea shells, shale or clay are the materials processed.

TABLE 1. DUST LOADING CHARACTERISTICS

Kiln Type	*Dust Loading*[a] gr/ft^3
Wet	1 to 13
	10 to 13[b]
Dry	1 to 12
	20 to 55[b]

[a] Before collection at stack conditions (400 to 600°F)
[b] New long kiln utilizing chains and/or lifters
SOURCE: Sussman, Reference 2

Processing controls vary with the age and size of the plant. Kiln tempera- 12
tures, which have a bearing on the firing rate and the raw mix feed rate, are indicated and usually recorded. Combustion equipment and controls will also vary; however, plant operators may be able to provide information even though data recording equipment may not be available.

A variety of dust control equipment is in use including bag-type collectors, 13
electrical precipitators, and multiclones. Generally, electrical precipitators are used to collect the dust entrained in the kiln effluent because of the high temperature of the exhaust gas. Baghouses using glass fabric bags have shown efficiencies of 99.5%+[4] and should find increasing use in this industry. Dusts resulting from storage (silos are under positive pressure due to air displacement in loading or from pneumatic conveying), grinding, cooling, and materials handling are usually controlled by bag-type collectors or mechanical collectors.

Figure 2 represents a flow diagram of a cement plant and indicates the 14
points of dust emission. The kiln, which is the primary dust source, is usually vented to an electrical precipitator or baghouse. Screens, mills, storage bins, and packaging facilities are generally vented to mechanical collectors and baghouses in series. Transfer points, drop points, elevator boots and loading stations should be hooded and materials collected. Depending on cross drafts and physical shape, hood design inlet velocities of 200 ft/min or more may be required.

In coal-fired plants, pulverizing systems are used to prepare the coal for 15
burning in the kiln. While the pulverizing system is "closed" and uses mechanical separators as part of the system, this ancillary process can also produce large quantities of dust.

D. INSPECTION POINTS

The inspector should survey the plant to determine the adequacy of dust pickup 16
at any location where dust is generated and, if necessary, request stack tests to determine the efficiency of the dust control equipment. The dust emission locations shown in Figure 2, except for the kiln, are at transfer points, size-reduction equipment, classifying equipment (screens), and bulk-loading facilities. Dust pickup efficiency of the hoods serving these areas is indicative of air pollution control performance. For example:

1. Good dust pickup means the system is operating at or near its design efficiency.
2. Poor dust pickup may mean plugged lines, air bleeding in through worn or damaged duct work, damaged fan, drive belt slippage or undersized motor.

The unmistakable sign of a cement plant is the plume from the kiln stack. 17
Visible dust can be seen in the carry-over after the steam has dissipated. By comparison with previous observations, when the control device is known to have been operating effectively, the inspector can make a qualitative judgment as to the efficiency of the control equipment.

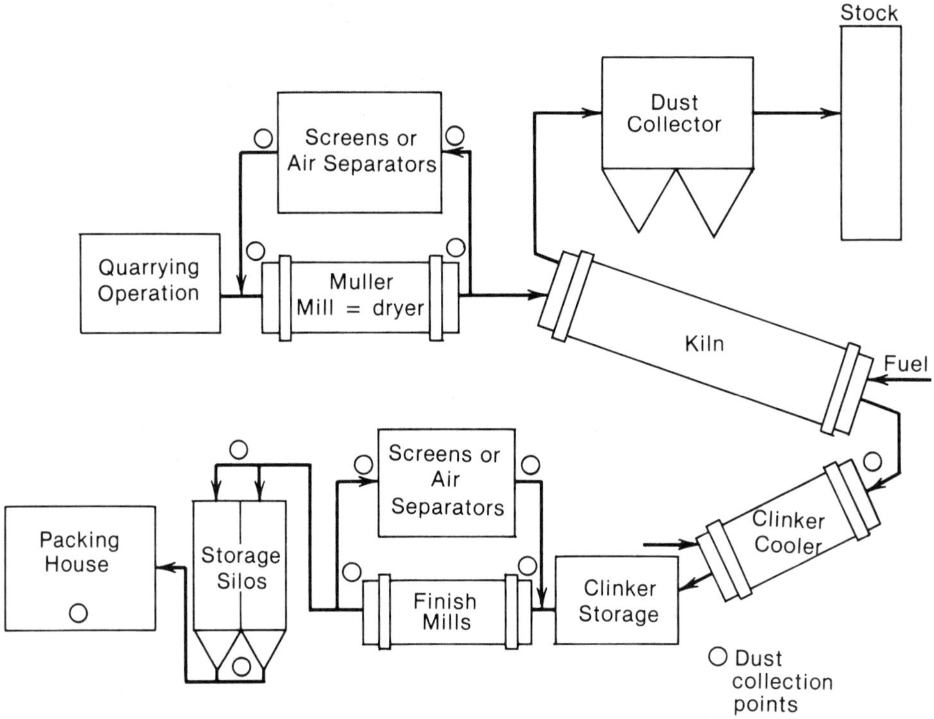

Figure 2. Flow diagram of cement plant operations (SOURCE: Sussman, Reference 2)

A check of the control room may provide evidence of malfunction of the equipment if there is a recording outlet opacity-indicator instrument. The inspector will find it extremely useful to make notes of the recording and nonrecording instrumentation, which can include kiln temperature, fuel flow measurements and gas analyzers for combustion performance. 18

Electrical precipitators, for example, are sensitive to the temperature and moisture content of the inlet gases. Resistivity (resistance to relinquishing negative charge to the collection electrode) for most particles increases sharply when the gases entering the precipitator are between 200° and 400°F. and less than 5% moisture. Collection efficiency is adversely affected by low resistivity. Thus the inspector should become familiar with the gas conditioning requirements of the precipitator and check the control instrumentation to see if the inlet conditions are in the proper range. The inspector should also check the rapping or washing cycle for removal of dust collected on the electrodes. Reentrainment of dust can occur if this cleaning function is not precise. Dust reentrainment can increase the opacity of the effluent and raise the exhaust grain loading. 19

Baghouses using glass fabric filters are extremely efficient air pollution control devices for cement kilns. The major operational problem is to maintain the temperature of the gases in the baghouse above their dew point. Insulation 20 of the baghouse enclosure, fan, and ducts is helpful in maintaining the proper temperature. Good operational procedures also require the exhaust fan to remain in operation during down time to prevent moisture buildup in the bags.

Inspection fundamentals apply regarding observations of equipment repair, 21
non-use of dust collection equipment, and housekeeping. Inspections should extend to roads and uncovered areas where dust has collected. Since the material is very fine, it can easily be disturbed by truck traffic and contribute to the background count of particulates for the area, as well as creating a public nuisance.

Dust collected from the kiln may be reinjected into the kiln, transported 22
dry by materials handling equipment to a waste disposal area where it is dampened before it is dumped, or mixed with water to form a slurry and pumped to a waste pond.

References

1. *Control Techniques for Particulate Air Pollutants.* Washington, D.C.: DHEW, PHS, NAPCA, January 1969.
2. Sussman, V. H. "Nonmetallic Mineral Products Industries." *Air Pollution,* Vol. III, A. C. Stern (ed.). New York: Academic Press, 1968.
3. *The Making of Portland Cement.* Skokie, Illinois: Portland Cement Association.
4. Kreichhelf, T. E., D. A. Kemnitz, and S. T. Cuffe. *Atmospheric Emissions from the Manufacture of Portland Cement.* Cincinnati, Ohio: DHEW, PHS, 1967.
5. Simon, H. *"Single-Stage Electrical Precipitators,"* in: *Air Pollution Engineering Manual,* J. A. Danielson (ed.). Cincinnati, Ohio: DHEW, PHS, National Center for Air Pollution Control and the Los Angeles County Air Pollution Control District, PHS No. 999-AP-40, 1967.

QUESTIONS FOR DISCUSSION

1. The section labeled "Description of Source" functions as an introduction. What kinds of background information does this section contain and how are these details useful to EPA inspectors?
2. How is the actual explanation of the cement-making process (types of details included) relevant to the EPA inspectors? What details does it include that might not be necessary for the following people?
   a. cement-plant workers
   b. homeowners
   c. general contractors
   d. decision makers
3. How does the section "Process Explanation" help the reader to understand the next two sections—"Emissions and Controls" and "Inspection Points?" How is process explanation used in these latter sections?
4. This model from the EPA manual pays a great deal of attention to equipment. Find examples emphasizing the importance of equipment to the processes here and explain why the manual includes this information for inspectors.
5. Visuals play an important part in the inspectors' training. How will the visuals included for the cement plant specifically help to prepare this audi-

ence for their inspections? How does Figure 1 differ from Figure 2? Why are both necessary? Why didn't the author of the manual begin the discussion with the flow diagram?

6. As the introduction to the chapter and previous models have shown, a process explanation contains many elements to be effective. To review these elements and to understand them better and use them in your writing, identify the following elements of a process explanation from this manual:

a. preview of the major steps
b. individual stages of the process
c. the role of the operator
d. advantages or disadvantages of the process
e. transitional words and phrases indicating temporal, spatial, and causal relationships

SUGGESTIONS FOR WRITING

1. Rewrite the process explanation of how Portland cement is made for one of the audiences cited in Question 2 above.

2. Write a brief explanation of a process people in your major field need to understand before starting a job. In your introduction inform readers that your explanation will help them complete one part of their orientation.

Part 3

Applications

chapter **6**

Instructions

Whereas a process explanation helps a reader to understand an operation (see Chapter 5), instructions describe a process that your reader will actually perform. He or she will use your instructions to get a specific job done. For that job to be accomplished safely and correctly, your instructions must be clear and easy to follow. Moreover, they have to contain everything necessary for a reader to complete the job. Missing or misplaced information can be confusing and possibly dangerous. Remember that when you write a set of instructions, you are writing for individuals who have less knowledge and experience than you do about the process they have to complete. And if trouble comes, your instructions will be like a life preserver keeping the reader from harm.

Instructions are found in many job settings. They show employees how to assemble, operate, maintain, or disassemble a piece of equipment. They may also inform workers how to follow a procedure, whether it is completing a job order or going through a security clearance. Written for consumers, instructions can describe an infinite number of tasks.

SPECIAL CONSIDERATIONS OF AUDIENCE

The most important part of preparing instructions is determining how much your audience knows. Instructions that ignore the needs of the reader are not only hard to use but can also be dangerous. For example, if you omitted an important detail (a piece of equipment should be assembled outdoors, not in the plant) because you thought a reader knew it and didn't, your instructions could cause serious injury. Remember that you are legally responsible for what you write. If a reader—employee or consumer—is injured because your instructions are incomplete or inaccurately written, you and your company might be held liable for damages.

Consider your readers' background, education, and purpose. Will you be writing for a do-it-yourselfer who has never done anything like this before or for a technician who has performed similar steps many times? Will your readers need a few basic explanations or will they require detailed technical descriptions throughout the procedure? The types of procedures completed by consumers do not often require the extensive precautions and regulations crucial to technicians who must repair sophisticated equipment. Whoever your audience is, never assume that they have your exact technical experience. If you are describing a procedure you have performed many times, you will undoubtedly need to remind yourself that every step is likely to be new to your readers.

No matter what your readers' background, you can meet their needs by following two important guidelines:

1. *Regard your readers as active participants in the process you want them to complete.* Consider them as your coworkers, even if they are not. Write as if you and they are working to accomplish a common goal.

2. *Consider your instructions an essential piece of equipment readers need to get a job done.* Give as much attention to your writing as you would to a precision instrument necessary for successfully performing the procedure you are describing.

One way to be sure that your instructions leave nothing out is to anticipate questions your audience may ask as they read. Following are some important questions readers may have:

1. What am I already expected to know? Are any particular skills required?
2. What materials, tools, equipment, or conditions are needed?
3. How long will it take to do each step of the procedure?
4. In what order must I follow those steps?
5. Are there any shortcuts or alternate steps?
6. Are there any precautions to take? Where and why?
7. What should I do if problems arise? What measures must I follow?

You must also consider the content, style, visuals, format, and safety notices.

CONTENT

The information you include will depend upon your audience. Instructions on installing home insulation for the do-it-yourselfer would correctly provide definitions and explanations of the "R value" for thermal resistance, the types of insulation available (fiberglass bats, cellulose, loose-fill fiberglass), and reasons for performing certain procedures, e.g., smoothing out plastic coverings. Such background information would be unnecessary for the professional home contractor doing the job.

You will also have to know when to explain a step for one audience and whether to delete that same step for another audience. Details that would be obvious to technicians might need elaborate explanations for consumers. For example, consider preparing instructions on how to administer insulin for two different audiences—the patient (consumer) and the nurse (expert). The patient would require detailed explanations describing how to clean the vial and how to withdraw and inject the insulin so that equipment, material, and the patient's skin are not contaminated. For the nurse, the direction "Maintain sterile technique" would normally be sufficient. For any audience, include only those steps that tell how to get a specific job done and that explain what to do if something goes wrong.

The best way to figure out which information to include in your instructions is to set about performing the task yourself, recording every step you take as you go. Following are some guidelines to prepare clear and useful instructions.

1. *Determine the exact task to be accomplished using your instructions.*

2. *Research the equipment or procedure to be employed.* If possible, speak with people knowledgeable about the topic. You may want to request a conference with an engineer or technician who is in charge of the equipment or with an executive who formulated the policy upon which the instructions will be based. Watch the product or equipment in operation or observe the procedure being performed. Obtain any available charts, models, or schematic diagrams. Review pertinent literature in professional journals, handbooks, technical bulletins, or specification sheets. By gathering this information, you will be better able to write about theory or mechanisms whenever necessary for readers.

3. *Collect the materials and equipment necessary to carry out the instructions.* Determine if the reader needs special sizes, models, or brands or if alternate tools or materials can be used. Where relevant, pay attention to any special conditions or handling of tools or materials. For example, note if something must be first cooled or heated. Keep detailed notes on what you gather so that you can record these facts—often in a separate section entitled "List of Equipment and Materials."

4. *Perform the procedure, step by step.* Be careful not to skip a step because you think a reader will already be familiar with the equipment or procedure. Take notes on the steps you perform. Be sure steps do not overlap or include too much information.

After you have compiled all the information necessary for the instructions and prepared a first draft, you will have to test your own instructions. First, try performing the steps yourself, following your own written instructions. Then, ask two or three members of the intended audience to try performing the task with your instructions. Revise your draft as long as readers have problems.

STYLE

Your audience governs the language you use. Consumers may not understand certain technical vocabulary or jargon, whereas technicians and often operators might not understand instructions without familiar jargon. Always consider whether you are writing for consumers, operators, decision makers, or experts.

Preparing instructions for different audiences requires skill. Professional health care personnel initiating cardiopulmonary resuscitation (CPR) can be told, "Find the carotid pulse," but this instruction has to be expressed in different, more understandable terms for the layperson: "Place your fingers on the Adam's apple and slide them into the groove in the side of the neck nearest to you." Consider how instructions on measuring the tension of a fan belt might be written for a mechanic as opposed to a consumer. The mechanic might be told, "Measure the belt tension and adjust it to 70 lbs. using a belt-tension gauge." Not having or needing a gauge, the consumer would be instructed, "Press down on the fan belt. If there is more than one-half inch of play, tighten the belt."

When you write instructions for a large, undifferentiated audience, you will have to define all technical terms. What exactly constitutes a "technical term"? Any word your readers might not know. Just because you may know a meaning of a word, do not assume your readers will, too. For example, in recent years there has been much research on the language of over-the-counter drug labels. The general public often misunderstands terms that seem very obvious to health researchers. Some college freshmen, in answering a questionnaire about over-the-counter drugs, thought that prostate trouble is a malady that afflicts elderly women. To avoid any possible confusion and legal trouble, choose your terms carefully, and define any that may confuse a nonspecialist.

Regardless of your reader's background, your instructions should be authoritative and easy to understand and carry out. Sure ways to make instructions authoritative and clear are to use imperative or active verbs, to give precise measurements, to use terms consistently, and to give short steps.

1. *Use verbs in the imperative mood and active voice.* Instructions are easiest when given as commands. Rather than writing "You might keep the contents under pressure for ten minutes," state "Seal top and keep the contents under pressure for ten minutes." In the first sentence, the word *might* raises a question in the reader's mind—should I or shouldn't I do this? Do I have a choice? Similarly, the passive voice—*might be kept*—does not clearly state *who* exactly is to do the pressurizing. The imperative "Keep the contents under pressure" leaves no such doubt.

2. *Include precise—and accurate—measurements.* Vague commands such as "Pour a little," "Adjust adequately," "Wait a while," or "Use the right size tool" offer little or no help to users who may question whether they have done that part of the job correctly. The following examples tell the reader precisely what to do:

> develop film at 68°
> fingerprint on 8 x 8 card stock
> place one drop of the patient's serum on a slide
> move bolt wrench ¾ of an inch

Exact language will help to lessen the readers' anxiety and assist them to get the job done right the first time.

3. *Use terms consistently.* To avoid confusing the reader, call a piece of equipment or a part by the same name each time you refer to it. For example, in a set of instructions where the reader has to use different types of screws (e.g., truss head, round head, hex head), don't refer to a "truss head screw" in one place and simply a "screw" or a "head screw" in another. The reader will not know what type of screw to use. Be consistent—always refer to it as a truss head screw.

4. *Break instructions into short, simple steps.* Your instructions will be easier for a reader to follow if they are separated into small, easy-to-perform steps. Be careful not to overload the reader by combining a number of separate steps into one step. By performing your own instructions ahead of time (see page 186), you should be able to separate a set of instructions into discrete steps. To further assist your reader, number each step so it is easy to find and to perform.

VISUALS

Visuals can be extremely helpful in assisting readers to understand your instructions. Photographs, drawings, sketches, and diagrams will considerably reduce the chances of a reader misunderstanding a direction. Visuals can show relationships and steps more economically and, many times, more clearly than words can. They also help reduce the number of words necessary. Moreover, a good visual can really help your readers to see what they must do. What would your reaction be to having to perform a set of instructions that contained no visuals as opposed to one that was abundantly illustrated? All well-designed visuals are effective, but exploded drawings are especially useful when you are showing readers how to assemble or disassemble a piece of equipment. These drawings, such as the one on page 195, show which piece goes first and often illustrate how to attach or detach one piece of equipment. The following guidelines will help you to incorporate your visuals effectively in your instructions:

1. Put visuals as close as possible to the text they refer to. Try to place them immediately after a step or at least on the same 2-page spread. Don't make readers search for a visual three or four pages away from the instruction to which it applies, because they are likely to be reading your instructions while performing the task.

2. Key visuals to the individual instructions to which they belong. Label all visuals clearly. Label any parts mentioned in the instructions.

3. Make sure your visuals clearly depict the exact task you want readers to perform. In a drawing or photograph, represent *only* those parts or sections

readers need to see for that particular step. Unnecessary detail can interfere with and make instructions more difficult to carry out.

Besides visuals, the layout of your instructions can help readers to follow them easily. Instructions that are easy to read are easy to follow. Don't bunch them together so that they look cluttered. Separate different parts of a set of instructions with plenty of white space and headings or subheadings to make them easy to follow and recheck if necessary. Different sizes of type (capital letters, boldface, italics, etc.) are helpful because they can emphasize especially important information, such as warnings or cautions. Color, too, can alert the reader to the different kinds of precautions and checks needed in performing a procedure.

FORMAT

The models in this chapter illustrate many formats and methods of organization. Format refers to the way instructions are printed on the page; organization refers to the order and number of parts they have. (See Chapter 1 for a review of the concepts *format* and *organization.*) Instructions vary greatly in their complexity, depending on the audience's purpose in using them and the specific job the instructions are to accomplish. Following are several common formats instructions can take:

1. Instructions may be given *solely or predominantly through visuals.* Visuals can be quickly understood by many readers regardless of their background. Visuals alone, however, should be used only for very basic instructions that do not have many complicated steps and that can be accomplished sequentially, such as opening a carton, putting on an oxygen mask, and logging on to a computer. For some more technical procedures visuals are used to clarify the written instructions for each step of the procedure. See the model on pages 195–196, where visuals help the reader perform an installation procedure accurately and quickly.

2. A *decision tree format* (also called a *flow chart*) is useful for instructions that require the readers to make, and see the results of, a number of decisions before or as they complete a job. Decision trees are especially effective because they allow the reader to see clearly and quickly the variety of options that can be exercised. As the name implies, a decision tree contains branches with connecting lines and symbols—boxes, triangles, squares, trapezoids. The life support decision tree model included on page 199 prepares the reader confronted with almost any type of cardiac or respiratory emergency. Decision trees are often used to demonstrate computer programs.

3. The *numbered steps format* is used for a variety of jobs, technical and nontechnical alike. This format is the mainstay of instructions in technical manuals. It is also found in cookbooks and model kits and is used with a variety of

consumer appliances. Each step is numbered sequentially and is further separated from other steps by white space to make it stand out and apart. The reader can thus keep track of what he or she has done. (Some instructions in numbered steps have boxes next to them for users to check once they have completed a step.) These steps are organized in an accurate chronological sequence—the order in which they must be followed for the user to accomplish the procedure. Some instructions in this format divide steps into related groups—e.g., Priming the Machine, Operating the Machine, Shutting the Machine Down *or* Assembly, Disassembly, Maintenance. Consumers using a can of spray paint or following directions on how to change the oil in their car may require only a few numbered steps. Installers and operators of complex equipment, on the other hand, would need detailed groups of steps divided into many sections and subsections.

4. A special type of instructions known as *troubleshooting instructions* is frequently formed into tables or charts at the beginning or end of a manual. As the name suggests, troubleshooting instructions pinpoint problems the reader may encounter and provide specific directions in checklists on how to solve these problems. Troubleshooting instructions are lifesavers for the reader suddenly confronted with malfunctioning equipment or unexpected outcomes in following a procedure. For the user's convenience, the tabular format lists the general steps that should be taken when trouble occurs—1. identify the problem, 2. isolate its causes, 3. find a solution. Unlike instructions in the numbered steps format, where all the steps are followed, troubleshooting instructions allow the reader to see quickly a specific step to follow when a particular problem arises. Since one problem may have many causes or solutions, the reader must be given all possible options. These options might include changing or discarding a part, repeating a step, turning off a piece of equipment, or calling a service representative. Often, each solution is numbered, with the most likely solution listed first.

Writing troubleshooting instructions requires rigorous audience analysis. Quite literally, you have to put yourself in the reader's predicament. To do this, anticipate the major difficulties a reader could encounter, determine how a reader will see or hear them, and describe the problem and the solution in easily identifiable terms. List symptoms and cures in highly descriptive and precise language. Types of noise, smoke, leaks, or interruptions need to be identified exactly so that the reader knows what to look for and how to correct it.

SAFETY NOTICES: WARNINGS, CAUTIONS, AND NOTES

Warnings, cautions, and notes give readers "emergency information," alerting them to procedures that have special significance. A warning informs readers that personal injury or even loss of life can result if they fail to exercise special care.

WARNING

DANGEROUS VOLTAGES ARE PRESENT. MAKE SURE THAT POWER HAS NOT BEEN CONNECTED TO MODEL 325 BEFORE REMOVING LOWER REAR PANEL.

Step 10. Connect sensor to electronics enclosure.

A caution notice tells the reader that damage to equipment can occur if its message is not heeded.

Step 6. Lubricate rear axle.

CAUTION

DO NOT USE EP LUBRICANTS IN UNIT EQUIPPED WITH BUILT-IN BACKSTOPS. ROTOR DAMAGE WILL OCCUR.

A note gives the reader information that makes the procedure easier to follow. Notes can help the reader avoid a problem, suggest an alternate way of doing something, or affirm that a present condition is not dangerous.

Step 8. Remove the shaft from the plug.
NOTE: Difficulty is sometimes encountered in this step due to an accumulation of deposits between the plug and the shaft. Application of a drop or two of lubricant to the shaft bore will facilitate removal.

Step 15. Replace the retainer coupling with capscrews in the hub of the handwheel.
NOTE: Two holes drilled in the handwheel hub are provided to accept the capscrews in the hubwheel.

Warning, caution, and note statements should be prominently inserted immediately above or below the instructions they qualify. If a particular step is especially dangerous, place a *WARNING* notation *before* the step to alert the reader to the potential danger. A number of distinctive symbols and typefaces to signal warnings and precautions are used—skulls and cross bones, exclamation points in triangles, the words *WARNING* or *CAUTION* in red. Use warning, caution, and note statements wherever they are necessary for the reader's safety and to protect equipment. A single set of instructions may use these techniques repeatedly, as the model on pages 209–211 shows. However, never use them indiscriminately. If you insert them without necessary cause the reader may disregard them when they are most necessary. Don't be like the boy who cried wolf.

INSTRUCTIONS IN MANUALS

Detailed instructions are included in manuals (brochures, booklets, book-length documents) published by manufacturers explaining how to use their product. Manuals are issued for everything from lawn mowers to generators and often use a combination of formats. Consumers receive such manuals when they buy home appliances. Operators and technicians rely on manuals to install, operate, maintain, and troubleshoot the technical equipment they use on their jobs. Instructions in a manual are often divided into the following sections: Introduction;

CONTENTS

Figure 6.1. Table of Contents for Filbert Manual.

Theory or Principles of Operation; List of Necessary Tools and Materials; Description of Mechanism(s); Step-by-Step Directions on the Installation, Operation, and Maintainence of the Equipment; Conclusion. Figure 6–1 reprints the table of contents from a detailed manual that accompanies a pump used in water pollution control. As you will see, this table of contents modifies and extends the parts of a manual listed above.

You may never have to write a complete manual by yourself. That is a job assigned to a company team of engineers, technical writers, and public relations personnel. However, in your own instructions you may have to include the following parts depending on your audience's background and purpose.

The *Introduction* gives the reader information on who your manual is for, how to use it, and how it is organized. It tells the reader what particular models your instructions refer to and what kind of tools and materials to use. The introduction may also mention benefits for consumers and stress safety features for all readers.

The *Theory/Description of a Mechanism* sections may be combined or listed separately. Adjust the technical detail to the needs of your audience. For example, consumers would not need or want elaborate descriptions and diagrams illustrating theories and mechanisms for simple home appliances, such as blenders, steam irons, and snow blowers. Technicians, on the other hand, would require such information to install and service an expensive piece of equipment. Sometimes technical details require physical description and process explanation (see Chapters 2 and 5).

The *Step-by-Step Procedures* are the core of your instructions. You should follow the guidelines listed earlier in this chapter (pages 187–188).

The *Conclusion* can vary from a brief wrap-up for consumers congratulating them on successfully completing a job (and indicating what something should look or sound like) to detailed inspection procedures for technical readers. Troubleshooting charts, parts lists, and service policies often conclude a technical instruction manual.

A CHECKLIST FOR WRITING INSTRUCTIONS

1. Determine how much knowledge and experience your audience has about the object of your instructions and use appropriate detail and terminology.
2. Watch the procedure in operation and perform it yourself, step by step if possible.
3. Select an appropriate format (primarily through visuals, decision trees, numbered steps, etc.).
4. Use the imperative mood and the active voice.
5. Give precise measurements.
6. List all necessary tools, equipment, and conditions.
7. Alert the reader to any necessary precautions and to all potential difficulties.
8. Indicate how long each step should take.
9. Use visuals when they are helpful in carrying out your instructions.
10. Write a draft and use it to try to perform the procedure yourself. Ask someone else to do the same. Revise the draft as necessary.

The following come from a service manual on installing and maintaining steelflex couplings used in kilns and paper mills. These instructions were written by the Falk Company, designers and manufacturers of gear drives and shaft couplings. Together with other instructions for Falk equipment, they are included in a large binder used for handy reference by installers, operators, maintainance crews, and cost engineers. This model is especially effective for its use of visuals to give instructions. Four different types of visuals—an exploded drawing with callouts, a line drawing, a table, and photographs—help the reader to accomplish the job quickly and accurately. The first page of this model primarily provides reference information about safety precautions, lubricant specifications, and maintenance schedules. The second page contains the actual installation steps, each one illustrated by an appropriate photograph.

Instructions for Installation and Maintenance of Steelflex Couplings

Falk Company

Instructions for Installation and Maintenance

SIZES 3 thru 190

STEELFLEX COUPLINGS

Horizontal and Vertical

Type F

Subject to change without notice.

438-110

SERVICE MANUAL

March 1980

Supersedes 2-73

PAGE 1

TYPE F STEELFLEX COUPLINGS

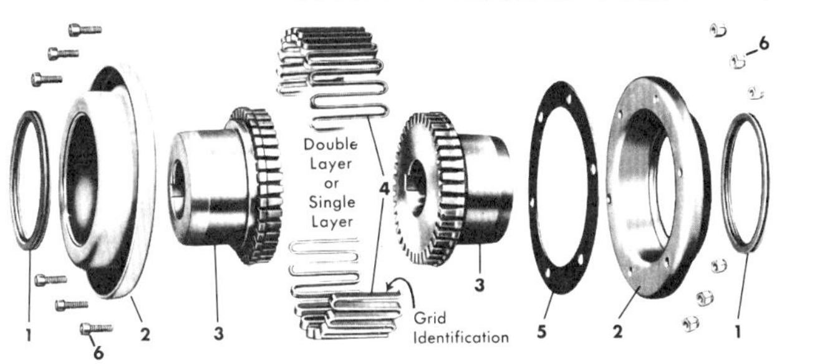

1. **Seal Ring**
2. **Half Cover**
3. **Hub (Specify bore and keyway)**
4. **Grid**
   INNER layer painted gray, stamped "IN"
   OUTER layer painted yellow, stamped "OUT"
5. **Gasket**
6. **Cap Screws and Nuts**

When ordering spare parts, specify hub bores and keyways and state coupling size as stamped on coupling cover and hub.

INSTALLATION — For best results, assemble standard couplings with minimum misalignment and to the gap shown in Table 1. Heat small interference fit hubs in an oven or in oil. Be sure that oil flash-point is 350°F (177°C) or higher. Apply flame heat evenly to large hubs, but DO NOT apply flame directly to the grid groove area. Also, DO NOT heat hubs over 275°F (135°C).

LIMITED END FLOAT — When electric motors, generators, engines, compressors ana other machines are fitted with sleeve or straight roller bearings, limited axial end float kits are recommended for protecting the bearings. Consult Falk for details.

GRID — DO NOT APPLY LOAD UNLESS complete grid is installed correctly. The number of segments and layers is listed in Table 1. Color coding is explained on Page 2 under "INSERT GRID." Grid identification is stamped at the end of the section as shown above.

Installation — It is necessary to spread the grid slightly so that it will pass over the coupling teeth at the outside diameter. To accomplish this with a minimum amount of spreading, start the grid at either end and tap the rungs only part way into the grooves. After all of the rungs are partially in their respective grooves, tap the grid all the way into place. When installing a two-layer grid, center the sections of the outer ("OUT") layer over the free ends of the inner ("IN") layer, see A above, and extend all cut ends in the same direction, see A and B above.

Removal — A round rod or screwdriver that will conveniently fit into the open loop ends of the grid is all that is required in the way of tools. Begin at the open end of the grid section and insert the rod or screwdriver into the loop ends. Use the teeth adjacent to each loop as a fulcrum and pry the grid out radially in even, gradual stages. Proceed alternately from side to side lifting the grid about half way out until the end of the grid is reached. By following the same procedure once again, the grid will clear the teeth.

LUBRICANT SPECIFICATIONS — Refer to Manual 428-010 for recommended lubricants. The following specifications apply to lubricants for Falk couplings which are lubricated annually and operate within ambient temperatures of 0° to 150°F (−18° to +66°C). For temperatures beyond this range, consult the Factory.

Dropping Point — 300°F (149°C) or higher.

Consistency — NLGI No. 2 with worked penetration value in the range of 250 to 300.

Separation and Resistance — Low oil separation rate and high resistance to separation from centrifuging.

Liquid Constituent — To possess good lubrication properties . . . equivalent to a high quality, well refined petroleum oil.

Inactive — Should not corrode steel or cause swelling or deterioration of neoprene.

Clean — Free from foreign inclusions.

Semi-permanent Lubrication — Up to eight years between lubrication checks may be obtained thru the use of "still-bottom" asphaltic based lubricants. Refer to Service Manual 428-012 for instructions.

MAINTENANCE — Adequate lubrication is essential for proper operation of the coupling. Lubricate at least once every twelve months. Lubricate oftener when the coupling is exposed to excessive moisture, extreme temperatures or rapid reversing or shock loads.

DISCONNECTING — To disconnect, clean coupling, oil hubs, remove cap screws and covers. Remove grid as shown at the left.

TABLE 1 INSTALLATION DATA*

Coupling Size	Max Speed rpm	Approx. Lubricant lb	Gap	Grid: No. of Layers	Grid: No. Segments Per Layer	Operating■ Alignment Limits: Offset (Max)	Operating■ Alignment Limits: Angular (Max)	Cover Bolt Torque (lb-in)
3	6000	1/16	1/8	1	1	.005	.005	100
4	6000	3/32	1/8	1	1	.005	.005	100
5	6000	1/8	1/8	1	1	.005	.005	100
6	6000	3/16	1/8	1	1	.005	.005	100
7	6000	3/16	1/8	1	2	.005	.005	100
8	5000	5/16	1/8	1	2	.010	.010	200
9	4500	3/8	1/8	1	2	.010	.010	200
10	3750	3/8	3/16	1	2	.010	.010	200
11	3600	1/2	3/16	1	2	.010	.010	200
12	3600	5/8	3/16	2	2	.012	.012	200
13	2700	3/4	3/16	2	2	.012	.012	200
14	2500	1 1/2	1/4	2	2	.012	.012	350
15	2400	1 1/2	1/4	2	2	.012	.012	280
16	2300	2	1/4	2	2	.012	.012	280
17	2200	2 3/4	1/4	2	2	.012	.012	280
18	2100	3 1/4	1/4	2	3	.012	.012	570
190	2000	8	1/4	2	4	.015	.015	570

* Refer to Bulletin 431-110 for maximum bores and Engineering Sheet 427-108 for reboring instructions.

■ Align couplings within "Operating Alignment Limits" specified above. Exceeding these limits reduces coupling life.

INSTALLATION OF TYPE F STEELFLEX COUPLINGS

Lock out starting switch of prime mover. Clean all metal parts using a non-flammable solvent.

COUPLING GAP

Maintain the coupling gap shown on Page 1 whenever possible.

CAUTION: Consult applicable local and national safety codes for proper guarding of rotating members.

1 MOUNT COVER, SEAL RING AND HUB ON SHAFT

Lightly coat seals with grease and install in covers. Place covers on shafts BEFORE mounting hubs. Mount hubs on shafts with hub faces flush with shaft ends. Tighten set screws when furnished.

2 GAP AND ANGULAR ALIGNMENT

Use a spacer bar equal in thickness to the gap specified in Table 1. Insert bar, as shown above, to same depth at 90° intervals and measure clearance between bar and hub face with feelers. The difference in minimum and maximum measurements should not exceed the ANGULAR limit specified in Table 1.

3 OFFSET ALIGNMENT

Align so that a straight edge rests squarely (or within the limits specified in Table 1) on both hubs as shown above and also at 90° intervals. Check with feelers. The clearance should not exceed the OFFSET limit specified in Table 1. Tighten all foundation bolts and repeat Steps 2 and 3. Realign coupling if necessary.

4 INSERT GASKET AND LUBRICATE

After coupling is aligned, carefully insert the gasket between hubs and hang it on either hub. Do not damage gasket. Next force as much lubricant as possible into the gap and grid grooves. Refer to Page 1 for lubrication specifications.

5 INSERT GRID

Insert grid as instructed on Page 1. Coupling Sizes 3 thru 11 have a single layer grid which is painted gray. Sizes 12 thru 190 have a two layer grid; the inner layer is painted gray and stamped "IN"; the outer layer is painted yellow and stamped "OUT." Use a soft mallet to seat grids.

6 PACK WITH LUBRICANT

Pack the spaces between and around the grid with as much lubricant as possible. Scrape or wipe excess lubricant off flush with top of grid. Lightly oil seal rings to facilitate the sliding of covers onto the hubs.

7 ASSEMBLE COVERS

Assemble covers so that lubrication fittings or NPT holes are 180° apart and align covers to prevent wobble. Tighten cover bolts and check seals for proper seating.

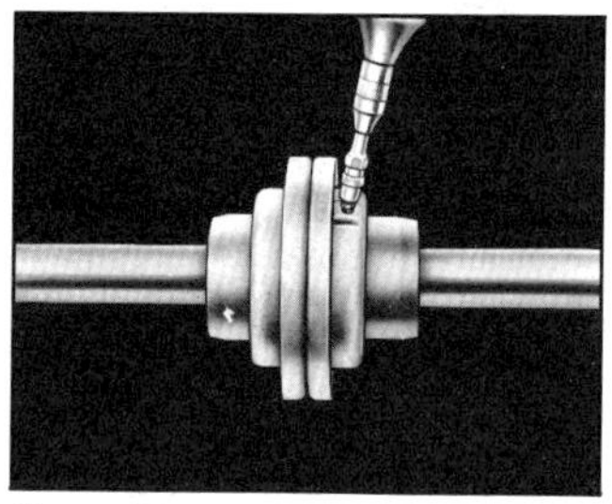

8 LUBRICATE EVERY 12 MONTHS

For couplings with permanent Alemite fittings, insert a smooth feeler gauge (.010) under seal 180° from fitting to vent cover. Use Alemite gun nozzle Z-737. For others, remove both lube plugs and insert your lube fitting. Lubricate couplings until grease flows through the vent or other opening. Then INSTALL LUBE PLUGS.

QUESTIONS FOR DISCUSSION

1. Why does this model begin with an exploded drawing of the steelflex couplings? How exactly will readers use this drawing? Is an exploded drawing the best type of visual to use for the kind of information the reader will need? Explain. What, if any, alternative visuals would you suggest?
2. Why is information about installing the grid listed both in the section labeled "Grid" and in Step 5? What are the differences between the two accounts?
3. Explain why Table 1 precedes the step-by-step instructions, how it is organized, and why it is necessary for the reader.
4. Could a reader install the steelflex couplings without the pictures included with the instructions? What advantages do the pictures offer to the reader?
5. How carefully are the pictures correlated with the written instructions? What kinds of duplication are there between written instructions and photographs?
6. Describe and evaluate the overall layout of this manual. Can you suggest any changes that might make the directions easier to follow? Explain.

SUGGESTIONS FOR WRITING

1. Prepare a set of instructions for a consumer or operator audience in which each step contains an appropriate visual together with written instructions. Identify your audience and its needs in a brief introduction.
2. Rewrite these instructions as a process explanation. (Review Chapter 5 for a detailed discussion of how to write process explanations.)

This decision tree was prepared by the American Heart Association for health care professionals familiar with the technical terms and procedures of emergency life support systems. Obviously, the tree does not show these individuals how to perform cardiopulmonary resuscitation (CPR). Its function is to provide an overview, a graphic summary, of the different choices these professionals might have to make during an emergency situation. As you read this tree, note how the format depicts the sequence of steps that might be taken.

A Life Support Decision Tree

American Heart Association

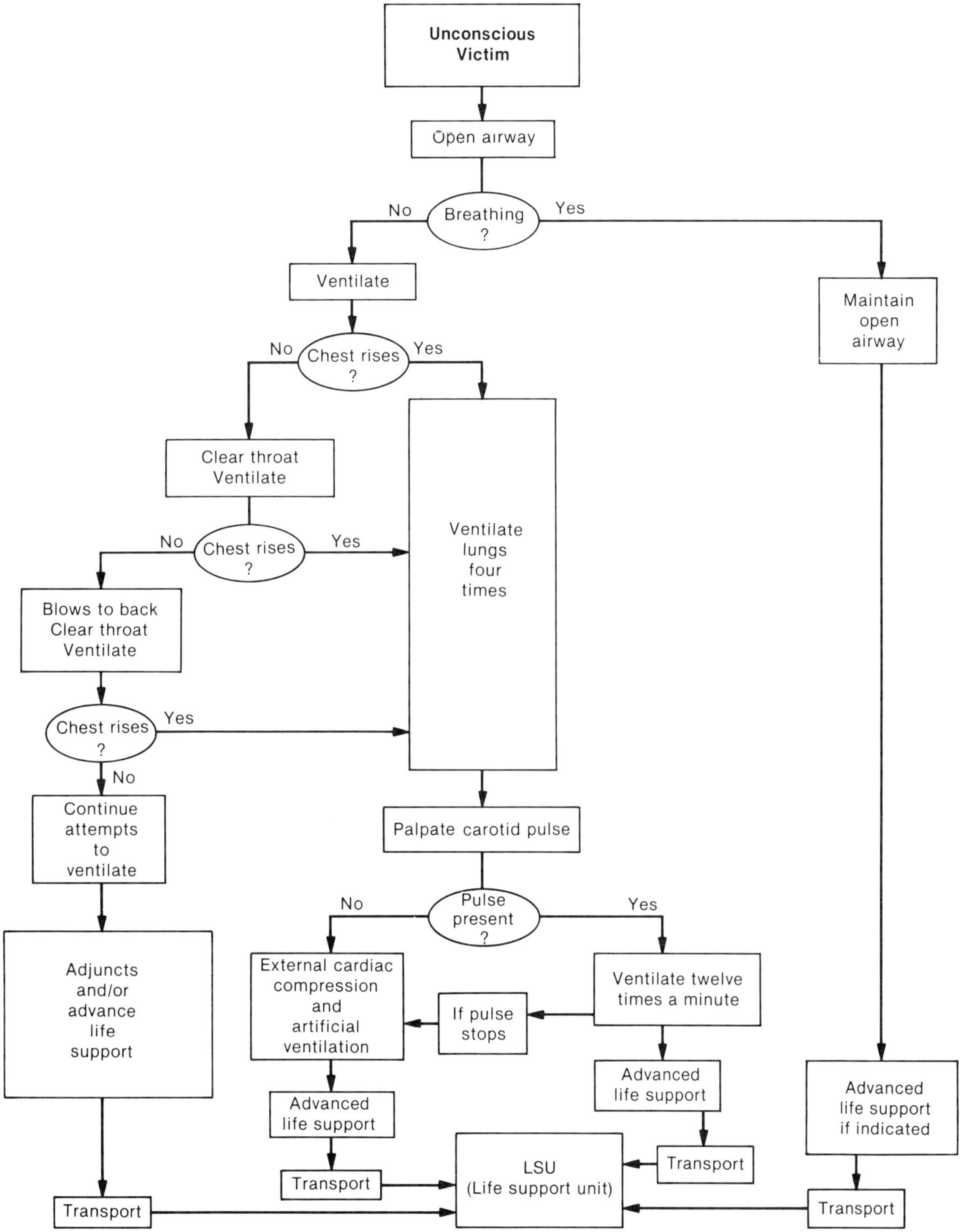
Unconscious Victim
Open airway
Breathing ?
No
Yes
Ventilate
Maintain open airway
Chest rises ?
No
Yes
Clear throat Ventilate
Ventilate lungs four times
Chest rises ?
No
Yes
Blows to back Clear throat Ventilate
Chest rises ?
Yes
No
Continue attempts to ventilate
Palpate carotid pulse
Pulse present ?
No
Yes
Adjuncts and/or advance life support
External cardiac compression and artificial ventilation
If pulse stops
Ventilate twelve times a minute
Advanced life support
Advanced life support
Advanced life support if indicated
LSU (Life support unit)
Transport
Transport
Transport
Transport

QUESTIONS FOR DISCUSSION

1. What different conditions and types of action do the main "branches" of the tree account for?
2. What function(s) do the different geometrical shapes serve for readers? How many different kinds of information do the arrows represent?
3. Evaluate the effectiveness of this decision tree for listing the different kinds of information contained in CPR instructions. How easy is the tree to follow?
4. What advantages or disadvantages does the decision tree format have over other formats discussed in this chapter? Would another format be more useful for giving CPR instructions to experts?
5. Name two or three other situations where instructions in a decision tree format might be used.

SUGGESTIONS FOR WRITING

1. Set up a decision tree for one of the situations you identified in Question 5. Clearly identify the audience for the decision tree.
2. Imagine you had to write instructions for a general audience on performing CPR. What format would be most appropriate? What terms from this decision tree would need defining? What procedures would need elaboration? What sort of visuals would be helpful? Prepare such a set of instructions.

The following instructions were taken from a U.S. Army manual written for the thousands of soldiers who operate troop carriers. These instructions explain a simple yet potentially dangerous process using technical terms familiar to this operator audience. Following a conventional format used by the military and numerous civilian organizations, these instructions include numbered steps in conjunction with many bold, carefully placed visuals. Pay special attention to the layout of these instructions, especially the relationship of the visuals to the numbered steps.

Starting an Engine with Outside Power

U.S. Army

STARTING ENGINE WITH OUTSIDE POWER

You can start your carrier when the batteries are dead by using a slave cable and an outside electrical source. The source can be charged batteries, or another vehicle. These instructions will tell you how to slave-start your carrier from another vehicle. You can also use your carrier as the power source to start another vehicle.

> **WARNING**
>
> **Do not allow personnel between vehicles during slave starting.**

1 Position the source vehicle next to your carrier so that the slave cable can connect the auxiliary power receptacles on both vehicles.

2 Make sure your carrier's master switch (1) is OFF.

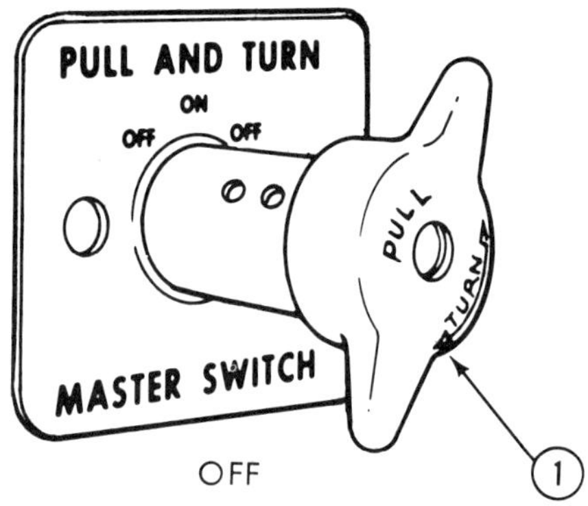

OFF

3 Unscrew the cap (2) from your auxiliary power receptacle (3).

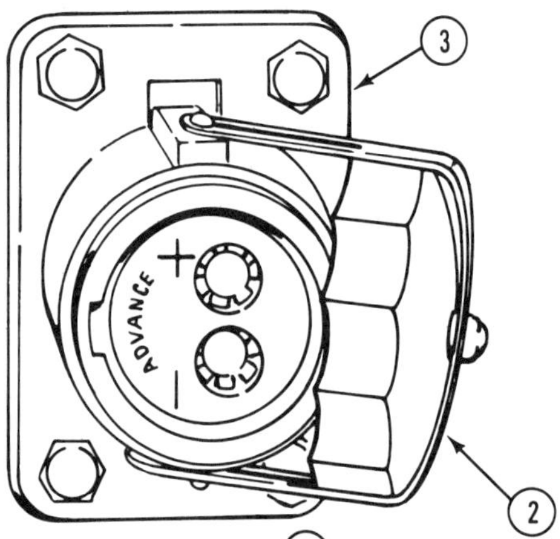

4 Connect the slave cable to the receptacle (3) on your carrier, then to the auxiliary power receptacle on the source vehicle.

> CAUTION
>
> Make sure cable prongs match the receptacle holes, + to + and - to -.

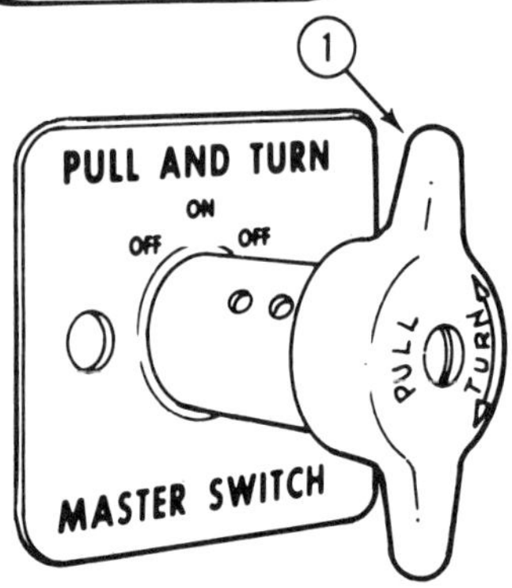

ON

5 Turn your master switch (1) ON and go ahead with a normal start (pages 2-8 through 2-11). Make sure the engine of the source vehicle is running fast enough to show charging on that vehicle's battery-generator gage.

6 After your engine starts, disconnect the slave cable, and screw the cap (2) on your receptacle (3).

QUESTIONS FOR DISCUSSION

1. Evaluate the appropriateness of these instructions for the intended audience. Concentrate on the language and content, based on the audience's specific job and background.
2. What information do the first two sentences give? Could this information be deleted without too much trouble for the reader? What heading might be used to set off these two sentences?
3. Why do the directions begin with a WARNING sign? Could the WARNING sign be placed between steps 1 and 2 or between steps 2 and 3? Why, or why not?
4. Why does the author switch, after step 3, to placing the steps on the right and the visuals on the left? Does this change make it easier or harder for readers to follow the instructions? Explain.
5. Why is step 6 necessary for the audience? Isn't the information it contains too obvious to include? Explain.

SUGGESTIONS FOR WRITING

1. Following the format of this model, write a short set of instructions using one of the topics below for a very large and diverse audience. Include appropriate visuals.
   a. cleaning a kitchen appliance
   b. removing stains from brick
   c. changing the oil in a car
   d. preparing a blood smear
   e. installing floor tiles
   f. cutting a piece of wood with a power saw
   g. planting a tree
2. Rewrite the instructions that you wrote for Question 1 as a process explanation. (See Chapter 5.)

Written for the do-it-yourselfer, these instructions on making a thermostat control timer appear in an information sheet available from a state Cooperative Extension Service. Correctly assuming that readers have little or no technical knowledge, the author emphasizes how following these instructions can increase a homeowner's savings, convenience, and comfort. The tone is friendly and the advice is practical. Every effort is made to make the reader's job easy and worthwhile. The model is especially interesting in the way that it adapts a number of the conventional sections of technical manuals (e.g., introduction, theory of operation, description of a mechanism, parts list) to meet the needs of nonspecialists. The actual step-by-step instructions are written in appropriate language and supply the right amount of detail for the audience.

How to Make a Thermostat Control Timer

Make A Thermostat Control Timer

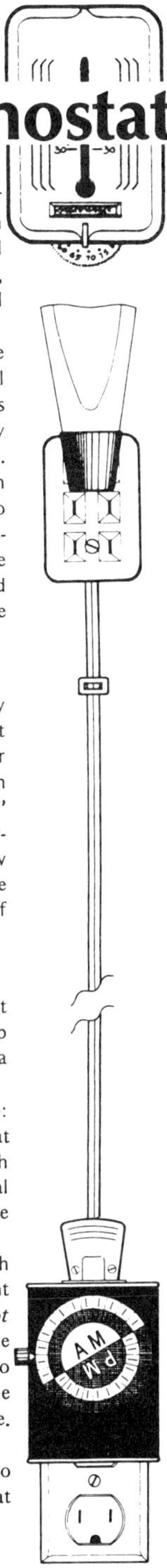

One way to cut heating costs and increase your comfort is to make a thermostat control timer which combines a night-light and electrical timer mounted on the wall below your thermostat. Most homes, apartments, and businesses with central-control thermostats can take advantage of this device.

The money saved on heating bills will pay for the parts needed to build your thermostat control timer -- probably within about seven months. Studies show you can get substantial energy cost savings by using a timer to activate your heater or air-conditioner.

Commercially manufactured thermostats with built-in timers are expensive and usually harder to install than the one described here. The main disadvantage of the homemade version for some people may be its looks. The cord, light fixture, and electrical boxes on the wall may prevent some consumers from trying it.

Operating The Timer

Your thermostat control timer will automatically turn up the heat in the morning and turn it down at night, at the times you have chosen. Just set the timer so the night-light below the thermostat comes on in the evening -- the bulb gives off heat that "fools" your thermostat into letting the night-time temperature in the home be maintained 5 to 8 degrees below your normal daytime setting. In the morning at the designated time, the timer turns the night-light off and the furnace comes back on.

Materials Needed

The only tools you need to build a thermostat control timer are: hammer, small screwdriver, sharp knife, pencil, ruler, and small nail (or a drill with a small bit, if you have one).

Purchase the following items at a hardware store:

- standard 4-watt *night-light.* Make sure that when it's plugged in, the night-light is vertical—with the bulb on top—and that the flat sides of its electrical prongs are parallel to the sides of the night-light. See figure 1.
- an *electrical outlet* that can be mounted with screws to a flat wall surface, called a "surface-mount convenience outlet." It must mount *on* the wall, *not in* the wall and have three pairs of slots where the night-light can be plugged in—this will allow you to experiment with how close the bulb is to the thermostat. The slots must all be on a flat surface. See figure 1.
- a 24-hour, plug-in *electrical timer,* designed to turn lights or appliances on and off automatically at pre-set times.
- *two-wire electrical lamp cord.* There should be enough to extend from your thermostat down the wall to the floor, and from there to the nearest wall outlet. Ask for standard 18/2 flat electrical lamp cord, type SPT-1.
- a two-prong *electrical plug,* commonly called a "quick-clamp plug" male adapter.
- a few small, *insulated electrical wire staples.*

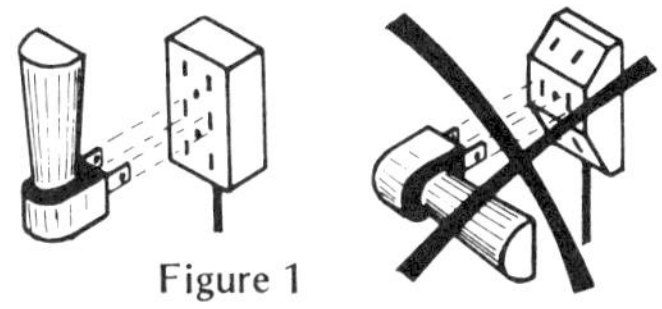

Figure 1

In developing the thermostat control timer, we used the following generally available parts. Although these specific parts were used, similar ones by different manufacturers may be found easily in most hardware stores.

We do not endorse any particular brand as working better than any other, as long as it fits the general specifications outlined in the preceding paragraph.

General Electric Hugger *night-light.* (No. GE 3967-41D-white).

Laviton Manufacturing Co. *surface-mount outlet* (No. 303, with mounting screws).

General Electric *quick-clamp plug* (No. GE 43002-white).

Sears *automatic plug-in timer* (No. 34H6442).

Step-By-Step Instructions

1. First plug the night-light (A) into the middle pair of slots on the surface-mount outlet (B). See figure 2.
2. Hold the complete unit (A and B) under your thermostat and measure approximately 2 to 2½ inches from the top of the night-light shade to the bottom of the thermostat. Then use your pencil to lightly mark the position of the outlet (B) on the wall.
3. Unplug the night-light (A) from the outlet (B). Holding the outlet against the wall where it was in Step 2, mark the positions of its screw holes on the wall.
4. Remove the outlet (B) and use a small nail or drill-bit to start the screw holes in the wall. These starter holes should be smaller than the screws included with the outlet, and about ½-inch deep.
5. Attach one end of the electrical lamp cord (C) to the outlet (B), according to the outlet manufacturer's instructions. Don't cut the cord to exact length yet -- you will adjust its length later.

Cooperative Extension Service   Mississippi State University

How To Build It, Step-By-Step:

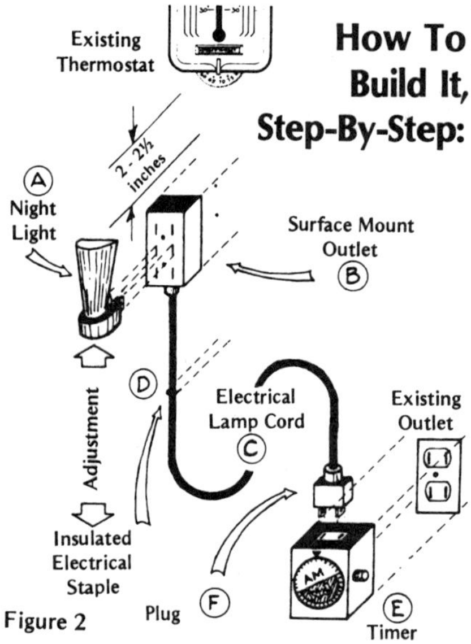

Figure 2

6. Attach the outlet (B) to the wall with the screws provided. Pull the cord (C) down gently to the floor or baseboard where it must change direction and fasten it there with an insualted electrical staple (D). Make sure the cord lies flat and is not twisted or kinked.

7. Plug the electrical timer (E) into the existing wall outlet closest to your thermostat.

8. Extend the cord (C) along the wall to the timer (E) you have just plugged-in. Staple the cord to the wall as necessary, spacing the electrical staples about 16 inches apart, to a point near the timer.

9. Adjust the length of the lamp cord (C), trim it and attach it to the quick-clamp plug (F), following the plug manufacturer's instructions. Then plug it into the timer.

10. Make sure the night-light's switch is on. Check to be sure the bulb is working by plugging it into any functioning house outlet. Then put it back into the center slots of the surface-mount outlet you have installed below your thermostat.

11. You will be able to find the best control of your thermostat by moving the night-light among the upper, middle, and lower levels of the outlet. The closer the bulb is to the thermostat, the cooler your home will be at night—and the more your energy savings will increase.

12. Your thermostat control timer is now ready to use. Just leave your thermostat on its normal daytime setting (68° F.). Now adjust your plug-in timer, following the manufacturer's directions, to the on and off settings you want to use.

For example, if you set the on tripper for 9:30 p.m. and the off tripper for 6:30 a.m., the night light will go on at 9:30 and your furnace or heating unit will be cut back starting around 9:45. The next morning your night-light will go off at 6:30, the furnace will start about 6:45 and the house temperature will return to a normal 68°.

Remember - your thermostat control timer does not shut off your furnace or heater completely. Any time the temperature falls below your adjusted thermostat setting, the heat will come on as usual, maintaining a lower, more energy-efficient but still safe temperature throughout the nighttime hours. Remember your timer is like any electric clock. It must be reset if the power is cut off for any reason.

Amount Of Savings

Savings of 1 percent for each degree your thermostat is set back for eight hours can easily be obtained during the cold season. Assuming you have a $40-amonth heating bill for seven months ($280 per year total), your savings of 5 percent with a five degree setback would amount to $14. Since your thermostat control timer can be built and installed for only about $10, it would pay for itself in just one heating season.

Modifications

The thermostat control timer can be modified to suit your individual needs. For instance, some households only need a comfortable level of heat in the mornings before work and in the evenings between work and bedtime. In those homes use a standard multiple-setting timer. Any automatic plug-in timer with multiple on-off settings, similar to the Sears model No. 34H6666, is suitable. It can adjust the furnace or heating unit several times a day.

Use With Air-Conditioner

You can also use the thermostat control timer for summertime savings if your air-conditioner uses the central wall thermostat in your home or office. To do this, you simply reverse the process used for controlling heat in the winter.

Installation is the same. The difference is that the night-light will be on when you want the house to be cool. For example, the timer can be set so that, when you get home after work, the air-conditioner has been on for about an hour.

Because the heat of the night-light bulb "fools" your thermostat into measuring a temperature about 8° warmer than the house really is, the thermostat can be set at 80° rather than the usual summertime 72°. When the night-light is on, your air-conditioner will operate and the house will be cooled, just as though the thermostat were set at 72°.

But, when the light is off, your air conditioner and thermostat will maintain a temperature no warmer than 80°—and you'll save cooling costs.

For maximum savings, convenience, and comfort you may want to experiment with various temperature settings on the thermostat.

Distributed in Mississippi by Joan Embree, Energy Extension Center. Adapted from information by Toby T. Benson, National Center for Appropriate Technology.

Information Sheet 1183
Extension Service of Mississippi State University, cooperating with U. S. Department of Agriculture. Published in furtherance of Acts of Congress, May 8 and June 30, 1914. L. N. WISE, Acting Director
(10M-8-81)

QUESTIONS FOR DISCUSSION

1. How effective are the first three paragraphs as an introduction to these instructions for the intended audience?
2. In the section "Operating the Timer," how has the author adapted the explanation of technical principles so that a consumer audience with no technical background can easily understand and apply them?
3. What kinds of additional details about tools and equipment might another, more technical audience (e.g., engineers, electricians, contractors) require? What types of details does the author give about these tools and equipment for the consumer reader?
4. Discuss the order and language of the numbered steps. Are steps 11 and 12 part of the instructions proper or could they be included elsewhere in the sheet? Explain.
5. Identify places in the numbered steps where CAUTION or NOTE signals might be placed? Does their omission here put the audience at a disadvantage? Would these signals have to be included if these directions were written for another audience? Explain.
6. Discuss the use of, and reference to, visuals in these instructions. Contrast the use of visuals in this model with those found in the instructions on installing a sensor on pages 194–195.

SUGGESTIONS FOR WRITING

1. Rewrite the section "Operating the Timer" for a more technical audience.
2. Using these instructions on a thermostat timer as your model, write a set of instructions for consumers on one of the following topics:
   a. installing an outside television antenna
   b. digging a well
   c. insulating an attic
   d. replacing a bicycle chain
   e. replacing a car battery
   f. installing a ceiling fan
   g. replacing a washer in a sink
   h. assembling a radio set
   i. finding buried wires on a lot or hidden nails in a house
   j. testing the chlorine level in a swimming pool
   k. making a small piece of wooden furniture

The following model is intended for a very technical group of readers. It comes from a technical manual published by the Foxboro Company to show how to assemble, disassemble, and maintain the sophisticated monitoring equipment they manufacture. The section of the manual we reprint here shows users how to install the sensor into a ball valve unit and shaft of a condensate unit, a water chiller system used at large plants for many types of industrial chemical processes. The sensor is an instrument that monitors the amount of direct current that is chemically produced in the water chiller. These instructions were written by a Foxboro service representative with years of experience in the field and a thorough working knowledge of the company's products. The audience is a group of engineering technicians and pipefitters who also have extensive training in the use of high technology equipment. The section begins with guidelines on mounting the sensor and a specification table and then presents the numbered steps. As you read this model, note that even though the subject of these instructions is highly technical, the directions on installing it are not—they are simple, clear, and easy to follow. Also note that numbering of tables and figures does not begin with 1 because the model is taken from the middle of the Foxboro manual. Figure 10 is not discussed in the model but rather is treated in an earlier portion of the manual.

How to Install a Sensor into the Ball Valve of a Water Chiller System

Foxboro Company

Output Wiring

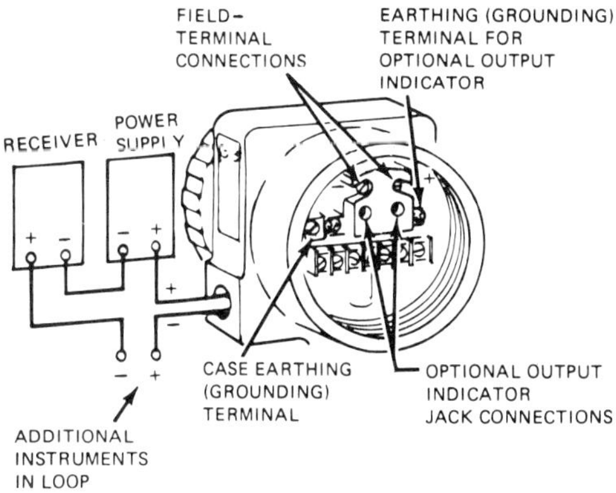

Figure 10. Transmitter Output Wiring

SENSOR INSTALLATION

Proper installation of the sensor is important for efficient and accurate operation.

For all applications and sensor configurations, mounting arrangements must be located so that:

1. Sample at the sensing area is representative of the solution.
2. Solution circulates actively and continuously past the sensing area.
3. Flow velocity at sensing area does not cause cavitation or cell damage.
4. Position and orientation of the sensor does not trap air bubbles within the sensing area.
5. Deposits of sediment or other foreign material do not accumulate within the sensing area.
6. If cable is installed in metal conduit (recommended), either flexible conduit should be used or some other provision made for removal of sensor from the process.
7. Lubricate all O-rings with thin film of suitable grease.

Ball Valve Assembly Installation

NOTE

A nonpurgeable ball valve assembly is illustrated, but these instructions are applicable to the purgeable ball valve assembly also. The purgeable ball valve assembly provides for two 1/4 NPT connections and shut-off valves (both user supplied) to allow purging of dirty solutions, high suspended solids, etc., before insertion of the sensor and during operation.

The ball valve assembly permits sensor body codes -PN and -TF sensors to be removed from a process stream or tank under rated temperature and pressure without draining the system or resorting to a bypass arrangement. The assembly consists of the ball valve unit and a shaft assembly. Standard flexible conduit (user provided) can be used for cable protection. For ball valve unit specifications, see Table 4.

A ball valve assembly installation must meet mounting arrangements specified under "Sensor Installation" section as well as the following requirements:

1. Direct the flow of the solution (as much as possible) toward the sensing area.
2. Provide a minimum of 660 mm (26 in) space for removal of the sensor in the ball valve unit (see Dimensional Print DP 611-122).

Table 4. Ball Valve Unit Specifications

PROCESS LINE CONNECTION (in)	MATERIAL OF PROCESS WETTED PARTS				RATED PRESSURE		MAXIMUM TEMPERATURE AT RATED PRESSURE		USED WITH 871EC SENSOR BODY CODE
	PROCESS CONNECTION AND HOUSING	BALL SEAT	BALL VALVE	O-RING	MPa	psi	°C	°F	
1 1/2 NPT	316 ss	ptfe	Glass-filled polypropylene	EPR	0.9	125	20	70	-PN
					0.5	70	65	150	
					0.3	50	95	200	
2 NPT	316 ss	ptfe	316 ss	EPR	1.4	200	105	225	-TF

Note: A complete assembly consists of one ball valve unit and one shaft assembly.

Assembly of Ball Valve Unit and Shaft Assembly

NOTE

This procedure is for a new installation. The process stream or tank is unfilled and unpressurized.

1. Disassemble the locknut and the housing nut from the installed ball valve housing (see Figure 11).

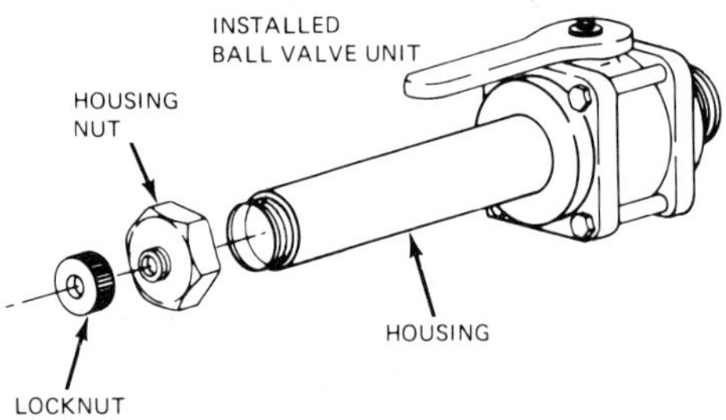

Figure 11.
Disassembly of Locknut and Housing Nut

2. Assemble the sensor cable through the shaft assembly. Insert and tighten the 3/4 NPT end of the sensor into the shaft assembly connection. Tighten the shaft assembly and shaft assembly connection (see Figure 12).

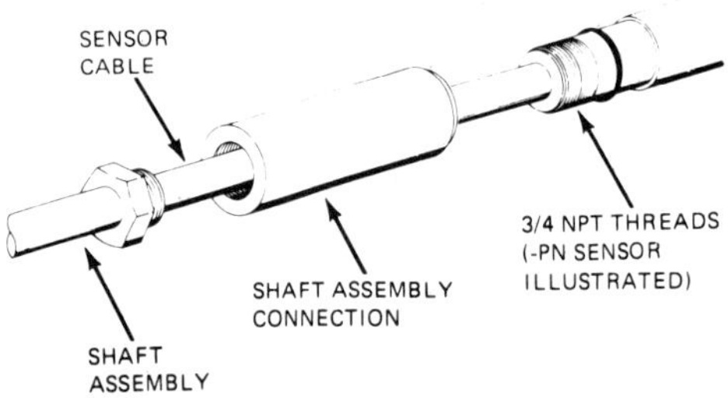

Figure 12.
Assembly of Shaft Assembly to Sensor

3. Assemble the shaft assembly as shown in Figure 13.

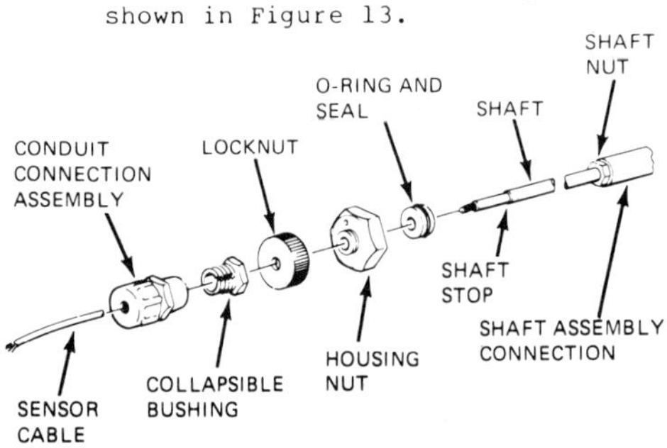

Figure 13. Assembly of Shaft Assembly

4. Lightly lubricate the shaft with silicone grease (or equivalent).

5. Slide the O-ring and seal against the shaft nut.

⚠ CAUTION

Do not nick the O-rings.

6. Slide the housing nut onto the shaft to a position just beyond the shaft stop.

7. Slide the locknut against the shaft stop.

8. Slide the collapsible bushing and the conduit connection onto the end of the shaft. (Do not tighten.)

Installation of Ball Valve Unit Shaft

1. Turn the handle on top of the ball valve unit (see Figure 14) to the close position (indicated by an arrow on the handle).

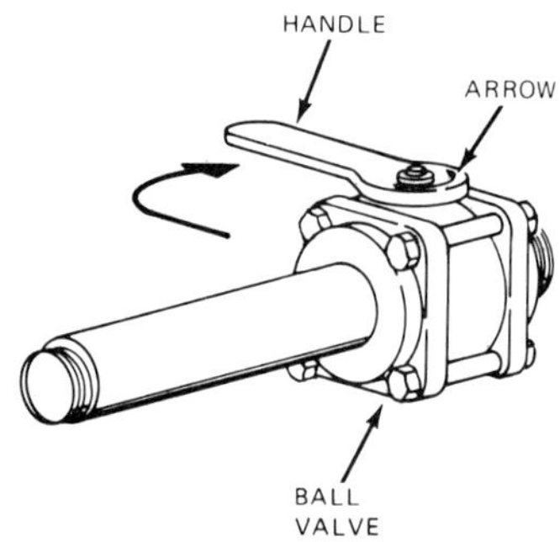

Figure 14. Closing Ball Valve

2. Fill and pressurize the process line or tank as required. Check for and eliminate leaks.

⚠ CAUTION

Do not exceed the sensor temperature and pressure limits.

3. Insert the sensor into the housing isolation chamber until the O-ring and seal are seated into the end of the housing (see Figure 15).

⚠ CAUTION

Do not nick the O-rings.

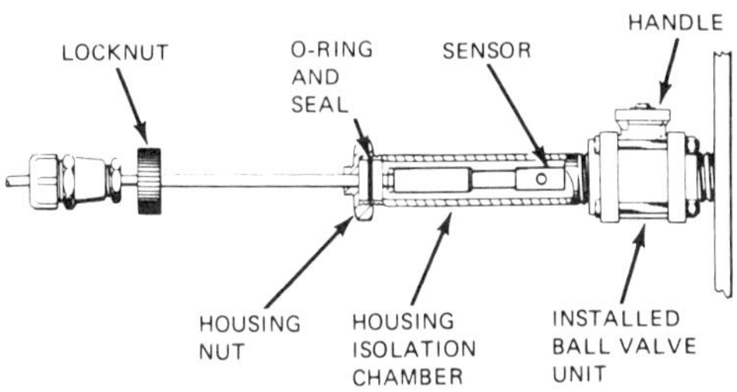

Figure 15.
Ball Valve Unit Shaft Installation

4. Tighten the housing nut onto the end of the housing.

NOTE
The housing nut is supplied lubricated and should be re-lubricated periodically to prevent galling.

5. Slowly turn the handle on top of the ball valve unit (see Figure 15) to the open position (indicated by an arrow on top of the handle).

6. Check for and eliminate leaks.

7. Tighten the locknut onto the housing nut.

8. Tighten the collapsible bushing against the locknut and install the conduit connection assembly onto the collapsible bushing.

Removal of Ball Valve Unit Shaft and Sensor

⚠ WARNING

Do not loosen the housing nut (see Figure 16) when the ball valve unit is open and the process line or tank is filled and pressurized .

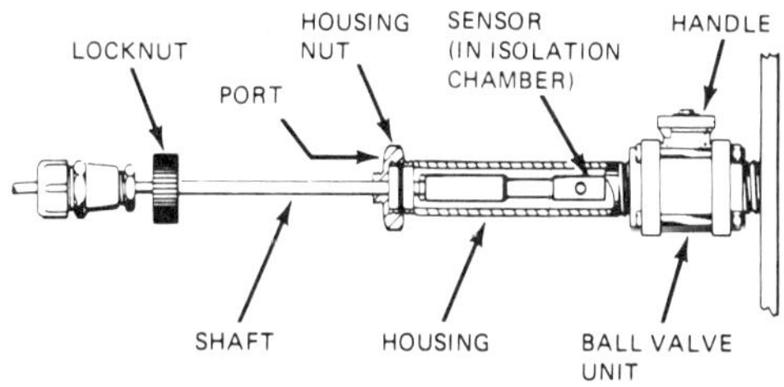

Figure 16.
Ball Valve Unit Shaft Removal

1. Carefully loosen and detach the locknut only.

2. Allow the sensor to enter into the ball valve unit housing isolation chamber (see Figure 16).

NOTE
Ensure that the sensor is against the housing nut mechanical stop (and has cleared the valve). Approximately 635 mm (25 in) of the shaft will be exposed.

3. Turn the handle on top of the ball valve unit (see Figure 16) to the close position (indicated by an arrow on top of the handle).

⚠ CAUTION

Do Not loosen the housing nut if a continuous leakage is observed through the port on the housing nut.

4. Slowly loosen housing nut.

⚠ CAUTION

There will be a release of pressure, and therefore some leakage. Wait until leakage stops before removing the housing nut completely.

Check to be sure the ball valve is completely closed. If the leakage continues, the ball valve is not closing. The process line or tank may have to emptied and depressurized

5. Remove the housing nut and withdraw the shaft assembly from the housing isolation chamber.

QUESTIONS FOR DISCUSSION

1. In what order does the writer give the reader information about the procedure to be performed? Explain the relationship of the different sections of these instructions to the overall process.
2. What types of information are included under the first two sections—"Sensor Installation" and "Ball Valve Assembly Installation"? Why does the audience need these details before progressing to the numbered steps beginning on page 210?
3. Describe the writer's sentences and word choices. How do they help the reader to accomplish the job?
4. How carefully are the visuals keyed to the numbered steps to which they apply? Explain with examples. Are the captions under each visual distracting or helpful? Explain.
5. Explain how, where, and why the author uses CAUTION and WARNING signals.
6. Describe the overall format of these instructions. How does it help the reader to understand and carry out the directions? Would you suggest any changes?
7. In what specific ways do these instructions reflect a keen awareness of the type of questions and problems the reader may have when installing the sensor?

SUGGESTIONS FOR WRITING

1. Write a set of instructions about a procedure that requires users to observe WARNING and CAUTION signals. Include information about safety precautions both in an introduction to your numbered steps and in the steps themselves. Indicate the audience you are writing for.
2. Select one section of this model and rewrite it as a process explanation. (See Chapter 5.)

The format for these troubleshooting instructions for an Apple II computer uses four columns moving from problem to solution. Most of the problems described here are indicated by the errors that users may see on the computer screen. Although not experts, users of this manual clearly know how the Apple II works. In their role as operators, they would appreciate the cabalese style.

Troubleshooting Hints for the Apple II

Problem	***Symptom***	***Possible Cause***	***Solution***
no data transfer	no sign of any com-munica-tion at all	cable wires not connected OK; jumper block facing wrong way	check all cable connections, then pin assignments; try reversing jumper block
characters garbled	jh2 3g%$Q	wrong baud rate	change SW1-1 to SW1-4 or use ⟨n⟩B command
		wrong data format	change SW2-1 (and SW2-2 in Comm Mode) or use ⟨n⟩D command to change format
		other device is off, out of paper, etc., off-line	turn on device, remedy its problems, put it on-line
paper not advancing	one line of smudge	printer needs line feeds from SSC	turn SW2-5 on or use L(inefeed E(nable command
printer is skipping lines	lines look like this	printer and SSC both generating ⟨LF⟩ after ⟨CR⟩	turn off SW2-5 in Printer Mode, or use L(inefeed D(isable command
missing characters	mssig caractrs	device buffer is overflowing	if device supports full RS-232-C handshaking, ensure all required cable wires are connected if device supports only ETX/ACK, set SIC P8A Mode if device supports XON/XOFF, set Printer Mode and use X(OFF E(nable cmd or set Comm Mode if device supports none of these, set delays with ⟨n⟩C, ⟨n⟩L and ⟨n⟩F cmds

Problem	*Symptom*	*Possible Cause*	*Solution*
device sticks at line's end going nuts	one long OK line, smudge at right end	device doesn't generate own ⟨CR⟩, and isn't getting enough from Apple	use SIC P8 Mode and ⟨n⟩N command, or Printer Mode and C command plus appropriate SW2-3 and SW2-4 have software send ⟨CR⟩ before right margin
Apple has occasional bad times	it works one minute & not next	ACIA interrupting the Apple when DCD or DSR changes	make sure that interrupt switch SW2-6 is off
Apple not working	dead kybd and screen	SSC in slot #3 under Pascal	Pascal expects external terminal to run the show
Apple kybd seems off	keystrokes all lost	echo off; keyboard zapped; IN# not Ø	use E(cho E(nable cmd; unzap with POKE; IN#Ø
screen seems off	nothing typed is displayed	device not echoing (half duplex) or ACIA not sending to screen	in Comm or Terminal Mode, use E(cho E(nable; in SIC or Printer Mode, use I command or SW2-3 & -4 on
screen is seeing double	eevveerryy tthhiinngg ttwwiiccee	device & SSC both echoing to Apple (full duplex)	use E(cho D(isable cmd in Comm Mode or use ⟨n⟩N cmd in Printer Mode
screen is spacing double	lines look like this	device generating and sending ⟨LF⟩ after ⟨CR⟩	use M(ask E(nable command to remove extra linefeeds
forced uppercase display	lowercase beCOMES UPPERCASE	Apple monitor changing letters in GETLINE routine	use ⟨n⟩T command to allow lowercase to pass through (not possible in Pascal)
Apple misses some characters at the beginning of lines	pple sses ome racters t the bgnning lines	screen scrolling too slowly, or BASIC or Pascal program running too slowly, and so ACIA overruns	turn off screen (⟨n⟩N or SW2-3 & -4 in Prtr Mode); reduce scroll window; use assembly language or faster program routines; use lower baud rate (3ØØ vs. 12ØØ); use ⟨n⟩C, ⟨n⟩L or ⟨n⟩F commands; in Comm Mode, chain (⟨n⟩S cmd) to 8Ø-column card with its own scrolling hardware

QUESTIONS FOR DISCUSSION

1. Categorize the audience for these instructions. How does the audience affect the content?
2. Is the second column, labeled *Symptom*, necessary? What information does it give the reader that is not available elsewhere in the table?
3. How are these troubleshooting instructions organized? Why do they begin with "no data transfer"?
4. Why do some of the solutions begin with an "if" rather than an imperative verb?
5. Why is the reader given two possible causes for some problems? Why doesn't the writer identify the major cause or explain why there might be more than one?
6. Why is the last solution so long? Could it be listed differently? How?

SUGGESTIONS FOR WRITING

1. Write a set of troubleshooting instructions on problems users might encounter with one of the following pieces of equipment. Divide your table into at least 4 major sections in the *Problem* category.
   a. lawn mower (starting and cutting action)
   b. oscilloscope
   c. television set (restrict your instructions to one major part, e.g., picture tube)
   d. vacuum cleaner
   e. microwave oven
   f. walkie talkie or pager
   g. sewing machine
   h. duplicating machine
2. Write a set of troubleshooting instructions that gives readers appropriate visuals (as the Apple II's section *Symptom* does) to help them solve their problems.

chapter 7

Investigative Reports

All types of technical writing depend on investigation, the careful collection and examination of information. Investigative reports are those that collect as well as evaluate empirical data for the reader. They are prepared after a writer has performed an experiment or test, made an inspection, or sampled an audience's perceptions and views. Such reports contain the results of the writer's original research—gathered by questioning, observing, measuring, comparing, ordering, and analyzing what happened under precisely described conditions. With analysis of this quantitative data, the writer is sometimes able to propose a solution to a problem or at least suggest areas for further study. Some investigative reports are written in response to specific questions (e.g., identifying or solving problems at a work site or with a piece of equipment). Because investigative reports provide the facts needed by researchers, they often precede proposals (Chapter 8) or feasibility studies (Chapter 9). And of course, investigative reports can also be taken up as the result of these other documents.

KINDS OF INVESTIGATIVE REPORTS

Investigative reports take many different forms—lab reports, reports of original scientific experiments, on-site inspection reports, time-and-motion studies, opinion surveys, and tests of the commercial viability of equipment or procedures. They can be based upon research conducted almost anywhere—in a pathology laboratory, in a forest, on an ocean atoll, in a shopping center, or in your own office or plant.

Investigative reports are written to answer a variety of questions, including the following:

1. How does the information in this report add to the body of information in a specific discipline (e.g., identifying a new species, determining the properties of a polymer)?

2. Does this report verify, extend, or refute the findings of a previous study?
3. How and why does this report compare or contrast subjects, specimens, equipment, or procedures (e.g., investigating whether a rotating 12-hour shift in a hospital is more cost effective than a fixed 8-hour shift, determining the tolerance of a given metal under different conditions)?
4. How does this report assess the effectiveness of a device or procedure (studying the performance of liquid helium to cool superconducting computers)?
5. Why and how does this report evaluate the health, safety, or environmental conditions at a work site or other location (e.g., determining the number of cotton dust particles in the air at a textile plant, assessing the noise level in an industrial plant)?

The model investigative reports in this chapter fall into one of three categories.

1. *Laboratory Research Reports.* For these reports the researcher studies subjects, specimens, procedures, or materials primarily in a laboratory setting that offers the benefits of highly sophisticated and sensitive equipment and a controlled environment. These reports often follow a conventional seven-part organization described on pages 219–222.

2. *Field Research Reports.* Here the researcher must travel to a site to collect and assess data that could not be replicated in a laboratory; information is included about the site, the time spent in the field, individuals present during the collection, the types and significance of data obtained, the continuity of visits, and possibly the need for future visits.

3. *Opinion Survey Reports.* For such reports the researcher determines the perceptions and opinions of a carefully selected audience, representative of a significant group and knowledgeable about the topic, gathered through questionnaires or interviews (in person or over the telephone). These questions must be precise, nonbiased, and relevant. Opinion survey reports summarize and assess respondents' views and offer specific recommendations for changes.

Investigative reports sometimes are a combination of any of these three types. For example, a geologist may go to a site to gather rocks (for a field research report) and later transport those rocks to a laboratory to analyze their chemical components (for a laboratory research report).

GUIDELINES FOR CONDUCTING THE RESEARCH

An effective investigative report demands careful collection and interpretation of evidence. Following are some guidelines to keep in mind when doing your research.

1. *Define and limit your topic.* In order to have a clear purpose that you can accomplish in the course of your research, be careful to restrict the problem

to be investigated. Stake out the right amount of territory to cover. Many writers start off with a topic too broad to investigate in the specified time and place. As a result, they confront many different topics, and solve none of them. Or they narrow their focus so closely that they pick an insignificant subject to study or only look at a part of a problem. To formulate your topic and narrow your field of research, there is certain background work you must do. Study the literature to make sure the problem has not already been solved and to identify relevant studies to help you solve your problem. You may also need to conduct preliminary tests or interviews to have a clear understanding of the significance and scope of the problem. When you identify the problem for readers, they will want to know what has been done and why you are conducting your present research.

2. *Use appropriate tools and methods.* To explore, measure, test, and analyze, choose the right equipment, materials, instruments, questionnaires, and environments. Sometimes you may need to construct your own tools or formulate your own methods. In any case, you must provide the reader with a precise description of materials, tests, procedures, and sites and the reasons you selected them.

3. *Gather data by careful observation and measurement.* To be able to tell the reader what took place and what was done, gather data carefully. Keep complete and objective records of times, conditions, subjects, and any changes observed. Be careful not to omit details from log books or lab notebooks. Although they may appear insignificant at the time, these details may be very important later. Also, never ignore or minimize relevant data, however difficult it is to gather. It would be misleading and sometimes unethical to base conclusions on incomplete data or untested assertions.

4. *Analyze the data.* Compare and contrast, noting patterns, habits, similarities. Proper use of scientific methods requires a control group (one not being varied or changed) and an experimental group (one being tested or changed) to judge one set of data against another and to eliminate error, chance, or misleading conclusions. In determining causation, decide which factors may be most important, which may be least important, and which have no effect at all.

5. *Assess the significance of your findings.* The reader will want to know what your results mean. Before drawing any conclusions, ask yourself if you have accomplished the goals of the experiment or test. Are there any areas of your study that need to be repeated or that will not support a generalization as readily as others? What can or should a reader do with your results? Do these conclusions have immediate application or will further research be necessary? Do the findings point to a better, more economical way of accomplishing something? Are there any unexpected beneficial results from your work—useful information that was not a projected part of your work?

THE STRUCTURE OF AN INVESTIGATIVE REPORT

Investigative reports are generally divided into seven sections: the abstract, the introduction, the materials and methods, the results, the discussion and conclusion, the recommendations, and the references. Not every report includes all these sections or gives them in this order; the exact organization can vary according to the writer's purpose, the audience's needs, the report's scope.

Often an investigative report will be divided into only two sections: introduction/methods and results/discussion. Sometimes the sections will be labeled with just those headings, but other times they may be replaced by specific headings relevant to the study under discussion—"The Use of a Spectrophotometer to Confirm Low Level Substances" instead of "Methods," for example. Short investigative reports—site-inspection reports or field reports, for example—frequently contain three sections: purpose, site description, and recommendations or actions taken. And investigative reports written for decision makers often conclude with a recommendations section.

The Abstract

Many investigative reports and studies begin with an abstract. An abstract summarizes the main points of your work, usually in a short paragraph of about 50 to 200 words, although abstracts for lengthy reports can be one or even two pages long. Abstracts enable your reader to determine whether to read, skim, or skip your report. The abstract must be able to stand alone and should not require reference to the longer work to be understood. It should give basic ideas, not a series of details. In business and industry, abstracts are often sent to many different departments of a company. To assist researchers in finding relevant scientific studies in an expanding world of scholarship, there are several journals that print abstracts in specific fields (e.g., *Chemical Abstracts, Computer and Control Abstracts, Engineering Index,* and *Psychological Abstracts*).

There are two types of abstracts—*informational* and *descriptive*. An informational abstract concisely tells readers about the problem you investigated, the methods you used to investigate it, and the conclusions you reached. It is a miniature version of the report itself. The following informational abstract was printed in the 1981 *Transactions of the American Society of Civil Engineers:*

> Many loessial soils are susceptible to structural collapse due to inundation or a rise in ground-water levels near hydraulic structures. These soils are often classified as unstable without adequate testing, as has been the case with a loess formation found in southeastern Washington. A study was conducted to determine if spread footing foundations on Palouse Loess would experience additional settlement due to increases in moisture contents. The physical and mineralogical properties were examined and compared to the corresponding properties of loess found in the midwestern states. Modified consolidation tests were performed on undisturbed

> specimens to examine the effect of initial water content on the settlement characteristics. Test results indicate that increases in moisture contents do not result in excess settlements. The type of clay cementing loessial soils appears to be the determining factor.[1]

Notice how the author gives the reader important details about the problem he studied and the conclusions he drew—the clay, not the moisture, affects the settlement of the soil. He includes only principal facts and omits background information, details about specific tests on properties, and specific information included from tables or charts.

A *descriptive abstract,* on the other hand, tells what topics a report covers without revealing its results or conclusions. It simply presents the subject of the report, almost like a table of contents in sentence form. A descriptive abstract of the article summarized in the preceding informational abstract might read as follows:

> A study was conducted to determine if spread footing foundations on Palouse Loess would experience additional settlement due to increases in moisture content. Physical and mineralogical properties were examined and modified consolidation tests were performed. Conclusions about causes for settlement are given.

Because of the abundance of research being done today, there is a trend away from descriptive abstracts and toward informational ones. After reading an informational abstract, readers will know the subject *and* conclusions of your work, even if they do not have the time to read it in detail.

The Introduction

This section informs readers about the purpose, scope, and background of your investigation. Since your purpose will often be to investigate a particular problem, be careful to tell readers why the problem is significant by emphasizing inconveniences, limitations, or risks and then citing particular examples of the problem. Specify for readers any limits you have placed on your investigation—this is a statement of the scope of your research. If something is to be excluded, tell readers precisely why. In a professional article or laboratory report, the introduction also contains a brief critical review of relevant studies on your topic to let readers know how your investigation will confirm, amplify, or refute earlier work. For an informal, on-the-job report, however, a literature review is usually unnecessary. For such reports, though, be careful to identify the particular business or industrial context in which the report was written—to solve a problem for a customer, to locate the most desirable site for a new plant, to determine the extent of a certain kind of pollution.

[1] Gary T. Lobdell, "Hydroconsolidation of Potential of Palouse Loess," *Journal of the Geotechnical Engineering Division,* Proceedings of the ASCE, 17, No. GT6 (June 1981): 733–742.

The Materials and Methods

In this section, you describe the overall procedures and kinds of equipment or tools that you used to conduct your investigation. Tell readers how you gathered the data in order to help them reproduce or verify your findings. To assist them, provide specific information on how samples or specimens were selected, obtained, and prepared, how informants for an opinion survey were chosen, where and how often a certain test was performed, what standard tests were used and when and why you varied from them, and what kinds of tools or analyses were involved (e.g., holographs, x-rays, pyrolysis). While scholarly laboratory and site-investigation research demands scrupulous descriptions of your methods, readers of an on-the-job report will be less interested in the fine details of your procedures than in knowing you have followed a standard and verifiable method. That is not to suggest that in doing the research for such reports you should be less conscientious or meticulous. It simply means that if your readers are decision makers with little or no experience in your field you should not overwhelm them with technical terms and explanations.

The Results

The results section presents the data—the hard facts—you have gathered. This information is usually listed through tables, graphs, and other figures. These visuals help you to present the data in an economical and orderly way and assist the reader to compare and contrast a large amount of quantitative information. Make sure that you identify these visuals for readers—pointing out the kinds of information they contain and correlating your comments in the results section with the specific references in these visuals. Be careful not to repeat in detail the information already contained in these visuals. Summarize key details to give readers the big picture. Some important points to emphasize include major differences or similarities in the data, numbers that point to an important pattern, and significant variations among specimens, metals, chemicals, or any other subject under investigation. Needless to say, your totals will play an important role in the results section.

The Discussion or Conclusion

The discussion or conclusion, which is sometimes combined with the results or the recommendation, gives the implications and impact of your results. This section tells readers what your facts mean by explaining the reasons for them. Throughout this section readers will ask the question why. Be sure that your explanations are concise and objective. Don't omit details that may seem hard to classify or appear out of place and don't overstate or understate. If you are unsure about outcomes, don't speculate. Your guess may be wrong and costly. When you explain for readers, isolate causes, separating, whenever possible, major from minor ones. Explain how and why a condition developed or a result occurred, and whether it is beneficial or disadvantageous. Elaborate on whether your results contradict or confirm previous explanations. Point out important effects and state their significance. In an on-the-job report pay special attention to how the results you are interpreting affect present or future business ventures.

The Recommendation

In this section you advise readers how to use the data you have gathered and interpreted for them. The recommendation may be brief—one or two sentences in an informal report—or a lengthy list of suggestions in a formal report. To be effective, recommendations must be precise and carefully based upon the evidence. In a piece of on-the-job writing, the author's recommendations are doubtless the most important part of the report for the decision maker. You might want to review the suggestions we make about writing a list of recommendations on pages 305–306.

The References

Not every investigative report contains references, or a list of sources (books, monographs, articles, papers) the author consulted in writing the report. Such a list would be out of place and unnecessary in a short, informal report done for a decision maker. However, following a scholarly article or government-sponsored laboratory report such a list is essential to prove to readers that the writer knows the discipline and that he or she has not misrepresented previous scholarship. This list can also be known as *Literature Cited, References Cited,* or *Bibliography.* Scholarly journals provide detailed instructions to authors on how to document information, as do the style manuals cited in the next section of this introduction. Make sure that you are familiar with the particular method of documenting research in your own field.

RESEARCH GUIDES AND STYLE MANUALS

Many disciplines publish their own style manuals to assist writers in preparing, writing, and presenting their research. These guides contain detailed instructions about punctuation, the use of visuals, formats, style, and methods of documentation. Ask an instructor in your major field if your discipline has its own guide or if it uses a guide from another area. Listed below are the titles of some of the most widely followed style manuals in science and technology.

BIOLOGY Council of Biology Editors, Committee on Form and Style. *CBE Style Manual.* 5th ed. Bethesda, Md.: Council of Biology Editors, 1983.

CHEMISTRY American Chemical Society. *Handbook for Authors of Papers in American Chemical Society Publications.* Washington, D.C.: American Chemical Society, 1978.

ENGINEERING Engineers Joint Council. Committee on Engineering Society Editors. *Recommended Practice for Style of Reference in Engineering Publications.* New York: Engineers Joint Council, 1977.

ELECTRONICS American Institute of Electrical and Electronics Engineers. *Information for IEEE Authors.* New York: American Institute of Electrical and Electronics Engineers, 1948; Supplement, 1965.

GEOLOGY U.S. Geological Survey. *Suggestions to Authors of Reports of the United States Geological Survey,* 6th Edition. Washington, D.C.: GPO, 1978.

MATHEMATICS American Mathematical Society. *Manual for Authors of Mathematical Papers.* 7th ed. Providence, Rhode Island: American Mathematical Society, 1980.

MEDICAL TECHNOLOGY *Handbook for Authors.* Bellaire, Texas: American Journal of Medical Technology, 1975.

METALLURGY American Society for Metals. *Instructions to Authors for the Preparation of Technical Papers.* Metal Park, Ohio: American Society for Metals, 1962.

PHYSICS American Institute of Physics. *Style Manual for Guidance in the Preparation of Papers for Journals Published by the American Institute of Physics and Its Members.* New York: American Institute of Physics, 1978.

PSYCHOLOGY American Psychological Association. *Publication Manual of the American Psychological Association.* 3rd ed. Washington, D.C.: American Psychological Association, 1983.

SOIL SCIENCE American Society of Agronomy, Crop Science Society of America, Soil Science Society of America. *Handbook and Style Manual for ASA, CSSA, and SSA Publications.* Madison, Wisconsin, 1976.

A CHECKLIST FOR WRITING AN INVESTIGATIVE REPORT

Conducting the Research

1. Define the topic for research.
2. Select the tools and methods appropriate for the study.
3. Gather data, keeping detailed records.
4. Analyze the data.
5. Assess the significance of your findings for readers.

Writing the Report

1. Select a suitable format, consulting when necessary an appropriate style manual.
2. Include an abstract that tells readers what your investigative report includes.
3. State the purpose, scope, and background of the report in the introduction.
4. Describe the methods followed and the tools used.
5. Present your results.
6. Discuss conclusions and make recommendations based on those conclusions.
7. Use appropriate formats for documentation.

LABORATORY RESEARCH REPORTS

In the following laboratory report, food chemists at the Food and Drug Administration (FDA) describe for other experts the composition of french fries obtained from ten chain restaurants. Conducted as part of an ongoing research program at the FDA, their work was originally published in the Journal of Agricultural and Food Chemistry. *The aim of FDA research is "to provide scientific support for the implementation of the laws provided by Congress for the protection of both producers and consumers of foods, drugs, and cosmetics. . . . Every research project can be directly related to specific provisions of the Federal Food, Drug, and Cosmetic Act of 1938 or an amendment thereto."*

Like many laboratory reports, this model was written by several authors who followed sophisticated laboratory procedures. Note that the results and discussion sections are combined, a common practice in laboratory research studies.

Individual Lipids and Proximate Analysis of Various Foods: French Fried Potatoes from Ten Chain Restaurants

Willard D. Hubbard, Allan R. Prosser, Alan J. Sheppard, David R. Newkirk, Theodore Osgood, Elias Tombropoulos, and Shirley T. Jones

French fried potatoes were obtained every third day from ten chain restau- 1
rants for a total of six samplings from each restaurant. The samples were analyzed for water, total fat, fatty acids, protein, ash, sterols, and cis,cis*-methylene interrupted polyunsaturated triglycerides. The data show that variations exist in the day-to-day selection of cooking oils used within a single restaurant. Also, the wide variations in data from one restaurant to another indicate that different oils were being selected. Values for total fat ranged from 14.0 to 28.4 g/100 g of product, and cholesterol values varied from near zero to 19.0 mg/100 g of product. Several restaurants appeared to use mixtures of animal and vegetable fats. One establishment used only a vegetable oil and others used only animal fat.*

More information is needed on the analysis of foods supplied by fast-food chain restaurants in view of the relatively large number of meals consumed by the public in these establishments. Of particular interest is the cholesterol and fatty acid content of foods. Samples of french fried potatoes were obtained from ten major fast-food chain restaurants. Water, total fat, fatty acids, protein, ash, sterols, and *cis,cis*-methylene interrupted polyunsaturated triglycerides were determined. 2

MATERIALS AND METHODS

Ten restaurants were each visited a total of six times in a sequence of every third day for collection of samples. The restaurants were Arby's, Burger Chef, Burger King, Gino's, Hardee's, Hot Shoppe Jr., Jack in the Box, McDonald's, Red Barn, and Roy Rogers. (The data in the tables do not reflect this order of listing.) Samples were homogenized in a Waring blender and stored at −2 °C for a few days before extraction. The extraction procedure, using chloroform-methanol, has been described by Sheppard et al. (1974). A sufficient amount of sample was taken for the extraction step so that approximately 1 g of fat was recovered. The methyl esters of the fatty acids were prepared by the procedure of the Association of Official Analytical Chemists (AOAC, 1975) as modified by Solomon et al. (1974). The method for the preparation of the butyrate derivatives for sterol analysis and the method for cholesterol have been described by Sheppard et al. (1974). Official methods of the AOAC (1975) were used for the determination of water, protein, and ash. 3

RESULTS AND DISCUSSION

All samples were analyzed in duplicate. The data (Tables 1 and 2) are mean values ± standard deviations for the 12 observations, i.e., six samples in duplicate from each restaurant. When standard deviation values are low, it is probable that the particular restaurant had been using the same type (source) of cooking oil during the sampling period. Conversely, changes in source of oils would appear as higher standard deviations. 4

The wide variations in data from one restaurant to another indicate that different oils are being selected and to a lesser degree that a single restaurant changes its cooking oil source from day to day. However, the samples from restaurants A, C, E, I, and J are reasonably similar in fatty acid methyl ester (Table 1) and total fat (Table 2) content. The results from restaurant D are unique for the series. The high percentages of C18:1 and total fat reveal that its fat source is quite different from the others. The finding of C12:0 fatty acid methyl esters (not shown in Table 1) in three of six samplings from restaurant H indicates that coconut or palm kernel oil may have been used as an oil source during the test period. 5

TABLE 1. FATTY ACID AND *CIS,CIS*-METHYLENE INTERRUPTED POLYUNSATURATED TRIGLYCERIDE (MIPT) CONTENT (G/100 G OF PRODUCT) OF FRENCH FRIED POTATOES FROM TEN FAST-FOOD RESTAURANTS[a]

	Fatty Acid Methyl Esters								
Restaurant	*C14:0*	*C14:1*	*C16:0*	*C16:1*	*C18:0*	*C18:1*	*C18:2*	*C18:3*	*MIPT*
A	0.5 ± 0.1	0.2 ± 0.1	4.3 ± 0.4	0.5 ± 0.2	2.7 ± 0.4	6.2 ± 0.7	0.5 ± 0.1	0.1 ± 0.1	0.3 ± 0.1
B	0.1 ± 0.1	ND[b]	3.8 ± 0.7	ND	1.1 ± 0.1	8.2 ± 1.2	1.2 ± 0.6	0.1 ± 0.1	0.8 ± 0.5
C	0.5 ± 0.1	0.2 ± 0.1	4.2 ± 0.5	0.5 ± 0.1	3.0 ± 0.4	6.9 ± 0.8	0.5 ± 0.1	Tr[c]	0.3 ± 0.1
D	Tr	ND	3.0 ± 0.4	ND	3.1 ± 0.5	17.0 ± 2.5	0.1 ± 0.1	ND	Tr
E	0.5 ± 0.1	0.2 ± 0.1	4.1 ± 0.5	0.4 ± 0.1	2.9 ± 0.4	7.2 ± 0.9	0.7 ± 0.1	Tr	0.5 ± 0.2
F	0.3 ± 0.1	0.1 ± 0.1	3.8 ± 0.5	0.4 ± 0.1	2.2 ± 0.3	6.4 ± 0.8	2.1 ± 0.7	0.2 ± 0.02	1.6 ± 0.4
G	0.3 ± 0.05	0.1 ± 0.1	3.4 ± 0.1	0.3 ± 0.1	2.6 ± 0.1	6.8 ± 0.5	1.2 ± 0.4	0.1 ± 0.1	0.8 ± 0.3
H	0.2 ± 0.1	Tr	2.5 ± 0.4	0.1 ± 0.1	1.5 ± 0.2	8.0 ± 0.8	0.8 ± 0.2	0.1 ± 0.1	0.5 ± 0.1
I	0.4 ± 0.05	0.2 ± 0.1	4.1 ± 0.2	0.5 ± 0.2	2.5 ± 0.3	6.5 ± 0.3	0.8 ± 0.3	0.1 ± 0.1	0.6 ± 0.2
J	0.5 ± 0.1	0.1 ± 0.1	4.4 ± 0.6	0.4 ± 0.2	2.3 ± 0.4	6.8 ± 0.7	1.0 ± 0.2	0.1 ± 0.1	0.8 ± 0.2

[a] Mean values ± standard deviation of six samples assayed in duplicate. [b] ND = none detected. [c] Tr = <0.1 g/100 g of product.

In Table 2, the similarities between restaurants A, C, E, I, and J are again 6
present, with restaurant D in a separate category. This restaurant may have been changing its source of oil frequently, since wide variations were observed in the data for campesterol, stigmasterol, and sitosterol on a day-to-day basis.

Restaurants F, G, and H appear to use mixtures of animal and vegetable 7
fats, since both cholesterol and sitosterol were found. Restaurant B used only a vegetable oil. The samples from restaurant D have the lowest water content, highest ash content, and highest total fat content of any of the samples. The ash content and total fat content of samples from restaurant D were significantly

TABLE 2. PROXIMATE ANALYSIS AND STEROL CONTENT OF FRENCH FRIED POTATOES FROM TEN FAST-FOOD RESTAURANTS[a]

	Proximate Analysis, g/100 g of Product				***Sterols, mg/100 g of Product***			
Restaurant	*Water*	*Protein*	*Ash*	*Total fat*	*Cholesterol*	*Campesterol*	*Stigmasterol*	*Sitosterol*
A	40.9 ± 3.9	4.2 ± 0.4	1.4 ± 0.2	16.0 ± 1.8	16.5 ± 2.5	ND[b]	ND	ND
B	41.5 ± 4.2	3.7 ± 0.3	1.6 ± 0.3	15.5 ± 1.3	ND	ND	ND	18.3 ± 4.4
C	35.6 ± 3.6	4.1 ± 0.3	1.7 ± 0.3	16.4 ± 1.6	18.2 ± 3.6	ND	ND	ND
D	32.1 ± 2.3	3.4 ± 0.2	3.0 ± 0.2	28.4 ± 4.9	ND	12.8 ± 6.9	10.8 ± 6.9	28.7 ± 6.7
E	40.2 ± 5.6	3.9 ± 0.2	1.7 ± 0.3	17.0 ± 1.9	13.1 ± 8.7	ND	ND	ND
F	38.3 ± 1.0	3.4 ± 0.2	1.5 ± 0.2	16.1 ± 1.5	13.7 ± 3.1	ND	ND	7.7 ± 8.2
G	38.2 ± 3.9	4.0 ± 0.5	1.9 ± 0.2	16.0 ± 0.9	12.1 ± 3.0	ND	ND	6.6 ± 9.9
H	43.3 ± 1.6	4.0 ± 0.4	2.0 ± 0.1	14.0 ± 0.8	4.2 ± 3.3	4.4 ± 5.0	2.8 ± 4.6	18.1 ± 1.7
I	35.9 ± 0.8	4.6 ± 0.3	2.0 ± 0.4	16.0 ± 1.2	19.0 ± 3.0	ND	ND	ND
J	33.0 ± 3.0	4.6 ± 0.5	1.9 ± 0.2	16.5 ± 1.7	17.8 ± 2.4	ND	ND	ND

[a] Mean values ± standard deviation of six samples assayed in duplicate. [b] ND = none detected.

different from those obtained from the other restaurants when subjected to statistical analysis using the outlier test.

Literature Cited

Association of Official Analytical Chemists, "Official Methods of Analysis," 12th ed. Washington, D.C., 1975, Sections 28.057–28.060.

Sheppard, A. J., Hubbard, W. D., Prosser, A. R., Interim Methodology Instructions No. 2 with modification No. 1 dated April 16, 1975, for Implementing Requirements of Section 1.18 of Title 21, Chapter 1, Subchapter A, Part I ("Labeling of Foods in Relation to Fat, Fatty Acid and Cholesterol Content") issued June 11, 1974, Division of Nutrition, Bureau of Foods, Food and Drug Administration, Washington, D.C.

Solomon, H. L., Hubbard, W. D., Prosser, A. R., Sheppard, A. J., *J. Am. Oil Chem. Soc.* **51,** 424 (1974).

QUESTIONS FOR DISCUSSION

1. Why do the authors supply an abstract for such a short study? What information about their investigation do the authors include in the abstract? What information from the article itself do they omit?
2. Is this abstract descriptive or informational?
3. Every reader of the article would like to know the identity of Restaurant D. Why don't the authors reveal this information?
4. Under the Materials and Methods section, the authors refer to other studies describing the procedures they follow (e.g., Sheppard et al., the AOAC). Why don't they explain these tests and procedures, or at least give the reader more background information about them?
5. How will readers use the information given in the Results and Discussion section? This section selectively repeats information recorded in the two tables accompanying the article. What facts from the tables do the authors explain and interpret and which facts do they not discuss? Why?
6. "Research must be objective both in its methods and in its presentation." Comment on how this article fulfills these objectives. Describe the authors' style as an objective means of presenting research.

SUGGESTIONS FOR WRITING

1. This article is clearly intended for specialists. Rewrite it so that consumers can grasp the main points.
2. Assume you are a food chemist working for a large fast food chain. Summarize the relevant sections of this study in a short memo (see Chapter 12) for (a) a decision maker and (b) a brief (200 word) article in your company's news magazine for employees.

The following laboratory report was written by an industrial chemist at an aluminum company. One of the company's products is sacrificial anodes, electrodes used to retard corrosion of metals, which it sells to manufacturers of hot water heaters. When one of its customers sent in a hot water heater that the customer said did not work because of a defective anode, the vice president of the aluminum company (an engineer with 12 years of management experience) asked the staff chemist to inspect the unit and prepare a report of the findings. The lab report that follows is written in memo form, an appropriate format for this type of in-house communication.

A Chemist's Laboratory Report for the General Aluminum Corporation

Virginia Owens

GENERAL ALUMINUM CORPORATION

Portland, Oregon

TO: Marshall T. Hirsch
Vice President
Corporate Relations

FROM: Virginia Owens V.O.
Staff Chemist

DATE: April 30, 1984

RE: Examination of Universal Appliance Corp.
50 gal. Water Heater, Serial # A2421152

Background and Purpose of Investigation

The 50 gal. water heater, #A2421152, with its last charge of water was shipped to us for inspection on March 31, 1984, by the Universal Corporation of Palos Verdes, California. Universal claimed that the tank had apparently failed in service because of our sacrificial anode. Based on the date stamp on the tank exterior, the manufacture date for the heater was January 1983. Thus, the unit falls within the necessary service agreement specified by the warranty. 1

In an attempt to determine the exact nature of the failure, you asked this office to inspect the tank and analyze its contents and its sacrificial aluminum alloy anode. The initial examination of the heater and its contents was carried out at the products support lab in Portland on Monday, April 25, 1984. 2

Chemical Examination

Six quarts of liquid were drained from the tank. The usual sediment, apparently mostly sand, was found in the first few quarts of effluent. No red rust was noted. 3

2

Analysis of water contained in the tank gave: 4

> Cl^-=40.4 ppm; total hardness (Ca&Mg) = 123.6 ppm;
> Na^+=41.7 ppm; $SO_4^{=}$=73.0 ppm; F^-=0.2 ppm; pH=7.5.

This may be compared with the following analysis for water 5
supplied to Palos Verdes on March 24, 1984, by California Water Services of Los Angeles:

> Cl^-=76 ppm; SO_4=98 ppm; total hardness = 186 ppm;
> total solids = 370 ppm; pH=8.5

Physical Examination

Anode Conditions

The tank's sacrificial anode was found to be more or less 6
uniformly covered with gelatinous hydrous oxides, presumably mostly those of aluminum. This condition is normal for anodes removed from service. Later x-ray diffraction analysis showed the corrosion products adhering to the rod were primarily AlO(OH) with a small amount of $CaSO_4$. The dry weight of the anode and adherent corrosion products after correction for cap weight was 642 grams as compared to 638 grams calculated for the weight of an anode of the same dimensions before installation. After necessary correction for the weight of adherent corrosion product (8.40 grams) the weight of the anode removed from service was 633.60 grams, indicating an alloy loss from the original anode of 4.4 grams. It is unlikely that the precision of this data is better than ± 1 gram.

Interior Surfaces

After the tank was drained, it was sawed in half 7
perpendicular to its long axis to facilitate examination of its interior surface. No anodic corrosion on the interior tank surface was seen. The brown porcelain liner was found to be in virtually new condition. There was a considerable amount of sand in the bottom of the tank.

An indication of protective activity by the sacrificial 8
anode was noted on several steel 'whiskers' which protruded through minute pinholes in the bottom liner surface. The whiskers were encrusted with adherent, white oxides--clear evidence of prior cathodic activity at the whisker sites.

3

No leaks in the tank were found by visual inspection or by application of fluorescent leak detection fluid. 9

Exterior Tank Bottom

Inspection of the exterior tank bottom showed fairly extensive rusting and scaling. Both black and red rust were present and considerable rust debris had fallen onto the burner and bottom pan. 10

Conclusions

No evidence of problems with the sacrificial anode was found. The amount of rust and scaling found on the exterior bottom of the tank seems greater than would be normally expected for a heater that was in service for less than 18 months. This condition may have been aggravated by incorrect burner adjustment and excessive condensate. Because our company does not manufacture the gas valve or thermostat, the operation of these parts was not investigated. It is possible that the failure reported by Universal is related to one or both of these elements. 11

The protective anode appeared normal; in fact, its action had succeeded in occluding three minor voids in the porcelain liner. 12

In summary, then, the problem Universal encountered must be attributed to a cause other than our sacrificial anode. 13

QUESTIONS FOR DISCUSSION

1. Although trained in the sciences, the reader of the report is not a chemist and has not worked in a laboratory for a number of years. Judging from the report, how much specific information about the topic does the reader possess? In what ways has the writer shaped her description of the investigation to meet the needs of such a reader? Has she given any explanations that would be unnecessary for another chemist? Which ones, if any?
2. How is the report organized? Discuss the order in which the chemist presents her information. In what ways is her report like other laboratory reports in this section of Chapter 7? In your answer discuss the function of each part of her memo. Would you recommend changes in the order? For example, should conclusions come first?
3. What methods does the chemist follow? Why doesn't she include a section labeled "Materials and Methods"?
4. Why does the chemist include information on the following:
   a. Palos Verdes water
   b. whisker sites
   c. porcelain liner
   d. weight of the anode
   e. amount of rust and scaling discovered
5. What conclusions does the chemist reach and how will they help her reader?

SUGGESTIONS FOR WRITING

1. Assume that you work for a large manufacturing company as a laboratory researcher. Write an investigative report in memo format similar in format and scope to the General Aluminum report on one of the following topics. Specify the intended reader.
   a. television tubes
   b. car batteries
   c. microcomputers
   d. pneumatic drills
   e. autoclaves
   f. centrifuges
   g. cameras
   h. car tires
   i. air conditioners
   j. animal feed
   k. food coloring or preservatives
2. Assume that you are the vice president of the aluminum company. Write a letter to the Universal Corporation explaining what your chemist found (See Chapter 12 on correspondence.) Assume that your reader is a decision maker with little or no background in the sciences or engineering.

FIELD RESEARCH REPORTS

Not all investigative research is done as part of a scientific study. The following field research report was written by an engineer for the plant manager and the plant superintendent of SMEPA, the Southeast Mississippi Electric Power Association, about his recent trip to inspect a company motor that was being repaired by another company. In his report the engineer describes what he found and discusses the implications of his investigations for the company.

A Field Inspection Report on 1A Mill Motor

Roger E. Smith

SOUTH MISSISSIPPI
ELECTRIC POWER ASSOCIATION

HIGHWAY 49 NORTH / P.O. BOX 1589 / HATTIESBURG, MS. 39401
TELEPHONE (601) 268-2083

TO: Marcus Ware, Manager of Production
Wayne Downs, Plant Superintendent

FROM: Roger E. Smith R.E.S.

DATE: April 9, 1982

SUBJECT: Inspection of 1A Mill Motor

On April 5, 1982, I traveled to the Westinghouse service shop in Mobile to inspect a 1A mill motor and its parts. The following people were present at this inspection: Wally Floyd, Westinghouse Corporation; Jim Webster, Westinghouse Corporation; Chuck Royal, Cleveland Electric. Westinghouse had originally brought the motor into their shop for overhaul and cleaning. 1

Wally Floyd, Manager of the Westinghouse Motor Shop in Mobile, had requested a SMEPA representative to inspect the motor because it required work beyond that specified in the original purchase order to correct the problem Westinghouse found. This motor had last been worked on by Cleveland Electric in the fall of 1981. 2

Below is a listing of the items I found that need to be corrected: 3

Inboard End:

1. The bearing dowel pin was recessed into its hole in the bearing cap so it did not lock the bearing in place. The dowel pin was fiber and not metal. The bearing had rotated slightly prior to motor disassembly.

2

2. The oil seal on both the bearing bracket and cap had previously been repaired in the fall of 1981 by Cleveland Electric and the repair had numerous voids apparently due `to improper welding.

3. The bearing bracket was cracked at the oil seal closest to the bearing. This crack is between the bearing and the motor, and if it is extensive, a new bracket may be required.

4. The motor shaft in the area of the oil seal had been "peened" and, because of several indentations, was not as smooth as a machined surface. Shaft repair had previously been performed in this same area (near the seal) and the shaft had voids where it had not been built up enough prior to final machining.

5. The end seal was loose.

6. The shaft-to-seal fit was bad.

7. Bolts were missing on the baffle.

8. Two (2) bolts were missing on the bearing bracket.

9. Oil had seeped into the rotor, but not more than would be expected for a motor used in this type of service.

Outboard End:

1. The bearing cap was worn on the top of the inner oil seal.

2. The bearing dowel pin was recessed into its hole on the bearing cap. This pin was metal and should have been fiber since this is the insulated bearing.

3

3. The bearing looked as if it had been hot at one time.

4. Two (2) bolts were missing from the end bracket.

5. Oil had seeped onto the rotor, but accumulation did not exceed normal levels.

Westinghouse is waiting for instructions from us on whether 4
to repair this motor or return it to our plant as is. Since Cleveland Electric may be responsible for correcting errors made during their initial servicing, the motor may have to be returned to them. They have offered to correct these problems at no cost to us.

Both Westinghouse and Cleveland Electric were told that they 5
would receive a reply from us after I submitted my findings to you. Please give me your instructions.

RES/tg
cc: Maynard T. Graves

QUESTIONS FOR DISCUSSION

1. Why does Roger Smith have to go to Mobile? Where does he tell his readers why he went? Why does he mention the names of others who were present for the inspection?
2. What role does Cleveland Electric play in Smith's report?
3. How does Smith gather and organize data for his readers? Would it have been appropriate for him to include sections entitled "Description of the 1A Mill Motor" and "Techniques of Inspection"? Why or why not?
4. Where does Smith give information that might in a formal field research report be labeled "Discussion" or "Results"? Provide examples from the memo in your answer.
5. Describe Smith's style. How would you characterize his diction? Judging from his language, to what kind of audience do his readers belong?

SUGGESTIONS FOR WRITING

1. Write a field research memo from Chuck Royal to *his* boss about the same inspection.
2. Using a memo format, write a short research study about (a) a trip you made for your employer or (b) a trip you made for a class for an instructor. as part of your classwork.

The following field research report first appeared in the prestigious New England Journal of Medicine. *Research took place at two hospital intensive-care units; the question posed was whether or not health care professionals washed their hands after contact with patients. The health care professionals were unaware that they were being observed. Handwashing is important to eliminate or to reduce nosocomial infections, infections transmitted to patients from hospital personnel. The mysophobia—fear of dirt and contamination—mentioned in the last paragraph of the study is one fear that hospital personnel must acquire. Although written primarily by physicians, articles in the* New England Journal of Medicine *are followed by a broad range of health care providers—nurses, technicians, hospital administrators, and consumers. The article reprinted here is divided into four sections—an unlabeled introduction, methods, results, and discussion.*

Hand-Washing Patterns in Medical Intensive-Care Units

Richard K. Albert and Frances Condie

Organisms that cause nosocomial infections can be transmitted by the hands of physicians, nurses, technicians, and other hospital personnel.[1-4] Hand washing is considered the single most important procedure in preventing nosocomial infections, and it has been recommended after contact with every patient by both the Centers for Disease Control and the American Hospital Association.[5,6] The risk of acquiring organisms, transmitting them to others, or causing a nosocomial infection may vary with the susceptibility of the patient and the type of patient contact.[7] However, because patients in intensive-care units are highly susceptible to nosocomial infections,[7,8] it has been suggested that personnel working with these patients should wash their hands more frequently.[7] Compliance with these recommendations has not been evaluated. Therefore, we determined the frequency of hand washing in two medical intensive-care units, using a method that would not bias the results. We found that, on the average, hospital personnel washed their hands after contact with patients less than half the time. Physicians were among the worst offenders. 1

METHODS

We recorded whether doctors, nurses, or other hospital personnel washed after direct contact with either patients or support equipment in the medical intensive-care unit of a university-affiliated teaching hospital. Observations were made 2 during 10 four-hour periods that were selected to include the ward teams' morning work rounds, during which maximal contact with patients was most likely. Recording was done under the guise of traffic-pattern observation, and none of the personnel knew that their hand-washing behavior was being observed. During each period, all patients in the intensive-care unit were observed, and all contacts were noted. Because house staff and attending physicians rotated every 30 days, we made observations during two successive months to maximize the number of participants being observed.

To determine whether the results applied only to the intensive-care unit 3
being studied, we observed another unit in a private hospital with completely different personnel, in an identical fashion for a total of 20 hours.

RESULTS

In the intensive-care unit of the university hospital, contact with 28 patients 4
was observed. All patients had vascular catheters, 13 (46 per cent) had urinary catheters, eight (29 per cent) had central venous or Swan-Ganz catheters, seven (25 per cent) had endotracheal tubes, and two (7 per cent) were receiving intravenous hyperalimentation. Forty different physicians or trainees (including first-year and second-year residents, medical students, and attending physicians) and 15 nurses had contact with these patients. Hand washing occurred after only 41 per cent of the contacts (Table 1). Physicians washed less frequently than did nurses ($P < 0.001$ by chi-square analysis). Multiple specific opportunities for nosocomial infection were observed: for example, urine bags were manipulated, intravenous dressings were changed, and respiratory equipment was adjusted without an intervening hand wash.

The hand-washing practices in the other intensive-care unit were similar 5
(Table 1).

TABLE 1. HAND-WASHING PATTERNS IN TWO MEDICAL INTENSIVE-CARE UNITS

	University Hospital		*Private Hospital*	
Group Observed	*No. of Contacts Followed by Washing / Total No. of Contacts*	*%*	*No. of Contacts Followed by Washing / Total No. of Contacts*	*%*
Physicians	90/326	28	7/52	14
Nurses	301/701	43	54/196	28
Respiratory therapists	91/120	76	13/27	48
Radiology technicians	7/16	44	1/4	25
Others	5/49	10	6/18	33
Total	494/1212	41	81/297	28

DISCUSSION

Compliance with hand-washing recommendations was poor in this unbiased 6
assessment. We found that basic concepts of infection control in the medical intensive-care unit were frequently ignored, particularly by physicians.

Hand contamination by a variety of bacteria has been documented,[1,2,4] and 7
contamination with some viruses is likely to occur. The hands of personnel may not only transmit organisms but may also become a reservoir for nosocomial organisms.[4,9]

We consider that hand washing is necessary after even minor contact with 8
patients or support equipment, because even the limited contact that occurs with taking the pulse, blood pressure, or oral temperature, or just touching a patient's hand, can result in transfer of organisms that can be recovered up to 2.5 hours later.[3] In addition, because the frequency of nosocomial infection is increased in patients with urinary catheters, endotracheal tubes, or vascular cannulas,[8] our patients were particularly prone to this complication. However, because we observed the activities occurring during morning work rounds, most of the contacts were extensive, as patients were examined, dressings were checked, and equipment was inspected and altered.

The low level of mysophobia that we observed in these intensive-care units 9
may be a major contributor to the high rate of nosocomial infection known to occur in this setting. Education of all personnel working in these areas is needed. Hand washing seems to be a procedure that is ignored or at least underemphasized in our present system of medical training.

References

1. Salzman, T. C., Clark J. J., Klemm, L. "Hand Contamination of Personnel as a Mechanism of Cross-Infection in Nosocomial Infections with Antibiotic-Resistant *Escherichia coli* and *Klebsiella-Aerobacter*." *Antimicrob. Agents Chemother.* 1967; 97–100.
2. Rammelkamp, C. H., Jr., Mortimer, E. A., Jr., Wolinsky, E. "Transmission of Streptococcal and Staphylococcal Infections." *Ann. Intern. Med.* 1964; 60:753–8.
3. Casewell, M., Phillips, I. "Hands as Route of Transmission for Klebsiella Species." *Br. Med. J.* 1977; 2:1315–7.
4. Knittle, M. A., Eitzman, D. V., Baer, H. "Role of Hand Contamination of Personnel in the Epidemiology of Gram-Negative Nosocomial Infections." *J. Pediatr.* 1975; 86:433–7.
5. Center for Disease Control. *Isolation Techniques for Use in Hospitals.* 2d ed. Washington, D.C.: Government Printing Office, 1975:19.
6. *Infection Control in the Hospital.* 3d ed. Chicago: American Hospital Association, 1974:111–3.
7. Steere, A. C., Mallison, G. F. "Handwashing Practices for the Prevention of Nosocomial Infections." *Ann. Intern. Med.* 1975; 83:683–90.
8. Schimpff, S. C., Miller, R. M., Polakavetz, S., Hornick, R. B. "Infection in the Severely Traumatized Patient." *Ann. Surg.* 1974; 179:352–7.
9. Maki, D. G. "Control of Colonization and Transmission of Pathogenic Bacteria in the Hospital." *Ann. Intern. Med.* 1978; 89:777–80.

QUESTIONS FOR DISCUSSION

1. How effectively does the introduction (a) provide a review of previous relevant literature, (b) establish a legitimate problem for field investigations, and (c) relate (b) to (a)?
2. The authors say that they use a "method that would not bias the results." Evaluate their work as objective field research based on this criterion. Specifically, comment on the use of (a) the traffic-pattern "guise," (b) the observations made over "two successive months," and (c) the two hospital units, one "in a private hospital."
3. Why do the authors describe the patients' treatment and appropriate apparatus in the Results section? Shouldn't this information be included in the introduction or Methods sections?
4. How carefully have the authors proved that their conclusions are justified? Where are these conclusions stated and how are they organized? What part does the table play in the authors' proofs?
5. Explain the way in which the authors, in the section "References," list page and volume numbers, years, titles, and places of publication.

SUGGESTIONS FOR WRITING

1. Assume you are the director of infection control at a hospital. Compose a memo to hospital personnel providing the rationale for handwashing and giving guidelines based upon the information in this study. Emphasize the importance of this field study in your memo, and cite this article specifically.
2. Write a field research report similar in format and length to this handwashing study aimed at experts on one of the following topics:
   a. the use of safety equipment by a group of factory workers
   b. the feeding habits and preferences of a particular species in a restricted environment, e.g., a zoo, a valley, a section of a forest
   c. the adherence to a company policy (e.g., dress code, identification procedure, security codes or procedures) by employees
   d. the amount and type of vehicular traffic using entrance and exit ramps on a four-lane highway
   e. the amount of noise emitted by trucks, airplanes, or machinery in a particular location
   f. the presence or absence of a particular mineral deposit at a site
   g. the speed (or success) of a new manufacturing or accounting procedure

The field research report that follows is a full-fledged formal report prepared by three researchers for the U.S. Bureau of Mines Coal Mine Health and Safety Program. The report investigates the use of a polymeric sealant to protect mine walls from deterioration. Because the U.S. Government does not endorse specific manufacturers' products, the proprietary names of the sealants are not given. This report was written for mining engineers, safety inspectors at the Bureau of Mines, decision makers at the coal company where the spray was tested, and executives at the chemical plant that manufactured the spray. As you read this investigative report, determine how its description and evaluation of the "field" meet the needs of these different readers.

Polymeric Sealant Used to Stop Shale Degradation in Coal Mines

J. C. Franklin, K. D. Weverstad, and R. F. Marquardt

ABSTRACT

The Bureau of Mines substantially reduced shale degradation in a coal mine by two spray-coat applications of a water-base epoxy sealant. More than 12,700 square feet of passageways were treated, and in 18 months only one failure occurred. This failure was due to improper spray application. 1

Material costs averaged 13 cents per square foot, and labor costs were 7 cents per square foot. Labor included moving all materials into the mine, the setup time for each of the areas coated, and application of coatings. 2

INTRODUCTION

Roof falls continue to be a main cause of accidents in coal mines (*2*). Some of these falls are caused by repeated cyclic wetting and drying of the shale by moisture condensing from or taken away by the mine air (*1, 6*). The sources of mine moisture are usually ventilation air, which is subject to daily and seasonal variations in temperature and humidity, and some open areas of mine water over which the ventilation air passes. Open mine water can be eliminated in some cases by piping, grouting, sealing, and covering drainage ditches. Ventilation air moisture can be prevented from degrading shale only by some method of sealing the rock to prevent the air-rock contact. 3

Tests have been performed on a variety of sealants including polyurethane foam, magnesium oxysulfate cement, and epoxy. Polyurethane foam is a good sealant, but presents a fire and health hazard (toxic gas release) when exposed 4 to fire (*7*). Magnesium oxysulfate-base cement, containing vermiculite (*8*), prevents shale sloughing and fireproofs urethane foam. Fraley (*3*) describes the process and difficulties of applying layers up to 1 inch thick by successive spray coating.

Neither of these sealants nor several others tried have been ideally suited 5
for coal mine use; however, sealants to control radon emission have been successful in recent Bureau of Mines studies. In its program of research on controlling radon and radon daughter hazards in uranium mines (*4*), the Bureau has identified a number of barrier coatings that appear to be good for both radon control and prevention of shale degradation. This report describes the field test of two water-based epoxies in a coal mine.

TEST MATERIALS

The two water-based epoxies selected for the mine test are generally used for 6
waterproofing concrete. One is used as a prime coat and the other as a top coat. For the coal mine tests the prime coat was diluted 25 percent with water. Thus, it penetrated the dust on the roof and adhered to the shale, bonding the loose dust to the shale. After curing, it was top-coated with the main sealant. Both coating materials are two-component systems mixed 1:1.

The manufacturer's test of the prime coat for water vapor transmission 7
(ASTM E-96) was 0.98 grams per square meter per 24 hours at 40° C and 100 percent relative humidity. The top-coat material was not tested. It is similar to the prime coat, except for added pigment, and has similar characteristics.

Before the test could begin, two water-base epoxies were submitted to 8
the Mining Enforcement and Safety Administration for fire testing. Samples of uncured components, along with cured materials applied to fireboard, were submitted. Data from pyrolysis of cured materials and smoke chamber test results determined by Hammon (*5*) were submitted.

Even though these materials were accepted for underground use and are 9
considered by industrial standards as nontoxic, some safety precautions should be followed. Protective clothing should be worn while spraying, and the user should wash with soap and water immediately after use. Personnel in the immediate area should wear respirators if they are exposed to the mist for any time.

An airless spray system is used to place the epoxy coat at a high pressure 10
(2,400 psi), for better penetration and crack filling.

FIELD TEST

The sites chosen for the test were in a Westmoreland Coal Co. mine that was 11
having a bad sloughing problem. The roof, which was exposed 18 months, deteriorated to the point of needing rebolting. Slabs up to 4 inches thick were falling,

Figure 1. Applying the top coat in intake air.

especially in intake and track areas. The Bureau applied the materials, and the company agreed to evaluate the sealant's performance for several months. The test included coating the roof in exhaust, track, and intake areas, plus some rib coating. Most of the test areas had been mined out for several months and required barring down of loose rock before coating. No other preparations were made. One section of drift in the track area, exposed for only a month, was coated without barring down.

In phase 1 the prime coat was applied to 190 linear feet of track area, 12
262 linear feet of intake airway, 100 linear feet of exhaust airway, and 120 linear feet of ribs. Almost 1 month elapsed between application of this prime coat and phase 2, top-coat spraying in the first three areas named above. Normally the first coat would cure for only a day or two before the second coat is applied; the 120 linear feet of ribs cured overnight. Also included in phase 2 was coating 63 linear feet near the face with both prime and top coats. The prime coat in this area cured only 18 hours before top-coat application.

Figure 1 shows the roof in intake air being coated with the top coat. As 13

Figure 2. Track area after top coat.

Figure 3. Tract area near the face before coating.

can be seen, the ribs were overlapped about 6 inches to stop moisture and air from penetrating the roof. A roof bolt that should have been removed can be seen near the upper center. Other bolts had rock around them so they were still supporting the roof.

Figure 2 shows the track area after application of the top coat. The floor 14
is covered with loose shale (a combination of fallen and barred down rock). Since little air moved through this area, the prime coat took more than 4 days to cure. This result is very important, as is shown later. As can be seen in both figures 1 and 2, the light-colored top coat helps in entry lighting.

Figure 3 shows the condition of the coated roof near the face, with some 15
loose thin slabs about ¼ inch thick beginning to peel off (to the right of the roof bolt on the right side). The roof was uneven and covered with rock dust. This track area also had little air movement. The grooves from the cutter prevented uniform coating and also cut down on the coverage. Additionally, a rib test was conducted for stabilization. The prime coat, diluted about 50 percent with water, was applied to both ribs in the intake airway. The prime coat was followed by a top coat after initial cure had taken place in 24 hours.

RESULTS AND ANALYSIS

Table 1 shows the coverage and cost breakdown for all areas coated. These 16
prices are based on 1,000-gallon purchases and do not include labor, which averaged 7 cents per square foot, including moving all materials into the mine, setup time, and application of coatings.

Inspected after 8 months, the original coated area was a virtually complete 17
success. The track area had one section about 4 feet square that came down in small pieces. There was a crack in the roof under the coating near the failure. This crack probably was not sealed in the area where the failure occurred. The return airway had one small failure at the extreme end of the coated section,

TABLE 1. TEST AREA COST ANALYSIS

Section	Area Sprayed, Ft^2	Primer, Gal	Water, Gal	Cost/ft^2	Top Coat, Gal	Cost/ft^2	Total Cost, Ft^2
Neutral	3,780	39	10	$0.05	40	$0.06	$0.11
Intake	5,250	24	6	.04	65	.06	.10
Return	1,953	24	6	.06	20	.05	.11
1 North tracks	1,260	12	3	.05	25	.10	.15
Ribs	540	13	6	.12	—	—	—
Total	12,783	112	31	—	150	—	—

a small piece about 3 inches square. The intake airway had several areas of fall. Most of these were very small and were at intersections.

The area where no roof preparations were made was almost a complete failure. It held for 7 months (November through June) and then started coming off. The rock came down in small sections (3 by 4 inches) and in most cases was less than one-fourth inch thick. In spite of the lack of special roof preparation, the most probable cause for this failure was that the prime coat was not completely cured before the top coat was applied. Only 18 hours was allowed for curing. The prime coat in the other track area was still tacky after 4 days of curing. When the top coat was applied, water, still present in the prime coat, could have been absorbed into the shale, causing failure of the rock. 18

Because of this failure, the coal and chemical companies conducted additional tests in the track area near the face. When the prime coat was given sufficient time to cure, there was no failure (8 months). 19

CONCLUSIONS AND RECOMMENDATIONS

The successful results and economical sealant costs indicate water-base epoxies to be an effective method for stopping shale degradation. Because the original areas are holding up well after 1.5 years' exposure, the coal company is in the process of coating about 6,000 linear feet of track area. It also is incorporating sealants as part of the roof-support plan. 20

From the test it is also concluded that all loose roof bolts should be removed, the holes plugged, and new bolts installed prior to coating. It is also necessary to prepare the roof adequately before coating, and to make certain that the prime coat penetrates well and is adequately cured before application of the top coat. 21

If a roof can be coated just after undermining, possibly only one coat would be needed. This would reduce material costs by half. Of course, spray equipment would have to be mounted on the continuous miner to spray the 22 roof as the miner retreats after making a cut. It is recommended that a one-coat test be made to investigate this possibility.

References

1. Aughenbaugh, W. E., and R. F. Dryewski. "Investigation of the Failure of Roofs in Coal Mines." University of Missouri—Rolla, USBM Contract H0111462, October 1973, 135 pp.
2. Barry, T., and Associates. "Fatality Analysis Data Base Development." *BuMines Open File Rept.* 22–73, 1972, 289 pp.; available for consultation at the Bureau of Mines libraries in Pittsburgh, Pa., Twin Cities, Minn., Denver, Colo., and Spokane, Wash.; at the Central Library, U.S. Department of the Interior, Washington, D.C.; and from National Technical Information Service, Springfield, Va. PB 219 250.
3. Fraley, J. E., and R. E. Simpson. "Magnesium Oxysulfate Cement Sealant in Coal Mines." *BuMines RI* 8150, 1976, 17 pp.
4. Franklin, J. C., R. C. Bates, and J. L. Habberstad. "Polymeric Sealants May Provide Effective Barriers to Radon Gas in Uranium Mines." *Eng. and Min. J.,* v. 176, No. 9, September 1975, pp. 116–118.
5. Hammon, H. G., K. Ernst, J. R. Gaskill, J. C. Newton, and C. J. Morris. "Development and Evaluation of Radon Sealants for Uranium Mines." Lawrence Livermore Lab., USBM Contract Report H0232047, May 1975, 67 pp.
6. Hartman, I., and H. P. Greenwald. "Effects of Changes in Moisture and Temperature on Mine Roof." *BuMines RI* 3588, 1941, 40 pp.
7. Mitchell, D. W., C. M. Murphy, and J. Nagy. "Fire Hazard of Urethane Foam in Mines." *BuMines RI* 6837, 1966, 29 pp.
8. Western Chemical and Manufacturing Company. "Sealant Systems Reduce Mine Fire Hazards." *Eng. and Min. J.,* v. 176, No. 9, September 1975, p. 119.

QUESTIONS FOR DISCUSSION

1. What specific problem does this report investigate and promise to solve? Point out the key sentences announcing this fact. Why do the authors discuss "a variety of sealants" before describing how they discovered a solution to the problem?

2. Why do the authors discuss "Test Materials" before describing the specific site where the investigations were made?

3. Two "failures" are cited in "Results and Analysis." Since the failure involving "the one section about 4-feet square" is so minor, did the authors really need to mention it? What explanations do the authors give to convince the reader that this failure was not the result of the proposed solution? Why do the authors want the reader to contrast this minor failure with the problems encountered "where no roof preparations were made"?

4. What conclusions do the authors reach and how are these conclusions linked to the results they report? Why do the authors combine the conclusions and recommendations sections? Could they have recorded this information in another way?

5. Not every audience for this report is equally interested in each section of the report. Identify the parts that would be of most interest to the following types of readers and explain why. Is there any overlap? Why?

a. decision makers
b. mining engineers
c. safety inspectors
d. officials at the Bureau of Mines

6. What information do the accompanying photographs give the reader that the text alone could not adequately supply? How well do the authors interpret these photographs for the reader? What other types of visuals could the authors have used? Support your answer by referring to specific examples from the report.

SUGGESTIONS FOR WRITING

1. Using this report as a model, write a field research study in the form of a long, formal report on one of the following topics.

a. the effectiveness of public transportation to your school, work, or some public area (e.g., a beach, a business district, a shopping center, a medical center)
b. the use of highway barriers or lights being used for rerouting, separating the flow of traffic, or closing off traffic on a local expressway
c. the performance of an alternate energy system (e.g., windmills, a photovoltaic system, solar energy collector) compared with a traditional energy system in a private residence or public building
d. the adaptability of a new computer language in a large office
e. the performance of a groundcover (e.g., Astroturf) at a ballpark
f. the behavior of a specific type of animal or plant relocated in a new environment
g. the wear resistance of a new fabric used by a specific occupational group (e.g., police officers, firefighters, food service personnel, underwater divers)

2. Compose a shorter, more informal report on the use of polymeric sealant for a less technical audience (e.g., the director of the public relations department at one of the mining companies or at the Bureau of Mines).

OPINION SURVEY REPORT

In this opinion survey the investigator mailed a 73-item questionnaire (included in the report) to carefully selected male and female agricultural engineering students. The survey was published in Engineering Education, *a professional journal concerned with "the study and improvement of engineering and technical instruction" and intended for "a broad readership of educators and engineers." The study was conducted to help engineering faculties and administrators design a curriculum that might increase "women's participation and success in higher education."*

It is notable that this report does not bury readers under an avalanche of data about each of the 73 questions. Instead, the author summarizes the students' opinions so that readers can see a pattern of responses emerging. Readers familiar with statistical significance testing (determining the percentage of responses that might be based on chance) can find meaningful information in the tables, though those without knowledge of this kind of testing can still obtain a lot of useful information from the report. The report is a model of how to study a problem, prepare a questionnaire for a restricted group of respondents, organize their responses, and then prepare a report based upon those responses.

Women in Engineering: Attitudes, Motivations and Experiences

Lee R. Erickson

Undergraduate men and women in agricultural engineering programs have much in common. Improved career counseling at the secondary level could help eliminate the few observed differences, providing the women with equivalent encouragement and preparation for an engineering career. 1

A nationwide survey of undergraduate agricultural engineering students (AE's) was undertaken in May 1980 as a portion of the Women's Educational Equity Act Program (WEEAP) at Montana State University. The project was funded to provide a model for curriculum review and modification for increasing women's participation and success in higher education. 2

THE SURVEY

A 73-item survey was developed to evaluate personal motivators, experiences 1
and effects of the college curriculum on men and women AE's. A nationally representative sample of the 1980 undergraduate AE population was selected with special emphasis on women AE's. Department heads from accredited agricultural engineering departments were contacted and asked to provide a list of female students and an "equivalent by class" male counterpart for each woman student. A random sampling technique was used to select about one-third of the approximately 200 women AE's enrolled in four-year programs for survey purposes. The same number of men were selected for study in a similar manner. Of the 130 questionnaries mailed (65 male, 65 female), 80 responses were received (41 male, 39 female) for an overall return rate of 62 percent. Of the total U.S. undergraduate AE population, approximately 16–20 percent of the women and 1–4 percent of the men responded.

The computer program SPSS[1] was used to generate response statistics and 2
to evaluate differences between male and female responses for each survey item. The survey was designed to test the null hypothesis that:

Men and women AE's are not significantly different in 3

1. Personal aptitudes related to AE job success,
2. Job-related personal motivators,
3. Career awareness at the high school level,
4. Their experiences in AE departments,
5. Their treatment by AE instructors,
6. Their reaction to AE text materials,
7. Their perceptions on why peers drop the AE curriculum.

The survey used a response scale of 5 (strong agreement) to 1 (strong dis- 4
agreement). Results are tabulated in the tables as follows: $\overline{x}$ = combined response for all respondents; $\overline{x}_m$ = average male response; and $\overline{x}_f$ = average female response. The significance level is based on the student's t test.[2] A significance level of .01 indicates a 99 percent probability that the difference in responses between male and female students is not due to chance.

SURVEY RESULTS

In general, men and women AE's were found to have quite similar personal 5
aptitudes and job-related personal motivators (Table 1). This finding is substantiated by the lack of significant difference in response between men and women in 20 of 32 survey items in these two categories and is consistent with other studies of men and women engineers.[3,4] Differences in aptitude and motivation noted in this study can be broken down along traditional sex role lines. Men were significantly more interested in mechanical devices, building and designing things, and in directing others in their work. They were also more interested

TABLE 1. PERSONAL DATA ON MEN AND WOMEN AgEng STUDENTS

Biographical Data				$\bar{X}$	$\bar{X}_m$	$\bar{X}_f$
1. Male 41 (51%) Female 39 (49%) (Total responses = 80 of 130)						
5. Class Level:	12 (15.0%)	Freshman	(1)			
	18 (22.5)	Sophomore	(2)			
	23 (28.7)	Junior	(3)	2.81	2.71	2.92
	27 (33.7)	Senior	(4)			
6. College GPA:	11 (13.7%)	2.0–2.5	(1)			
	23 (28.7)	2.5–3.0	(2)	2.75	2.73	2.77
	21 (26.2)	3.0–3.5	(3)			
	25 (31.3)	3.5–4.0	(4)			

Personal Aptitudes and Motivations

I am good at or I really enjoy working with:	$\bar{X}$	$\bar{X}_m$	$\bar{X}_f$	*Signif. Level*
7. people	4.06	4.00	4.13	
8. animals	3.97	3.90	4.05	
9. plants	3.88	3.66	4.10	.050
10. mechanical devices	3.90	4.33	3.47	.001
11. ideas and concepts	4.01	4.13	3.90	
I like to:				
12. build things	4.25	4.39	4.10	.200
13. design and analyze	3.96	4.10	3.82	.200
14. direct others in their work	3.45	3.63	3.26	.050
15. be directed in my own work	3.14	3.07	3.20	
16. have active, outdoor employment	4.36	4.44	4.28	
I wanted to:				
17. follow a family tradition	2.30	2.44	2.15	
18. do something unusual	3.22	3.15	3.29	
19. be associated with agriculture	4.17	4.24	4.10	
20. have high job satisfaction	4.46	4.46	4.46	
21. be able to live in a rural area	3.95	4.15	3.74	.050
22. have a high salary	3.61	3.73	3.49	
23. have a personal independence in my work	4.05	4.12	3.97	
24. have a secure job	4.06	3.98	3.95	
25. be competitive with peers	3.29	3.34	3.23	
26. learn more about specific areas	3.96	3.98	3.95	
27. work outdoors	4.19	4.39	3.97	.050
28. utilize my creative talents	4.15	4.20	4.10	
29. be able to teach others	3.44	3.46	3.42	
30. help society	3.76	3.71	3.82	
31. work with my hands	3.97	4.15	3.80	
32. use ag. engineering as a step to another career	2.84	2.98	2.69	.200

in working in a rural environment, working outdoors and working with their hands. Women were more interested in plants than were men. Men may be more inclined to use AE as a step in their career, although neither sex is strongly motivated in this direction.

Men had a significant advantage in preparation for engineering careers 6
(Table 2). High school counselors made them more aware of engineering as a possible career and gave them more encouragement to take necessary preparatory math and science courses. Men were also more encouraged in developing manual skills. Math and sciences were encouraged more than technical skills for both groups. Neither group had a strong understanding or awareness of men's or women's contributions to engineering, and were not particularly aware of career opportunities for agricultural engineers.

At the college level (Table 3), men and women AE's had similar experiences 7
with AE instructors and departments. Instructors were considered fair and impartial. Texts used were considered appealing and encouraging by both sexes. Women students would like to see more reference to women AE's in the technical literature, and are interested in role models as well. Departments were most persuasive in acquiring and keeping students through their promotional efforts and course offerings.

The final questions (Table 4) are a general inquiry into the causes of attrition 8
in the AE program. Information here is generally second-hand and is in a form that is difficult to evaluate statistically. Some trends emerge: Most losses (about

TABLE 2. CAREER AWARENESS AT HIGH SCHOOL LEVEL

I was encouraged to:	$\bar{X}$	$\bar{X}_m$	$\bar{X}_f$	*Signif. Level*
33. take math/science classes	4.16	4.32	4.00	.20
34. improve technical/work skills	2.71	2.93	2.47	.05
I was encouraged to try engineering by:				
35. parents and family	3.29	3.22	3.37	
36. male peers and friends	2.70	2.83	2.55	
37. female peers and friends	2.47	2.49	2.45	
38. teachers and counselors	3.22	3.59	2.82	.01
I was made aware of:				
39. engineering as a possible career	3.62	3.88	3.34	.05
40. subjects that would help prepare for engineering	3.28	3.50	3.05	.05
41. ag. engineers and what they do	2.41	2.55	2.26	
42. contributions made to engineering by women	2.19	2.20	2.18	
43. contributions made to engineering by men	2.77	2.73	2.82	
44. a practicing engineer in the community	2.45	2.53	2.37	
45. my high school was considered "good" for college prep students	3.67	3.73	3.60	

TABLE 3. COLLEGE EXPERIENCES OF MEN AND WOMEN AgEng STUDENTS

	$\bar{X}$	$\bar{X}_m$	$\bar{X}_f$	*Signif. Level*
College AgEng Department				
46. An introductory course in engineering or ag. helped make me aware of AE and influenced my decision to study in this area.	3.14	3.17	3.10	
47. The college catalog description was helpful in making me aware of opportunities in AE.	3.36	3.39	3.33	
48. Scholarship availability is important in attracting men and women students in AE.	2.75	2.83	2.67	
49. Promotional efforts and materials put out by the AE department are important and should show men and women students in typical AE activities.	4.05	4.05	4.05	
50. Courses offered in the AE curriculum appeared to meet my personal educational goals.	3.85	3.80	3.90	
51. The ASAE Student Branch Club encouraged participation of all students and was a consideration in my becoming an AE.	2.95	2.98	2.92	
52. Other AE students encouraged my presence in AE.	3.25	3.23	3.28	
Agricultural Engineering Instructors:				
53. Encouraged my presence in AE	3.92	3.88	3.97	
54. Treated men and women in a similar manner in classroom interactions	3.89	3.95	3.82	
55. Are impartial and unbiased by student gender in class assignments and grading	3.85	3.85	3.85	
56. Arranged physical aspects of class and lab work to accommodate everyone	3.86	3.90	3.82	
57. Avoided embarrassing students or putting them "on the spot"	3.66	3.74	3.58	
58. A woman instructor is necessary as a role model for women students and would be an asset to the department.	3.00	2.63	3.39	.01
Text Materials				
59. Technical materials were presented in a manner appealing to both sexes.	3.52	3.53	3.51	
60. Text materials were designed to encourage both men and women in AE.	3.29	3.39	3.18	
61. Women's contributions to AE were apparent in required texts and references.	2.25	2.53	1.97	.01
62. It is important to have women's contributions to the AE literature pointed out in texts and references.	3.17	2.92	3.41	.05
63. Presence of women engineers in textbook photographs and visual aids is important and should be encouraged.	3.33	3.05	3.61	.01

TABLE 4. ATTRITION AMONG AgEng STUDENTS

Students I know who dropped AE as a viable career objective did so, to the best of my knowledge, because of: (check those that apply)

		Percentage Responses (rounded to nearest whole %)			
		Men	*Women*	*Men & Women*	*None*
64.	inadequate technical background	24	14	21	41
65.	inadequate manual skills	4	14	6	76
66.	not interested in technical studies	26	13	20	41
67.	social pressures	3	4	5	88
68.	family and peer pressure	1	4	1	94
69.	negative faculty/staff attitudes	6	1	3	90
70.	biased text materials	1	0	1	98
71.	financial limitations	9	1	9	81
72.	unknown reasons	8	1	8	83
73.	other (specify)	11	3	1	85

half) of both sexes occur because students are not interested in or capable of technical studies. Men appear to have more problems in the areas of finance and staff pressure; women have more problems with manual or physical capabilities and with peer pressure. Text materials were a very minor cause for attrition. The "other" category (item 73) was dominated by male attrition, with a tendency to switch to another engineering program that better met the student's needs.

IMPLICATIONS OF THE FINDINGS

A successful program for encouraging and retaining women agricultural engi- 9
neers (or any engineering students) must begin at the high school level. The present study indicates that women are not strongly encouraged to prepare for entry into a professional engineering curriculum. Counselors are not doing an adequate job in pointing out alternative or nontraditional careers such as engineering. Better counseling could significantly affect the numbers of women enrolling in engineering and substantially improve their chances for success in this nontraditional career area. Women in secondary schools need more hands-on experience at building and designing things to promote manual and technical proficiencies and self-confidence. Such training could also lead to more interest and greater success in nontraditional employment areas.

The study also indicates that both sexes were relatively ignorant of the 10
contributions to engineering by women, in part because very little work has been done by women in traditional agricultural engineering areas. The few professionally active women AE's need to be encouraged to be as visible as possible in the technical literature. Contributions of all agricultural engineers need to

be made more visible to high school students. Greater visibility of role models in industry and at the college level may help and encourage women and men AE students.[5]

CONCLUSIONS AND RECOMMENDATIONS

A program to set up a procedure to inform students about agricultural engineering and the necessary entry level preparations at the high school level has been developed. An educational packet targeted at high school math and science teachers has gathered information on local and national sources related to agricultural engineering career opportunities, women in engineering, academic requirements, names of additional resources and other materials. 11

The educational packet will inform educators about agricultural engineering as a career and a field of study, and will give students access to literature on the field through their high school instructors or counselors. 12

The overall program is intended to serve as a model for student education and recruitment. The model will be of use to agricultural engineering departments that wish to develop a similar approach to encouraging and retaining women in their curriculum. 13

References

1. Nie, N. H., et al., *SPSS: Statistical Package for the Social Sciences,* 2nd edit., McGraw-Hill, 1975.
2. Ostle, B., *Statistics in Research,* 2nd edit., Iowa State University Press, Ames, Iowa, 1963.
3. Durchholz, P., "Women in a Man's World: The Female Engineers," *Engineering Education,* Jan. 1977, pp. 292–299.
4. Friedman, S. M., *Women in Engineering: Influential Factors for Career Choice,* Lehigh University, Bethlehem, Pa., 1978.
5. Gardner, R. E., "Women in Engineering: The Impact of Attitudinal Differences on Educational Institutions," *Engineering Education,* Dec. 1976, pp. 223–240.

QUESTIONS FOR DISCUSSION

1. Why did the investigator send the questionnaire to men as well as to women? Couldn't Lee Erickson have accomplished the goals of the study by mailing the questionnaire just to women AEs?

2. In what ways does the part of the report entitled "The Survey" correspond to a "Materials and Methods" section in a laboratory or field research report? In your response, discuss the significance to the survey of the following:
   a. the role of the department heads
   b. a random sampling technique
   c. the computer program from the SPSS collection

d. the use of the null hypothesis (the determination of which responses are the result of chance)

3. How is the questionnaire organized? What types of questions does the investigator include and how do these questions help Lee Erickson to accomplish the goals of the study?
4. Discuss the ways the investigator summarizes responses and highlights important points. What kind of information from the questionnaire is omitted? What is included? How does the author use the tables, and why do these tables help the reader?
5. Is the part of this report labeled "Implications of the Findings" comparable to a "Discussion" section in a field or laboratory report? Explain your answer by assessing the kinds of information Erickson presents and why the reader needs or wants to know these facts.
6. How are the conclusions and recommendations tied to the overall objectives of the study? Elaborate on how the students' responses justify the conclusions and recommendations Erickson offers.

SUGGESTIONS FOR WRITING

1. Design a short (5-question) questionnaire on some school or work policy—schedules, fringe benefits, parking, meals—and distribute it to 10–15 people. Write a report to a decision maker on the results of your survey and include a recommendation.
2. Prepare a statement of a problem that has occurred at school or at a store or plant or office where you work. Then prepare a short questionnaire (8–10 questions) and give the questionnaire to two separate but related groups (e.g., male and female majors in agricultural engineering). After gathering the groups' responses, write a 5–7 page informal report to a decision maker (a chairperson, dean, personnel director, manager) explaining the purpose of your study, the type of questionnaire you used, the way you interpreted responses, and the conclusions and recommendations you can offer based on those responses.

chapter **8**

Proposals

A proposal is a document that suggests a specific course of action for consideration. Proposals are made to solve a problem, to offer a service, to procure goods, to gain approval for a study. Proposals are written for many different reasons and for many jobs, large and small. They can be written to a professor to gain approval for a paper topic or research report. Sent up the administrative ladder in your own company, a proposal can demonstrate to decision makers that there is a need for a change (to improve communication, to purchase new equipment, to increase productivity, to expand markets) and that it is both feasible and wise to make it now and in the way you specify. Proposals can also convince a federal or private agency (e.g., the Department of Defense, Rockefeller Foundation) to support your scientific or technological research.

Proposals must be persuasive. A good proposal not only points out benefits but also alerts the reader to the disadvantages of delaying, refusing to make a change, or deciding to award a contract to someone else.

To persuade your readers to authorize your work or adopt your plan, you need to have a clear idea about how your proposal is likely to be read and what exactly readers want from it. Readers of a proposal are generally busy decision makers whose desks are piled high with papers, reports, and possibly other proposals competing with yours. Their chief job is to evaluate— to judge the merits of the work you want approved. Anything you can do to bring your readers to a favorable decision quickly obviously benefits you. These readers are primarily interested in how your proposal will meet their needs— make a job easier, make the company more successful, enhance the reputation of the agency. Your proposal must promise future benefits. Your readers-evaluators will have very clear, strict criteria by which your work will be judged. Among the most important considerations for such readers is cost. After all, most proposals ask for or require some type of funding, whether fees for work, support for research, or funds to purchase new goods or services. You might expect your reader to be more than a little skeptical and perhaps a little stingy. Executives who decide the fate of proposals are no pushovers.

Writing for such an audience requires you to supply persuasive and complete answers to some tough questions. You can expect the reader of a proposal to ask the following types of questions:

1. What problem does the writer propose to solve?
2. What is the actual work the writer plans to accomplish?
3. What benefits does it promise?
4. What procedures will be followed to get the job done?
5. What are the company's (or the writer's) qualifications to do the job?
6. Are there any potential difficulties with the proposed project?
7. How long will it take to do the job?
8. How much will it cost?

Convincing answers to these and related questions will help you to get your proposal approved. This chapter will show you how to prepare a variety of proposals that will provide the right answers and thus win the reader's approval of your proposal.

SOLICITED AND UNSOLICITED PROPOSALS

Proposals can be solicited or unsolicited, depending upon who calls for the plan of action. When a company or government agency has a problem that it wants solved or needs to purchase goods or services, it solicits solutions and offers through a Request for Proposals (RFP), or a Request for Bids (RFB). A *solicited proposal,* therefore, is one written in response to a request. An RFP describes the problem the company wants solved or goods or services it needs, outlines the specific work that must be done, and provides detailed instructions on how the proposal is to be prepared. These instructions include the types of information (especially specifications regarding size, weight, color, height, components, and capability) the writer must supply and the format in which it must be given. The small print in RFPs can be formidable, often dictating how the proposal is to be typed, duplicated, and even mailed. To win or bid for the contract, the proposal writer must faithfully comply with the RFP, for proposals that do not follow the guidelines in the RFP are returned for revision or even rejected. RFPs spur competition. Each year companies spend millions of dollars tailormaking proposals to obtain jobs. By winning a large contract from the government, a company can keep its work force busy for years. The U.S. government awards billions of dollars each year to companies and individuals who write successful proposals.

Fulfilling the requirements of an RFP can take days or years, depending on the scope of the job. A proposal supplying equipment (installing air conditioning units) or a service (decorating an office) can be handled easily in a few paragraphs. Proposals for complex projects—building a data processing system, testing aircraft—may require several volumes, each covering a different dimension of the job (engineering, management, costs).

A single solicited proposal can have one or many readers. A proposal to a small business may be read only by the manager, whereas one to a large company may be studied by experts in engineering, safety, transportation, and accounting.

An *unsolicited proposal* is developed at a company's or individual's own initiative and not in direct response to a specific request made by a potential customer or government agency. When you prepare a report to your employer, without being asked to, on installing security cameras in the company parking lot, you are writing an unsolicited proposal. If your company wishes to offer its technical services (e.g., assessing the strengths and weaknesses of a manufacturing process, providing training programs) because it has done similar work for one firm and thinks another firm might be interested in the same expertise, it will do so in an unsolicited proposal. Unlike writing a solicited proposal, where the employer or customer is aware of the problem or need, you have to convince the reader of an unsolicited proposal that there is a problem and a great need to solve it. To sell the reader, you have to define and describe the problem as it applies to the specific set of goals established by the reader's company. When you sell the problem, you are likely to sell the customer on your solution for it.

Preparing unsolicited proposals is both challenging and risky. Just as potential customers may award you a contract immediately, so they might ask for more details and then decide your idea is good but unaffordable. It is risky to invest time and money in a proposal that someone may refuse even to consider. For these reasons always regard the reader of an unsolicited proposal as a highly skeptical individual who will question your conception of the problem and may even deny its existence. Before expending a lot of effort, therefore, it is wise to do some preliminary, if informal, research. Try to speak with members of the company or agency, as Llewellyn Toulmin, the author of the proposal on pages 290–299, did. Make exploratory telephone calls and send query letters before launching an unsolicited proposal. If you are writing an unsolicited proposal to your employer, it is prudent to discuss your idea briefly first to find out whether or not he or she seems interested.

INTERNAL AND EXTERNAL PROPOSALS

In addition to being classified as solicited or unsolicited, proposals can further be categorized as (a) internal or (b) external. Each of these types is discussed in detail below.

Internal Proposals

An internal proposal is usually written informally as a memorandum for a decision maker within your own company, and it can be solicited or unsolicited.

The preceding example about installing security cameras in the company parking lot and Rannie Fox's memo to Les Axelrod in Chapter 1, pages 48–50, are internal proposals. Internal proposals often suggest changes or improvements in the organization. When you have an idea that might save the company time, money, or labor; improve employee morale; or enhance its public image, you may present your idea in an internal proposal. Such proposals frequently cover such suggestions as buying a new kind of equipment, hiring new employees, training personnel, changing schedules, or performing a procedure in a new way. If an employer requests you to submit a proposal on possible ways of cutting production costs without decreasing quality, you would be writing a solicited internal proposal.

An internal proposal is usually sent to your immediate superior, although it may eventually be sent higher up the organizational ladder for approval. Never assume that your audience will enthusiastically embrace your suggestion and implement it at once, for change is slow in all organizations. Moreover, company egos may be at stake: No employer likes to be told that his or her way of doing something is not the best way. Consider the human factor.

When you write an internal proposal about a problem or a need the reader may not know about, you must first convince the reader that the problem exists. The worst thing you can do is to assume that a reader will automatically agree with you and will know as much about it as you do. Document the existence and scope of the problem or need by supplying specific dates, places, and costs. If you see your company losing money because of an inefficient communication system, spell out precisely where, why, and how much in dollars and cents. Tell the reader where you obtained your figures; prove that you have done your homework. Management will not supply the figures and cost analyses for you. General remarks such as "as you know from current practice" or "we have present problems because of communication difficulties" will persuade no one. It is also unwise to shift the burden for supplying evidence to another operation in the company—"The business office or the maintenance department can supply you with the details."

Another important element in an internal proposal is to demonstrate that current equipment, personnel, and facilities are insufficient to solve the problem you are bringing to the decision maker's attention. If you were proposing that the company institute a training workshop on a new telecommunications system for employees, you would have to prove that current guidelines or policies were not sufficient to train and advise employees in this crucial area. Explain why a new workshop is important to the company now. Show how such a workshop would improve certain practices in the company by emphasizing how its absence is causing difficulty.

After identifying and explaining the problem, give the specifics of your plan. In the telecommunication workshop example, state who will conduct it (an outside consultant? someone from another division of your company?), who is to participate in it, what they will learn and what materials will be required, where and when it will be held, and, finally, how much it will cost the company in time and money. In brief, do not take anything for granted. Your reader surely will not.

One last point about internal proposals. Submitting them can mutually benefit you and your immediate superior who, seeing a good idea, will recommend that your plan be implemented. Just make sure that what you propose is carefully researched and diplomatically written.

External Proposals

An external proposal is sent outside your organization to a government agency or other potential user of your product, service, or expertise. Most external proposals you write will be sales proposals.

The goal of all sales proposals is to win a contract. They can be solicited or unsolicited. If they are responding to an RFP, they do not have to convince the reader there is a problem. If they are unsolicited, sales proposals may have to sell the reader on the problem as well as on the solution.

Sales proposals can be long or short, formal or informal, depending on the size of the job and the requirements of the audience. Contractors bidding on a large construction job—a twenty-story office building—submit proposals in a long formal report. The parts of that report—transmittal letter, numerous sections for explanation and justification of different phases of the job, appendices—are necessary to supply the detailed technical information the customer needs and will demand. For a smaller job—paving the parking lot in a small shopping center—a letter or short report would be appropriate and sufficient.

In evaluating sales proposals, the reader will be looking for the offer that best accomplishes the job for the least amount of money. Your task is to demonstrate that you and your firm are right for the job. Do that by emphasizing how your company has the experience and technical qualifications needed. But do not forget to discuss the client in your presentation. Spending all your time boasting is just so much sales talk. Bring your client into the proposal by showing how he or she will benefit specifically from your company's product or service. Spell out the advantages in investing in you and your organization. If you are responding to an RFP, read and reread the technical requirements the client wants bidders to meet. If your proposal is unsolicited, formulate the problem you are identifying for readers so that the solution is consistent with the company's goals and objectives.

PROPOSAL FORMATS

Proposal formats may be informal or formal. Which one you choose will depend on (a) your readers' needs and requirements, including whether they want you to follow the strict guidelines of an RFP, (b) the size of the job you propose, and (c) the cost of that job.

Informal Formats

An informal format—usually a letter or a memo—is suitable for a short internal or external proposal. A short, unsolicited internal proposal, for example, generally includes an introduction defining the problem or need, a body outlining the proposal solution step by step, and a conclusion summing up the benefits. An external solicited sales proposal for a relatively small job might suitably include the following parts:

1. *Qualifications for the Job*—your or your company's track record—number of years in business; types of work successfully completed; professional capabilities. Mention special features your firm offers, e.g., equipment or sophisticated techniques or training programs.
2. *Nature and Scope of Your Services*—the kinds of work you propose doing. Carefully detail the types, number, quality, and frequency of goods or services you intend to provide. Also spell out any services that you will not supply.
3. *Procedures and Methods*—the ways in which you will do the work.
4. *Schedule*—a breakdown of tasks or jobs according to the time they take. In addition, you may include a guarantee that the jobs will be performed as promised.
5. *Costs*—sometimes listed as a flat fee or computed according to an hourly rate.
6. *Feasibility*—a justification that the work can and should be done. Emphasize why it is possible to perform the services you have outlined in #2 above. Explain the means the customer has to judge the progress and quality of your work, e.g., progress reports or site inspections.

These six parts can be arranged, combined, or abbreviated as you see fit. For example, it might be possible to include a time schedule in the section on scope of services; or feasibility might be taken up with procedures and methods.

The plan above would, of course, have to be modified if your sales proposal were unsolicited. You would want to begin with a statement of the problem and how you learned of it. Then you would discuss your or your company's unique qualifications to solve it.

Formal Proposals

A long formal proposal is like a long report. It contains many different sections to give readers the detailed technical explanations they will need when you are bidding on a large job. A long formal proposal can contain the following parts.

The Letter of Transmittal. The letter of transmittal introduces your proposal. This letter may be a separate cover letter or may be attached to the proposal itself. Start by telling readers why you are writing. If responding to an RFP, indicate where you learned of it. If you are submitting an unsolicited proposal, state where you found out about the problem. Next briefly describe your firm's qualifications for the job. In a subsequent paragraph you might also highlight the major advantages of the plan you are submitting. Acknowledge any help you have received in preparing your proposal, i.e., conferences with the reader or employees at his or her firm. Finally, offer to answer any questions or to supply additional information should the reader require it. See the letter accompanying the Toulmin proposal on page 289.

The Abstract. This is a brief, informative summary of the proposal. Since it will be read by all audiences—scientists and laypersons alike—keep it clear and as free from technical jargon as possible. Emphasize the problem, briefly describe your plan to solve it, and conclude with your qualifications. You may want to review the discussion of abstracts on pages 219–220.

The Introduction. This section includes a statement of purpose, a description of the problem or need, and an explanation of the scope of your work. In the statement of purpose tell the readers why you are submitting the proposal. If solicited, refer to the RFP; if unsolicited, mention any previous contacts you have had with the company or agency. In describing the major area you plan to study, demonstrate that you understand both the background and the significance of the problem. If, for example, you were writing about the need for a scientific study, you could present the problem as a lack of scientific information and the need of further research. In explaining the problem, show the reader that you are familiar with the organization, agency, or company and its needs and goals. Document the problem by concentrating on important factors such as financial losses, decreased productivity, delays, or ineffective training programs. In the scope section tell the reader how your proposal is structured, what each part includes, and how the reader can use each to determine the benefits your proposal promises.

The Plan. This section, also called Methods, Approach, or Tasks, lists the specific procedures, step by step, phase by phase, that you plan to follow to solve the problem. Inform the reader of what will be done, by what methods, by whom, when, and where. For very complex jobs, each task (or job objective) can be broken down according to the following organization: Purpose, Activities to be Accomplished Phase by Phase, Desired Outcome, and Anticipated Problems and Solutions.

Evaluation Procedures. Included in this part of the proposal are the tests, or quality checks, you will perform to determine that the job is being done satisfactorily. Tell readers you will validate progress and results by performing additional experiments, or distributing questionnaires after the job is finished, or by sending periodic progress reports.

Qualifications. This section, also labeled Competency or Capabilities, gives the firm's track record. It provides the reader with the credentials of the people who would be working on the project. The emphasis should be on the skills and the experiences these individuals have acquired on similar jobs and would bring to the proposed job. For large projects, companies include the résumés of key employees as well as a list of references—previous clients who will verify the quality of work a firm has done for them.

Available Resources. Spell out the existing available facilities (laboratories, test sites), and equipment (computers, microscopes, word processers) both on-site and away from your institution or company that will be used. This section also informs the reader if additional equipment must be purchased or rented.

Costs. One of the most important parts of the proposal, this section often contains an itemized account as well as a narrative explanation of all costs to the client. Line items in that budget include expenses for salaries (and rates of pay) and fringe benefits, equipment, travel, supplies, operating expenses (e.g., the use of computer time), and overhead. Notify the reader about your intended fee schedule—e.g., will you receive three equal payments or be paid half before the project begins and the other half when the work is completed.

Conclusion. The conclusion offers the last opportunity to persuade your reader. Briefly summarize the main advantages your proposal offers. Explain also how accepting your proposal will bring the reader the results stipulated in an RFP. Where appropriate, stress how your services or goods are superior to those of competitors. Distinguish yourself from the competition in the reader's eyes.

Appendices. This section contains the supporting documentation for your project. It might include blueprints, graphs, tables, drawings, statistical data, copies of technical reports, and descriptions of complex equipment too detailed for inclusion in one of the major sections of the proposal.

A CHECKLIST FOR WRITING A PROPOSAL

1. Identify your reader and try to determine his or her concerns and preferences.
2. For a solicited proposal, study the RFP closely for proposal requirements.
3. For an unsolicited proposal, try to talk with the reader informally to determine whether there is any potential interest. Select the appropriate format for your proposal.
4. Identify and document the problem or question you intend to study.
5. Spell out clearly the work you propose to do.
6. Supply details about procedures, methods, materials, personnel involved.
7. Propose a realistic schedule.
8. Estimate the cost, showing why it would be a good investment.
9. State the qualifications of yourself or your company to do the job.
10. Consider any potential competition and tailor your proposal to show the superiority of your plan.

11. Emphasize the benefits of your proposal to your reader.
12. Be especially diplomatic in an unsolicited proposal to convince readers of the problem—you don't want your boss to think you're criticizing the way he or she has been running things.
13. Present your proposal in as clear and attractive a format as possible. This is especially important with sales proposals, where competition can be stiff.

The following memorandum is an unsolicited internal proposal written by an employee of a small factory that manufactures small tools. In length, amount of technical detail, and perhaps even proposed costs, this proposal is typical of many internal proposals employees in business, government, and industry write each year. The writer is proposing a plan to reduce losses of tools and equipment through employee negligence or theft. Her audience is the busy plant manager who is not fully aware of the gravity of the problem. Thus, the writer's goal was to convince the manager that the firm's current policy of preventing losses was not working and that a new, and stronger, plan had to be adopted. Because that new plan involved some important changes at the factory, the writer knew that the reader would demand convincing and complete answers to the questions he would surely raise.

A Plan to Reduce Supply Loss at Willis Machinery

Andrea Blair

WILLIS MACHINERY

INTEROFFICE MEMORANDUM

TO: Arthur Farnsworth
Plant Manager

FROM: Andrea Blair A.B.
Shift 1 Supervisor

SUBJECT: Plan to Reduce Supply Loss

DATE: April 19, 1984

Purpose of Memorandum

I am writing to obtain your authorization for a proposed 1
plan to reduce tools and supply losses now occurring through negligence or theft.

The Problem

The recently completed inventory for the first quarter of 2
1984 indicates a sharp 25% increase in the disappearance of tools and supplies. It appears that the current policy of posting signs and issuing memos to encourage employees to reduce losses of parts and tools seems to have had little positive effect.

First quarter losses of tools, motor oil, gaskets, and 3
transmission fluid amounted to nearly $1200. Periodic shortages of such hardware items as cotter pins, wedge bolts, and spacers have also been noted and attributed to employee pilferage.

In short, Willis Machinery is losing a large amount of the 4
types of tools, supplies, and hardware items that employees can use for home repair projects. Most of these items are impossible to trace once they have been removed from the plant.

Arthur Farnsworth
April 19, 1984
Page 2

A Practical Solution

In view of our belt-tightening policy for this year, we need 5
to make every effort to diminish these unreasonable losses. Therefore, I propose the following plan for your consideration.

1. All parts and equipment should be relocated into the single large storage room in the rear of Work Area 3. This room is centrally located, has large enough doors to accommodate fork lifts, and has more than enough space for the supplies currently stored near Areas 1 and 5.

2. Remove the employee smoking area from the rear of Work Area 3 to the rear of Work Areas 1 and 5. This change will not only create more storage space but will provide employees with additional, more convenient rest areas. In addition, this move will increase the physical separation of employees from supplies during employee rest periods.

3. Initiate a signout policy for all tools and equipment. Require employees to sign for all tools and supplies requested. Requests will be initiated with a numbered 3 X 5 request slip which can be prepared at a very low cost (less than 2 cents per slip) on the copier in the Administration Building. Use of request slips will allow a machinist to send a helper for supplies. This proposed system will provide both a record of supplies requested and supplies actually issued. Employees signing for equipment will provide their employee identification number and the number of the supply request form submitted.

4. Employ a parts/supply room clerk to administer the sign-out policy. Instead of hiring and training a new employee, I recommend reassigning one of our more experienced loading dock employees (e.g., Sally Morran, Bill Scheffers, Gerry Klowick). Work there is seasonal. In busy seasons, we could make temporary reassignments or hire less expensive part-time employees. I estimate that

Arthur Farnsworth
April 19, 1984
Page 3

we would need to use additional people in the dock area no more than 4 weeks a year.

Costs

The only expenses I see the company having will be for the following items: 6

Transferring Tools and Supplies to Work Area 3 (20 Hours of Labor @ $8.00 per Hour)	$ 160.00
Use of Part-Time or Reassigned Employee for Docks for 4 Weeks per Year	800.00
Year's Supply of Sign-Out Slips	100.00
	$1060.00

The system I propose shoul· cut losses by 77%. Considering that first quarter losses .re $1,200, the annual savings created by this plan shoul.l amount to $2,636 (77% of $4,800 annual losses = $3,696. Subtract the cost of the proposed plan of $1,060: $3,696 less $1,060 = $2,636 savings for the first year alone.) 7

Although I do not expect this plan to eliminate pilferage and losses compietely, I do believe that it will provide obstacles for spur-of-the-moment thieves and make them reconsider. The plan certainly will allow the company to keep better track of supplies and tools. 8

Please let me know your decision as soon as possible. I am ready to put the plan into operation within two or three days of your authorization. 9

QUESTIONS FOR DISCUSSION

1. How is the description of the problem intended to demonstrate to Farnsworth that it is serious and that Blair has a clear understanding of it and the company's interests?
2. Is the order in which Blair lists the four steps of the plan effective? Why? Would another order have been more convincing? Explain.
3. Where in the proposal has Blair done an exceptionally good job of anticipating and answering possible objections Farnsworth might have about the plan?
4. Blair writes about a potentially very emotional topic—employee theft—without injecting emotion into the proposal. How is this accomplished?
5. Why does Blair add the paragraph about annual losses immediately after the itemized account of the expenses the company would have if it accepted the proposal?
6. What function(s) do the last two paragraphs serve?

SUGGESTIONS FOR WRITING

1. Write a short internal proposal similar to Blair's on one of the following topics or on one assigned by your instructor:
   a. writing and designing a handbook for new employees
   b. buying a van for company trips (explain why a company car is no longer useful)
   c. purchasing additional software packages for the computer used in your department
   d. redecorating a reception area or an employee lounge
   e. creating a day-care center for children of employees
2. Using the information in Blair's proposal, write a memorandum from her to one of her subordinates about enacting the proposal. (See Chapter 12 for details about memorandum writing.)

The next proposal is an unsolicited internal proposal submitted to the management of the Lockheed Company of California by the Lockheed Employees' Recreation Club (L.E.R.C.) that tries to persuade the company to construct a new physical fitness facility at the Burbank plant. The authors, directors at the L.E.R.C., did not present their plan in a simple memo format, for the scope and cost of the project required a longer, more formal proposal. The authors of the Lockheed proposal did substantial research before submitting their work, surveying employees' opinions, securing estimates for construction costs, and preparing a detailed plan of operation for the proposed facility.

A Proposal to Management by the Lockheed Employees' Recreation Club

Robert A. Pindroh and Kenneth C. Wattenberger

INTRODUCTION

In recent years the employees and management of Lockheed have shown increasing interest in health and physical fitness. The Lockheed Employees' Recreation Club (L.E.R.C.) staff and council representatives have received more than 1,000 inquiries regarding a physical fitness program and facility. 1

Because of the employees' great interest in such a facility and the advantages it presents for them as well as Lockheed, the L.E.R.C. staff has developed the following physical fitness facility proposal for consideration by the Lockheed management. 2

AIM

The aim and purpose of the L.E.R.C. physical fitness program is threefold: 3

1. To promote employees' understanding of the need and value of physical fitness in maintaining good health.
2. To provide a fitness program designed to increase the participants' knowledge of physical fitness and improve their fitness and productivity through individualized programs.
3. To encourage the participants to maintain a lifelong fitness program.

Through regular exercise programs, participants can expect to increase muscular strength and endurance, improve flexibility and posture, reduce blood 4 pressure, lower resting pulse rates, and control body weight more easily. All of this, of course, means increased work performance and increased productivity to Lockheed.

NEED AND JUSTIFICATION

The creation of fitness facilities with shower and locker rooms has been a long- 5
range objective for L.E.R.C. for several years. Studies have shown that companies are justified in their concern about decreasing productivity, increasing medical costs, and poor morale because employees are not in good physical shape. Unquestionably, a company benefits from a healthy and fitness-conscious work force. The benefits that a company derives from healthy employees include bottom-line cost effectiveness with increased productivity, reduced company health costs, and a more positive employee attitude toward the company.

An on-site fitness facility could also be valuable in recruiting highly qualified 6
and skilled personnel to meet the company's expanding needs.

The L.E.R.C. Health Club at the Palmdale plant gives proof of the popularity 7
and usefulness of an on-site health facility. Over 200 men and women employees use and benefit from the facility monthly. To date the L.E.R.C. has received over 1,100 unsolicited signatures petitioning for a fitness center in Burbank. Therefore, we know that a fitness center at Burbank would also be highly used and have a sizeable impact in increasing employee health and morale.

Easily accessible shower and dressing room facilities would also encourage 8
employees to participate in such health promoting programs as jogging, volleyball, cycling, softball, and basketball.

ON- VERSUS OFF-SITE FITNESS CENTER

An on-site fitness center is highly recommended over an off-site program for 9
the following reasons:

1. The company would demonstrate a very concrete interest in and commitment to employee fitness.
2. Employees would be able to use the facilities at times convenient for them, e.g., before and after working hours.
3. A fitness program sponsored by Lockheed would emphasize the company's real commitment to lifelong health improvement programs for employees far more than could off-site health centers with their commercial sales approach.
4. The employees' cost would be considerably less than at off-site centers, and yet the employees would receive more individual attention.
5. The on-site facilities will motivate employees to participate in fitness programs when they hear of their colleagues' experiences with the equipment

or programs. In addition, the employees will take greater pride in their company when they can participate in their "own" fitness improvement program.

ESTABLISHING THE FITNESS CENTER

Establishing an on-site physical fitness facility raises questions about operating 10
procedures, location, costs, and staffing. The next sections of this proposal provide answers to these and other relevant questions management may have.

Membership Fees

Membership fees would be $104.00 a year which would be collected by payroll 11
deduction at $2.00 per pay period. Payroll deduction authorization cards will be provided if L.E.R.C. is granted this service.

Operating Rules

Current planning is to offer membership on a first-come, first-served basis with 12
maximum participation limited to 2,000 members. When it is necessary, we will prepare a waiting list with new members added to the program at the beginning of each quarter.

Registration Forms

Before any employee can be admitted, three forms will need to be completed: 13

1. The Physical Fitness and Health Questionnaire,
2. Release From Liability,
3. Payroll Deduction Authorization (Appendix A not included).

Program Attendance

All participants entering the program should have had a thorough medical exami- 14
nation during the past 12 months, including an exercise stress test. Any costs involved must be borne by the participant.

Employees will be able to use the facility at their convenience. If the facility 15
becomes overcrowded at any one time, then sign-ups for exercise periods will be offered.

Participants would be encouraged to attend at least twice a week. A daily 16 sign-in log and personal exercise record sheet will be used to evaluate attendance and program results.

Membership Promotion

An intensive publicity program will go into effect when construction begins. 17
This program will include ads in the Lockheed in-house publication, the *Star*, bulletin board posters, and an open house at the facility.

At the Open House, a two-step process will mark our lead-in to formal 18
registration for program participants:

1. Five days (Monday through Friday) of Open House with continual demonstrations of equipment, educational and instructional talks, and educational literature.
2. Registration at the door for all visitors, a button or stick-on patch being provided to publicize the Center.

All employees on the open house visitors' list will be sent a letter and a 19
Payroll Deduction Authorization Card.

Proposed Site, Size, and Construction Schedule

The recommended space to meet the minimum needs of the center is approxi- 20
mately 5,400 square feet. Attached is a sample design of the proposed facility. Included in the attachments is an outline of construction materials and finishes (Appendix B not included). The location for the proposed facility is adjacent to Kelly Johnson Hall at Robert Gross Park (Appendix C not included). This area and its surrounding functions is the most compatible with the facility. The recommended site and plans will meet immediate needs and provide for modest expansion. The construction would take approximately 295 calendar days.

Cost Analysis

• Site preparation: modification to existing building (blood bank), and walks, planters and picnic area	$ 15,620.00
• Construction of the building including interiors, but excluding lockers and exercise room equipment	315,400.00
• Final Design, working drawings, specifications, calculations, and normal supervision during construction	18,261.00

• Lockheed Facilities Field Coordinator (4% of construction cost)		12,616.00
• Contingency (15% of construction cost)		47,310.00
• Equipment		38,459.00
300 Double Tier Lockers	$13,055.00	
Women's Exercise Room		
1 Treadmill	$ 1,950.00	
1 Testing Equipment	558.00	
2 Stationary Bikes	400.00	
3 Mini-Trampolines	600.00	
1 Multi-Press Station	795.00	
1 Seated Leg Press	1,250.00	
1 Lat. Pulley Station	750.00	
1 Wall Pulley	825.00	
1 Butterfly Machine	1,695.00	
1 Leg Extension and Curl	750.00	
2 Abdominal Board Racks	300.00	
2 Flat Abdominal Boards	295.00	
1 Double Twister	375.00	
1 Rack and 6 Pair Beauty Bells	598.00	
1 Rack and 6 Beauty Bars	790.00	
1 Scale	185.00	
1 Flat Bench	125.00	
1 Incline Bench	220.00	
1 Stationary Bike	200.00	
Total Women's Exercise Room	$12,661.00	
Men's Exercise Room		
1 Treadmill	$ 1,950.00	
1 Multi-Station Machine	5,000.00	
1 Olympic Barbell	425.00	
1 Olympic Bench/Squat Rack/Plate Holder Combo	595.00	
3 Dumbbell Racks	660.00	
15 Pair Dumbbells 5-75's	1,325.00	

Item	Cost	Total
2 Stationary Bikes	400.00	
2 Adjustable Flat Benches	270.00	
1 Incline Bench	220.00	
2 Bent Knee Abdominal Boards	780.00	
1 Roman Chair	375.00	
1 Scale	185.00	
Testing Equipment	558.00	
Total Men's Exercise Room	$12,743.00	
TOTAL COST		$447,666.00

Alternate Energy Source

The use of solar energy is being considered as an alternate source of energy for the structure. This would add approximately $34,000 to the construction cost of the facility. However, Lockheed would receive Federal and State tax credits of approximately $17,000. In addition, the use of solar energy would save L.E.R.C. 80% on the cost of natural gas. To our knowledge, this facility would be the first of its kind in the United States to use solar energy as an alternate energy source. 21

Fitness Center Staff

Fitness Specialists. These staff members (one male, one female) will have the responsibility for planning, developing and implementing a fitness program which will provide participants with (a) a minimization of coronary risk factors, (b) development of joint flexibility, (c) enhancement of muscular strength and endurance, and (d) relaxation training. 22

Two Student Interns. Under the direction of the fitness specialists these staff members will assist in the instruction and demonstration of fitness techniques and aid in the overall operation of the Center. 23

Operating Hours

Projected operating hours are from 6:30 A.M. until 9:00 P.M. Monday through Thursday. On Friday the hours would be from 6:30 A.M. until 5:30 P.M. On Saturday and Sunday the facility would be closed. 24

We also suggest that the facility be made available for specialized rehabilita- 25 tion classes, lectures, a cardiac therapy program and other uses by Lockheed's managerial and medical staff.

FINANCING

It is projected that membership dues and the revenue from the sale of health food items, vitamins, and clothing would provide the necessary income to enable the program to be self-sustaining. 26

With near total capacity use it is anticipated that L.E.R.C. could generate approximately $200,000 per year which would offset operating expenses and allow us to amortize the original investment. 27

The L.E.R.C. proposes that if the Lockheed-California Company constructs the physical fitness building, L.E.R.C. will provide the necessary exercise apparatus, testing equipment and maintenance in addition to conducting a comprehensive physical fitness program. 28

The L.E.R.C. Board of Directors and Council may be receptive to any cooperative effort in constructing the fitness facility and financing the operation. 29

1981 Operating Plan

Income	
Membership Dues (2,000 x $104/yr.)	$208,000.00
Miscellaneous (vitamins, clothes, etc.)	8,000.00
Total income	$216,000.00
Expense	
Instructors and Fringe Benefits at 50%	$ 51,000.00
Maintenance	3,300.00
Material	7,000.00
Utilities (present energy sources)	25,000.00
Insurance	5,000.00
Equipment Repair	1,000.00
Miscellaneous and Contingency	5,000.00
Total Expense	97,300.00
Balance Income Over (Under) Expense	$118,700.00

1982–1984 3-Year Operating Plan

Income	
Membership Dues (2,000 x $104/yr.)	$624,000.00
Miscellaneous	30,000.00
Total Income	$654,000.00

Expense	
Instructors and Fringe Benefits at 50% and Cost of Living Raise	$235,620.00
Maintenance	12,000.00
Material	25,000.00
Utilities (present energy sources)	112,500.00
Insurance	18,000.00
Equipment Repair	4,500.00
Capital Improvements	20,000.00
Miscellaneous and Contingency	15,000.00
Total Expense	442,620.00
Balance Income Over (Under) Expense	$211,380.00

At the projected rate of income, the fitness center could be amortized before 30
1992.

CONCLUSION

The L.E.R.C. believes the construction of a physical fitness facility and the subse- 31
quent establishment of a fitness program is a sound investment in human resources. Such an investment will promote greater employee health, morale, and loyalty which should result in increased productivity.

The L.E.R.C. hopes to receive a favorable response from the Lockheed- 32
California Company. In providing this facility and encouraging physical fitness for its employees, Lockheed will be building a healthier work force for the future.

QUESTIONS FOR DISCUSSION

1. What problems do the authors bring to the attention of the Lockheed management? Why did these problems prompt the authors to write the proposal?
2. Where and how in the proposal do the authors indicate that this is an unsolicited proposal? Comment on the authors' strategy and tone in explaining their purpose.
3. The authors outline three aims for a physical fitness facility. Is there enough evidence from the proposal that these aims will be achieved by establishing the physical fitness facility and program at the Burbank site?
4. Why do the authors include the section "On- Versus Off-Site Fitness Center"? How does this section help the authors to convince management to fund the proposal? Why does this section appear early, rather than later, in the proposal?
5. What kinds of assurance is the Lockheed Company given that, if funded,

the facility and program will be carefully operated for the benefit of employees and the Company?

6. Why don't the writers provide information about their qualifications? Whose qualifications do they give? Why?
7. The writers suggest numerous ways to finance the physical fitness facility. What alternatives are given and why? Would the authors of an external proposal have suggested alternate ways of financing a plan?

SUGGESTIONS FOR WRITING

1. Based upon information contained in this proposal, write one of the following documents:
   a. a letter to contractors in the Burbank area asking for estimates on the physical fitness facility (assume it will be 5,550 sq. ft. and have 4 walkways, 18 planters, and a half-acre picnic area)
   b. a memo to the director of the physical plant at the Burbank site asking for information about available sites
   c. a letter from the Lockheed management accepting the proposal but wanting the L.E.R.C. to pay for half of the construction costs; supply a reason for Lockheed's decision
2. Write an unsolicited internal proposal in memo format about one of the following topics. Specify your immediate audience.
   a. instituting split shifts in your office or factory
   b. purchasing new equipment (e.g., microwave ovens, scanners)
   c. adding a new course in your major
   d. improving communication between one department and another (nursing/pharmacy; purchasing/transportation)
   e. extending the library hours
   f. improving your company's image (e.g., advertising, public service, change of product)
   g. using student workers in your company (e.g., interns)

Here is a solicited sales proposal written in letter format. It responds to an RFP sent out by the regional manager of a large insurance company who must make sure that her company receives the best bid on a service contract for its office equipment. The RFP specifies the quantity and type of equipment requiring maintenance. The author of the proposal is the manager of an office supply and maintenance firm who must convince the reader that his firm can service the equipment more efficiently and cheaper than the competitors can. To increase his chances of getting his proposal accepted, the writer has related each of his services to the reader's specific needs.

A Proposal to Supply Maintenance Service for National Home Life's Office Equipment

Jerry Golden

HIGH TECHNOLOGY CORPORATION

706 BROADWAY DRIVE • P.O. BOX 1528
HATTIESBURG, MISSISSIPPI 39401
(601) 544-7711

November 25, 1983

Ms. Regina L. Gordon-Muñoz
Manager
National Home Life Insurance Company
P.O. Box 67842
Hattiesburg, MS 39401

Dear Ms. Gordon-Muñoz:

In response to your RFP of November 10, 1983, I am pleased to submit our bid for a maintenance contract for your 1984-1985 fiscal year. We would like to use our 50 years of experience in office equipment sales and service to assist you as we do 37 other local firms. Because of this previous experience, HTC can give you superior service at a cost lower than our competitors'. 1

Our goal is to keep your office functioning smoothly with little interruption. Since our service department has the latest diagnostic equipment, any problems with your office equipment can be discovered and corrected quickly and efficiently, resulting in less downtime for your machines. 2

For our client's benefit, we keep a history, similar to a doctor's patient chart, on every one of the machines we service. That way any unusual problems can be identified and their cause eliminated. The enclosed sample chart [not included] will give you an idea of the care we will take in maintaining records on your equipment. 3

You will find our service contract especially cost effective. Unlike many of our competitors, when replacing worn or damaged parts we supply new ones at no additional expense to National Home Life. You will be spared the extra expense and inconvenience, therefore, of having to negotiate separate service and sales contracts for the same piece of office equipment. 4

These are just the highlights of our service agreement. Let me outline other advantages you would have in accepting our service agreement. 5

Ms. Regina L. Gordon-Muñoz
November 25, 1983
Page 2

Our Qualified Personnel

Because our service department of seven technicians is the largest in the area, you will be assured of receiving the attention you expect. These technicians have a total of 64 years of on-the-job experience. The department is headed by Wayne Hurst, who has 23 years of experience in servicing office equipment, nine with IBM as southeastern Regional Manager. All our technicians are factory trained and have earned high marks for proficiency. At least every two months they update their knowledge by learning, through in-house programs, about the very latest modifications and retrofits in office repair techniques, specifically those on Savin copiers, A. B. Dick equipment, Olivetti typewriters, and all makes of dictaphones. Whenever necessary, our technicians receive special training as well. For example, last March the Nashua Corporation flew in an engineer to instruct us on the changes which needed to be made when one of our customers, the Lange Coal Company, switched from Savin to Nashua parts on its Savin copiers. 6

Fast Response Time

Our service technicians are radio-dispatched to respond faster to your call. When a call is made to our main office, we send the technician nearest you. Because of our radio-dispatch system, most of our calls are handled within an hour, some within as little as fifteen minutes. As a further service to you, all our vehicles carry 90% of the parts most often replaced so that each technician is ready to respond to your emergency calls at any time. 7

Zero Defects Program

This program was devised to alert us to repeat calls from a customer. Should we receive a call on a machine we have serviced during the previous seven days, the call is logged onto a special analysis board. The service supervisor will then check the initial call to see if both service calls were made because of the same or a similar problem. If the original problem has persisted, both the service supervisor and the original technician will respond to correct it. At that time we will determine if the problem resulted from customer error or a defective part. This return service is 8

Ms. Regina L. Gordon-Muñoz
November 25, 1983
Page 3

needed on fewer than 1.8% of our calls, but our objective is to reduce that percentage to zero.

Availability of Parts

As an area Savin dealer, we have the largest supply of Savin 9
parts in Southern Mississippi. We also stock parts from such manufacturers as A. B. Dick, Olivetti, IBM, and Masters. Many times, however, we find that other supply and engineering firms have modifications superior to those you may be using. Although these parts may cost us more, we will place them when necessary on any of your machines under contract. Through precise records and a thorough knowledge of these specific parts, we are able to offer less downtime and fewer service calls. In addition, we can frequently determine if a part will fail after a certain volume is reached and replace it before this occurs. This practice will keep your machines functioning trouble-free and sustain high productivity.

Procedures

As part of our service contract for 1984-1985, we will 10
perform the following procedures on each of the machines listed.

Savin Copiers

We will routinely inspect each Savin copier a minimum of 11
once a month to be sure that it is giving you maximum copy quality. We will also completely examine each of your Savin copiers as often as outlined in the Savin Service Manual and will do the following:

1. replace any worn or damaged parts (e.g., scraper blades, reverse rollers, drums, and the venturi)

2. rebuild the motor bearings, hoses, and scraper blades where necessary

3. make adjustments necessary for proper use of the toner.

Ms. Regina L. Gordon-Muñoz
November 25, 1983
Page 4

Mimeographs and Duplicators

We will pick up all mimeographs and duplicators at least 12
once during the contract year and examine each machine. As part of our service to you we will:

1. replace the feed wheels in the ink pad, the drum wick, and any other worn or damaged part

2. make all adjustments recommended by the A. B. Dick Service Manual and its updates

3. replace any other parts as required.

Typewriters, Mechanical Calculators & Adding Machines

These machines will be periodically inspected, and as part 13
of the service contract we will:

1. provide chemical cleaning in our shop for any typewriter requiring it

2. recondition and change all worn or damaged parts on typewriters brought in for chemical cleaning

3. replace any parts on calculators and adding machines as needed at no additional charge.

Norelco Dictation Equipment

We will periodically inspect all dictating machines as part 14
of our guaranteed maintenance program.

Costs

The rates for these services for one year are listed below: 15

8 Copiers @ $600.00 each	$ 4,800
15 Mimeo machines @ $50.00 a machine	750
75 typewriters @ $50.00 a machine	3,750
30 dictaphones @ $28.00 a unit	840
	$10,140

Ms. Regina L. Gordon-Muñoz
November 25, 1983
Page 5

I believe that you will find these rates are more reasonable 16
than, and less likely to change as often as, those of our competitors. While many other companies have increased their charges from 50% to 150% in the last eight years, HTC has increased an average of less than 28% for the same period.

Thank you for considering this proposal. I hope it has given 17
you the information you need about the quality and quantity of the work we want to perform on the office equipment at National Home Life. Should you accept our proposal, please sign a copy of this letter in the space below.

If you have any questions, please call me at 564-4600, 18
extension 23.

Sincerely yours,

Jerry Golden

Jerry Golden
General Manager

I accept HTC's maintenance agreement for the fiscal year 1984-1985.

____

Regina L. Gordon-Muñoz
Manager
National Home Life Insurance Company
Hattiesburg, MS 39401

QUESTIONS FOR DISCUSSION

1. What specific problems does the HTC proposal claim to solve? Since he responds to an RFP, Golden realizes that Gordon-Muñoz is aware of these problems. But how does his reference to them reveal that he has carefully assessed Gordon-Muñoz's needs as a manager of the insurance company's office?
2. What functions do the first four paragraphs serve? How might they help to persuade Gordon-Muñoz to give the service contract to HTC? How do they favorably dispose the reader to accept the rest of the proposal? Why and how is the information in these four paragraphs repeated elsewhere in the proposal?
3. How does Golden link his company's competence and resourcefulness to customer needs and benefits? Provide examples from the proposal as part of your answer.
4. Where and how does Golden inform Gordon-Muñoz without sounding boastful that she will receive benefits from HTC that its competitors will not give?
5. Discuss the format and order of Golden's section on Procedures. How effective are they in presenting both factual details and persuasive commentary?
6. Why are costs discussed last? Should they be discussed earlier in the proposal (e.g., before Procedures)? Explain.

SUGGESTIONS FOR WRITING

1. Assume you are a general manager for a company. Write a sales proposal in letter format to one of the following individuals on the subject specified below:
   a. to the owner of a shopping center on providing cleaning services for outside areas
   b. to the manager of a large apartment complex on maintaining lawns and shrubbery
   c. to the city engineer on conducting tests for noise, water, or air pollution in a specific part of the city
   d. to the administrator of a local hospital on securing the maintenance contract on the hospital's television and radio sets
   e. to the head of a computer/data processing center on servicing hardware
   f. to the director of your college library on improving study areas (e.g., building carrels, installing soundproofing)
   g. to the chair of the biology department at your college on supplying laboratory animals
2. Write a sales proposal that responds to the following RFP:

   Appleton Police Dept., 30 Jan., wants to receive bids on supplying and installing computerized data processing system capable of storing and calling up data by name, number, and arrest/docket type. Training program for operators should also be included in bid. System must be in operation by 1 March.

The author of the following external proposal, Llewellyn Toulmin, is an experienced and successful proposal writer. We asked him to comment on the context, audience, and organization of his work.

The unsolicited proposal example provided below was written after discussions with city officials revealed that one of their regular, pressing management (and political) problems was the snow control program. Snowstorms and poor snow control programs have had severe, negative political consequences in New York, Chicago, and Washington, D.C. Similarly, the City of "Riverton, Delaware" had occasionally been criticized for its apparent inability to provide "bare pavement" arterials and residential streets within a few days after a major snowstorm.

Federal, state, and local agencies generally have mechanisms for dealing with unsolicited proposals such as the Riverton study that follows. However, it is always advisable to meet with the client and discuss at length his or her subjects of interest, and the approximate budget available, before submitting an unsolicited proposal. Sometimes it is wise to spend only a person-day or two and submit a very brief "concept paper" or letter proposal. Then if the concept is rejected, little has been lost. This proposal took about 70–80 hours of a professional's time to write. A letter proposal was not used first since we were fairly confident from our discussions with city officials that the proposal would be favorably received.

The primary audience was local government administrators at the level of department head, assistant head, and division chief with long experience in snow control. Possible secondary audiences included city council members with little technical experience in snow control, and officials in the City's budget, accounting, and other staff offices who might be asked to approve or comment on the proposal.

The proposal format follows a much-used pattern:

- Brief statement of the problem
- Background and history of the problem; delineation of its main issues
- Technical approach to dealing with studying and/or solving the problem broken down into sequential, separable tasks
- The consultant's ability to perform the task, including general corporate background and structure (omitted in this proposal for reasons of space), organizational approach to this project (teams, task forces, etc.) and "blurbs" on each staffer, tailored to the project
- Person day breakdown by task
- Line item budget

The proposal was largely accepted as written, although several additional study issues were added to the list of issues in Task 1. The project generally proceeded as planned, and apparently satisfied the client, since "Acme Consultants, Inc." was subsequently hired to do a small follow-up contract to monitor and analyze the city snow program's performance during a major snowstorm.

An Unsolicited Proposal to Provide Assistance in Upgrading a City Snow Control Program

Llewellyn Toulmin

ACME CONSULTANTS, INC.

111 River Road, Riverton, Delaware 19709
Phone: (302) 673-4796

March 15, 1982

Mr. Tom Jones, Director
Department of Transportation
City of Riverton
City Hall
Riverton, DE 19711

Dear Mr. Jones:

Acme Consultants, Inc. is pleased to present the enclosed 1
proposal to provide consultant services to the Snow Emergency Task Force. Execution of the work plan outlined will result in both a significantly more cost-effective snow control program and increased public awareness of salient emergency procedures and applicable municipal regulations.

We feel that this proposal represents a challenging 2
assignment and look forward to being of assistance. We also would like to thank you and your staff and representatives of the Department of Environmental Services for meeting with Acme on March 2. Those discussions were very helpful. I will be contacting you shortly to learn your reactions to our proposal.

Sincerely,

John Jackson

John Jackson, Chairman
Board of Directors

JJ/dlr
Enclosure
cc: Bill Tucker, DES

I. PURPOSE AND BACKGROUND

Purpose

This proposal responds to the identified need by the departments of Transportation and Environmental Services for technical assistance and analytic support in reassessing and upgrading the Riverton Snow and Ice Control Program. Completion of the scope of work and tasks described below will result in: 1

- A snow and ice program that maximizes productivity of current City personnel and equipment
- An improved overall response that resolves interagency and intergovernmental issues and clarifies responsibilities
- Higher levels of public awareness of policies and procedures, thereby minimizing both overall disruption during inclement conditions and the number of citizen complaints

Background

Primary responsibility for snow removal within the City resides with the departments of Transportation and Environmental Services (DOT and DES) under the direction of the official in charge of the long-established interdepartmental Task Force, whose mission it is to organize and respond to snow emergencies. Directions for implementing the City's overall response as embodied in guidelines, policies, and procedures are recorded in the *Program For Snow and Ice Control.* That document was originally prepared in the early sixties and has been updated recently only on an ad hoc basis in response to significant policy and procedural changes. 2

The status of the City's resources and capabilities to respond to snow emergencies recently has changed so dramatically that many previously enforced plans and policies are no longer operative. Consequently, the program manual no longer accurately reflects current City responses to snow emergencies. These changes have been necessitated by substantial reductions-in-force within DOT and DES and the merger of several Bureaus to form DES. In DES alone, staffing has declined by 1,700 employees over the last eight years, including 128 street cleaners in 1980 alone. Budget authority for snow removal in the current year (estimated at $1.2 million) is also significantly lower (in absolute terms) over levels reached in the 1960's. Unfortunately, however, the amount of snowfall is not regulated by the amount of appropriated authority. This winter, the task force has had to implement various responses on 21 snow-related days—generating expenses of more than $2.2 million through February alone. 3

In addition to being a highly volatile budget item from year to year, snow control is one of the most highly visible public services. This visibility is of a 4 very special nature because snow control activity is intensive and controversial, but for only limited times. Public attention is gauged by the City through the number of recorded complaints/requests for service. While this measure can be a good management indicator of service delivery, at times it can be a misleading statistic. Indeed, the public's lack of awareness of the complexities of snow removal procedures (and municipal regulations) can prompt hundreds of complaints and requests for service. For example, complaints about icy sidewalks refer to a problem over which the City has no jurisdiction.

In light of these realities, it is an appropriate time for the Department to 5
reassess fundamentally snow emergency policies and procedures. It should be emphasized, however, that the fundamental elements of the plan remain valid. The supporting elements of the plan require a fresh look. And of equal importance, the public should be informed of the updated plan.

II. TECHNICAL APPROACH

Task I. Resolve Pending Technical/Policy Issues

A number of complex and interrelated policy/technical issues have been identi- 6
fied by Task Force representatives. These issues have been generated primarily from the recent substantial changes in available personnel- and non-personnel-related resources. The resolution of these issues (which are identified below) will flow logically from the proposed reassessment of the fundamental aspects of the snow program. Such basic questions as *how, when,* and *where* the City's resources should be deployed need to be re-examined at this juncture based on a careful policy assessment.

A preliminary list of the key issues to be addressed in updating the new 7
Snow Control Program Plan follows. These issues will need to be refined and ranked before developing a project Work Plan that details the approach and scheduling for each project activity.

1. *Should the Department move to an alternative to wet pavements in snow conditions?* Applying abrasives to roadways and repairing the resulting potholes is becoming more costly while the available vehicle fleet is diminishing. Can the City implement less costly alternatives, at least for some streets, that do not increase risk unduly and also do not meet with unacceptable citizen resistance? Where and when, if ever, should abrasives other than sand and salt be stockpiled and applied?

2. *How should the Task Force respond in residential areas?* Response in these areas needs to be re-examined in light of the lack of consensus regarding how extensively to plow and spread abrasives. Enforcement of parking regulations is an important related issue. Also included here are such subissues as whether or where barrels should be placed in residential areas, which hills

should be plowed, how alleys should be dealt with, what kind of tradeoff there should be between plowed streets and plowed-in cars, and how extensively prowl trucks should be used.

3. *Are resources deployed most effectively and efficiently?* Resource issues, in addition to those inherent in (1) and (2) above, include whether manning can be adjusted to reduce the amount of overtime and especially the frequency of shifts well in excess of 12 hours during snow emergencies, whether the nature and level of contractor use is the most cost-effective possible, and whether any changes are needed in operating procedures, logistics, or the snow command center.

4. *Capital Fund Related Questions.* Several issues relate to capital fund items. Specifically, what are the implications of the administrative directive to reduce the fleet size; what special and multiple-use equipment should the Department purchase, especially as the bulk of the DOT fleet comes due for replacement, and what needs to be done to improve storage facilities?

5. *Should resources be deployed based on ward and neighborhood boundaries?* Recently, members of the City Council have requested redesign to allow more rigorous legislative oversight of service delivery. This, however, may result in certain inefficiencies and financial problems because of the geographic locations of DES and DOT facilities and of salt stockpiles, differences in equipment types maintained by DES versus DOT, and the introduction of more routes that are not defined solely based on practicality and compactness.

6. *Can any legislative/budgetary changes be implemented to permit better planning and resource utilization?* Appropriate levels of general fund appropriations for snow removal are difficult to forecast, which results in dramatic swings from "shortfalls" to "surpluses." In years of "shortfall," readiness may drop and capital investment plans may be disrupted.

7. *Intergovernmental responsibilities.* Several intergovernmental issues have been identified. One is what agency should be responsible for clearing subway parking-lot access roads and bus stops and also fringe parking lots. A second is whether the subway system should continue to be responsible for clearing certain streets, and whether other city departments should be included, thus creating a comprehensive, coordinated snow plan. Another issue is whether efforts to include the local National Guard's resources in certain contingency situations should be renewed. Finally, the tie-in between snow emergency plans and general emergency plans needs reassessment.

The Acme team will assist the Task Force in addressing these issues by: 8

- Gathering and analyzing appropriate Departmental records and data available from other sources on previous emergency efforts
- Conducting in-depth consultations with key members of the Task Force
- Examining lessons from the experience of other cities

- Modelling the impacts of alternative decision options (sometimes using the routing model described in Task III)

For example, analysis might be performed of the capital and life-cycle costs 9
of alternative equipment and of the labor costs and snow control levels achievable with each type of equipment. This analysis also would have to examine issues of maintenance needs and capacity, and equipment capability, requirements and cost with multiple applications (e.g., trash removal, street cleaning, or road maintenance and plowing/spreading).

Key policy questions will be examined in Option Papers that recommend 10
and justify a position or examine the pros and cons of alternative decisions for each relevant issue. These papers will help to structure DOT/DES decision-making, starting with the issues that require the earliest decision in order to meet the plan development and implementation schedule. DOT/DES policy decisions will be incorporated, as appropriate, into the upgraded Snow Control Program Manual.

Task II. Update Snow Control Manual

Acme will produce a new Snow Control Manual. The manual will integrate 11
new/revised policy and modified procedures, instructions, and route maps (as required). Acme will work in conjunction with DOT/DES staff in defining the contents and organization of the new manual. Adequate review time for designated City officials with sign-off authority will be provided before a "camera-ready" version is prepared. The new manual will be organized to allow easy future updating as required (e.g., the text will be prepared on a word processor and the plan will be designed for insertion in 3-ring binders). Furthermore, all modifications to the current manual will be consistent with the basic purpose of the document to inform field and operations staff quickly of applicable guidelines and vehicle routes and assignments. It is anticipated, however, that the new design will enhance the document's utility to non–Task Force individuals.

Task III. Design and Implement Routing Model

DES currently uses a computerized routing model to improve the productivity 12
of its trash collection effort. A similar approach is feasible for snow removal. An optimum routing algorithm will indicate the best routes for trucks to follow for different depths and rates of snowfall. It should be able to accommodate several types of equipment with different capacities and suggest routes that permit refilling of trucks' salt and sand supplies with minimum productivity loss. A routing model can suggest the territory to assign to each garage location in order to maximize productivity, or it can set routes within assigned territories. Specifically, likely benefits from a model include:

- Maximum plowing of side streets while clearing arterials at the desired frequency, or minimization of force needed to carry out desired snow clearance and abrasive spreading
- Matching of equipment types with suitable routes to maximize spreading/plowing with the force available
- Productivity-based division of DES and DOT territories

The first step in developing a routing model is to decide on the objectives 13
of the model. For example, one could develop a model that minimizes the workforce/equipment needed to achieve a specified level of snow control or that maximizes the vehicle miles of roadway cleared (i.e., the segment length times the traffic volume for each road segment).

Next, desired constraints should be defined. A constraint, for example, can 14
assure that a desired level of clearance occurs on arterials, give certain streets priority, define time-dependent priorities (this latter can be tricky), accommodate coffee breaks, or assure that certain streets will be kept cleared to a minimum standard.

Once the model's objectives and constraints are defined (or even a few 15
alternative sets of objectives and constraints), they must be translated into mathematical form. The appropriate formulation will, to some extent, depend on the solution method contemplated. In this case, we expect that a set-covering heuristic driven by an integer programming algorithm is probably most appropriate, but we will have to verify this once the model's objectives and constraints have been specified.

Next, the model must be set up on the computer. We would expect to 16
use the road network in DES's trash collection routing system (or, if this proves difficult, a road network tape available from the Council of Governments that our proposed Project Director, Dr. O. R. Econ, has used before).

When the model first runs on the computer, it must be tested and the 17
logic of its outputs examined. The model's route plan should be reviewed by DOT and DES operations supervisors. In some cases, routes may not make sense to them or their experience may suggest changes that would make the routes easier for crews to adapt to with little loss in longterm productivity. To the extent possible within the project timeframe and the proposed level of effort for this task, the model will be revised and routings found where appropriate. Finally, the model will be documented.

Task IV. Design and Implement Public Information Campaign

Especially with substantial changes in City snow policies and procedures, it is 18
important for the public to be informed of relevant snow response-related information. Acme will design a public relations campaign that will increase public awareness of City policies and procedures and related municipal regulations.

One essential component of the campaign will be a brochure or flyer that 19
summarizes the key points of information, such as which areas get plowed in

what order; the snow emergency number, etc. Acme will investigate low-cost ways of distributing the brochure (for example, an insert to telephone bills). The brochure distribution probably will be complemented by an electronic public-service announcement and print media "blitz" in the fall that will highlight critical information.

The major purpose of the snow information campaign, and indeed, of 20
the entire snow control program, is to make the transportation system serve people better during and after winter storms. Even a casual observer in the City knows that our most serious snow-related transportation breakdowns arise when heavy snowfall occurs on a weekday afternoon. It is Acme's observation that the problem is a result of fear as much of breakdowns/shutdown of the transportation system or of the difficulty of travel in the snow. This suggests the need for a second kind of public information—namely, information on the status of roadway clearance operations during weekday afternoon storms. Unlike when the snow plan was first prepared, the City transportation system is no longer just a roadway system. Weather-related problems in subway operation affect roadway problems (and the fear of problems) and vice versa. The solution we suggest is the issuance of press releases every half hour on roadway clearance plans and progress, subway operating status and prognosis, and detailed weather information, whenever a serious afternoon storm is in progress while most workers are downtown.

III. MANAGEMENT AND STAFFING PLAN

Management Structure

The study will be managed by a Project Director and a Deputy. These individuals 21
also will lead much of the technical effort on the study. Task work will be organized on a team basis as shown in Exhibit 1. Note that staff generally are involved with the efforts of several teams, thus assuring continuity, communication, and coordination. In addition, project staff meetings will be held weekly.

Proposed Staffing

The paragraphs that follow briefly describe the relevant skills and experience 22
of the staff members whom we propose to assign to this effort. Résumés of these individuals are appended [not included here].

Dr. O. R. Econ, a Vice President of Acme, is proposed as Project Director 23
and Routing Model Team Leader for this effort. Dr. Econ directs our Economic and Information Systems Group and has led more than 60 public sector contract efforts over the past 12 years, most of them planning-related, systems analysis,

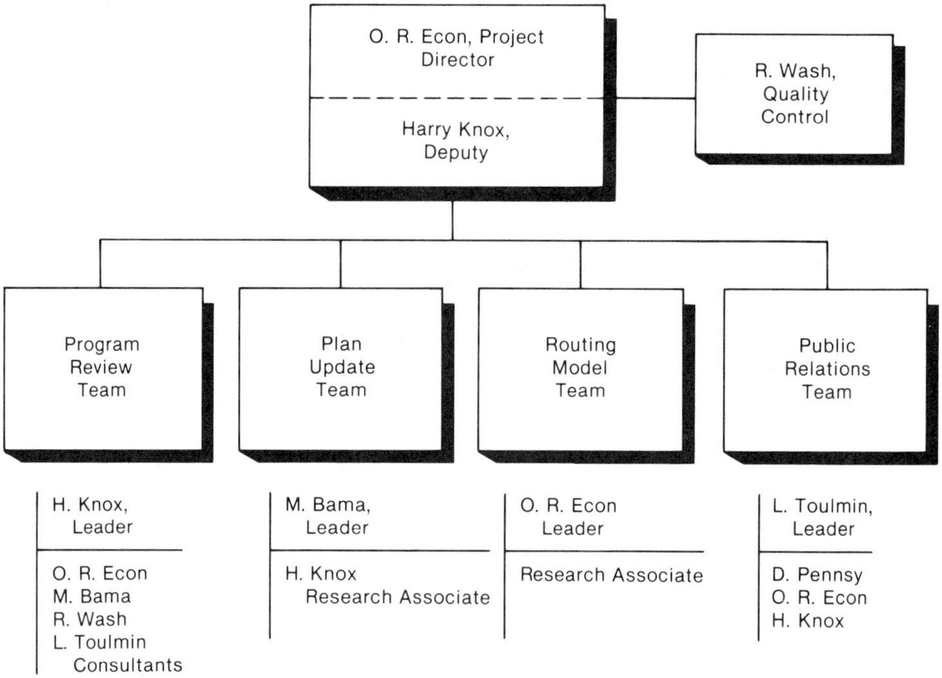

EXHIBIT 1. PROJECT ORGANIZATION CHART

or modelling studies. His educational background is ideal for the present assignment. He holds a B.S. in Urban Systems Engineering, an M.S. in Urban Planning, an M.S. in Operations Research, and a Ph.D. in Regional Science with concentrations in urban public facility location modelling and transportation modelling. He is a skilled cost-effective analyst, statistician, and computer analyst. He has worked on such varied problems as pollution incidence, housing and economic development policy analysis, transportation modelling, insurance availability and adequacy, evaluation of Federal health promotion media campaigns, economic impacts of infrastructure development, recreation trends analysis, elderly and handicapped transportation program development and operations, and ambulance station location. Dr. Econ will devote 80 percent of his time to this study.

Mr. Harry Knox is proposed as Deputy Project Director and Program Review 24
Team Leader. Mr. Knox, who is a Deputy Director of our Economic and Information Systems group, holds an M.S. in Urban Policy Analysis and has 10 years of experience in the analysis and improvement of urban management systems. Currently, he is about to complete a feasibility and cost-effectiveness study of a mandatory direct bank deposit program to replace the City payroll system and is developing an Adopt-a-Park program and a Gifts Catalog for the Riverton Department of Recreation. He also has performed program review or procedural documentation and training for the City financial management system, and park-

ing violations, water-meter billing, low-income-housing management, and land-use review systems for the City. Mr. Knox will spend 80 percent of his time on this study.

Dr. Robert Wash, a Principal of Acme, will serve as Quality Control Monitor 25
for the project and also lend expertise in intergovernmental relations and program analysis. An economist, Dr. Wash has over 15 years of government contract management and oversight experience. His assignments as Deputy Director of the Office of Policy Analysis for the U.S. Conference of Mayors and National League of Cities and as Assistant Administrator for Program Planning and Evaluation of the Social and Rehabilitation Service at HHS involved him particularly deeply in the analysis and resolution of intergovernmental relations problems at all levels of government. Dr. Wash is also a skilled and experienced management analyst. He will devote 20 percent of his time to this study.

Ms. Madeliene Bama will serve as a senior analyst and Plan Update Team 26
Leader on this study. She holds an M.S. in Transportation Planning and will receive her J.D. in May 1982. Her 10 years of work experience is split between transportation planning, job training, urban planning, and systems analysis efforts, including two and a half years as a Transportation Planner with the Council of Governments. Ms. Bama will devote 90 percent of her time to this study.

Mr. Daniel Pennsy will serve as a senior analyst on the study. Mr. Pennsy 27
has 15 years of experience as a trainer and developer of human relations and personnel management systems for state and local governments and community organizations. He is an expert in promoting smooth organizational change and sound public relations. His studies at Acme include an investigation on the problems of black alcoholism and provision of technical assistance to developers in meeting the affirmative action subcontracting and employment goals set by the Main Street Development Corporation. Mr. Pennsy will devote 25 percent of his time to this study.

Mr. Llewellyn Toulmin will serve as a senior analyst and Public Relations 28
Team Leader on this study. He will provide expertise in public relations, labor relations and budget analysis. Mr. Toulmin holds a Masters in Public Administration with a concentration in Public Finance and currently is enrolled in a Public Administration Ph.D. program. His 10 years of experience include three years as a local budget analyst and three years as a labor economist and budget analyst for the American Federation of State, County, and Municipal Employees. He has extensive experience in the analysis and costing out of productivity improvement plans, including a major management study of a waste-vehicle maintenance program. An excellent writer, Mr. Toulmin also has served as Public Affairs Director for the United Farm Workers in Florida. Mr. Toulmin will devote 40 percent of his time to the study.

A Ph.D. or M.S. candidate from a program that offers a concentration in 29
public facility location and vehicle routing systems will be hired to do the bulk of the work, under Dr. Econ's supervision, in developing the routing model. This individual will have strong statistical, mathematical, and computer skills and a strong background in urban service systems analysis and planning. We

have started recruitment of this individual and anticipate no problem in having a full-time person onboard in early May.

Personnel Assignments and Time Schedule

Exhibit 2 indicates our planned personnel assignments by task. The project will be completed in six months; the new Snow Plan should be ready for use by October 1, 1982.

EXHIBIT 2. PERSON DAYS BY PERSON AND TASK

	Resolve Issues	*Update Manual*	*Develop Routing Model*	*Develop PR Campaign*	*Total*
O. R. Econ	45	20	10	10	85
H. Knox	45	30		20	95
R. Wash	17	2	1	3	23
M. Bama	50	50		10	110
D. Pennsy	5			30	35
L. Toulmin	15			40	55
Research Associate	25	15	66		106
Total Technical	202	117	77	113	509
Secretary	45	35	6	30	116
TOTAL	247	152	83	143	625

IV. BUDGET

Direct Labor	$ 73,710
Labor Overhead @ 97%	71,545
Computer	9,400
Telephone, Messenger, Supplies, Materials and Local Travel	690
Printing, Xerox, and Map Preparation	4,000
Expert Consultants on Technical Issues, Media Campaigns	4,500
Day Rate Data Coders	1,500
General and Administrative Expenses (@ 12%)	19,840
Total Costs	185,186
Fee @ 8%	14,815
TOTAL	$200,000

Overhead and G & A Are In Accordance With March, 1982 DCAA Audit

Payment Schedule

A firm fixed price contract is proposed, with an initial progress payment billing 31
of $20,000 two weeks after contract award, and ten progress payment billings of $16,000 each, submitted on the first and fifteenth of each month thereafter. A final payment of $20,000 will be due upon final Department of Transportation acceptance of the final study products.

QUESTIONS FOR DISCUSSION

1. Identify discussions in the proposal that characterize it as an unsolicited proposal. How would this proposal be written differently if it had been solicited? What kinds of changes would have to be made, especially in Part II on Technical Approach?
2. What specific problems with the snow control program does the proposal identify for the city of Riverton? Summarize these problems in a paragraph or two.
3. How much knowledge about these problems does the Task Force have? How does the proposal use the Task Force to sell the plan advocated by Acme?
4. In a paragraph or two, summarize the methods the Acme firm will use to solve the city of Riverton's problems with snow removal. Then summarize, in a paragraph or two, the expected results and effects of following the Acme plan.
5. Mr. Toulmin's headnote provides information about the different readers the proposal will have. Which parts of the proposal will be of most interest to (a) members of the city council, (b) snow removal experts, (c) accountants? How does the proposal make technical concepts easy for nontechnical readers to understand?
6. If the proposal appends résumés of all Acme employees who will work on the project, why are the "blurbs" necessary? How is each "blurb" tailor-made to prove that Acme Consultants can do the job as promised?
7. Discuss the importance of the layout of the proposal to convince readers to accept Acme's plan. Comment on visuals, headings, spacing, etc.
8. Identify Mr. Toulmin's use of the following rhetorical techniques in the proposal: (a) definition, (b) comparison/contrast, (c) classification, (d) process description.

SUGGESTIONS FOR WRITING

1. Write a letter of acceptance to Acme Consultants but ask for more information, as the city of Riverton did according to Mr. Toulmin's headnote.
2. Write an external proposal modeled after this Acme model on any topic your instructor approves.

chapter 9

Feasibility Studies

A feasibility study is written to help management decide on whether a project can and should be carried out. The other names by which a feasibility study is known point to its function (*recommendation report*) and its method of interpreting the evidence (*analysis report*). A feasibility study weighs various options in order to recommend the most suitable course of action. In Chapter 1, the feasibility study in memorandum format from the landscape engineer assessed the advantages and disadvantages of three types of possible ground covers that could be used at an industrial park (pages 22–25). Your employer may ask you to investigate which types of word processors would be best for the company's text-editing needs, or which sites would be the most promising for a new hotel, or which chemical techniques would produce the best results in your company's plant. A feasibility study does not always investigate multiple options. In some cases you may compare a single new option with an existing facility, plan, technique, or product. Every feasibility study, however, uses convincing supporting documentation to justify the alternative or option the writer thinks is best.

A feasibility study answers the following types of questions for the decision-maker reader:

1. Do we need to undertake this project?
2. Is it possible or likely at this time?
3. What are all the realistic options?
4. Which of these options best meets our needs? Why?
5. How will this one option, as opposed to the others, benefit our company or agency (e.g., expand markets, increase productivity, improve maintenance, make the environment safer, maximize use of equipment or personnel)?
6. What capabilities are necessary to implement this project (personnel, equipment, facilities, transportation, services)?
7. How much will this project cost to implement?

8. When should this project begin (immediately, next week, next month, next quarter, indefinite period)?

THE RELATIONSHIP OF FEASIBILITY STUDIES TO PROPOSALS

Feasibility studies are similar to proposals in that they often begin with a problem to be solved. Unlike proposals, however, they do not presuppose that there is only one good way to solve that problem. Instead of concentrating on a single plan, feasibility studies present the advantages and disadvantages of several different alternatives. Often, a feasibility study results in a call for proposals; and sometimes proposals are made to do a feasibility study. And, suitably enough, proposals may contain a feasibility study to show that what is being proposed can be accomplished.

THE GOALS OF A FEASIBILITY STUDY

A successful feasibility study needs to accomplish the following four goals:

1. Identify the project or procedures under consideration.
2. Establish, classify, and discuss the various alternatives (or options) and the criteria by which each alternative will be judged.
3. Draw conclusions about the capabilities and limitations of each option.
4. Recommend which, if any, option best meets the established criteria.

THE PARTS OF A FEASIBILITY STUDY

These four goals are generally discussed in four sections of the study: Introduction, Body, Conclusion, and Recommendation. As the models in this chapter show, however, feasibility reports can follow a number of different formats and use a variety of parts.

The Introduction

The introduction serves a number of essential functions. In general, it gives an overview of the study, telling the reader the purpose of the study, what the options are, how you evaluated these options, and what your recommendation is. It is wise to write the introduction last, after you have written and assessed the body of the study.

The Purpose. This section previews the major points of your study. Begin with a clear statement of what you have investigated and what your recommendation is. Do not keep the reader in suspense until the end of the study to find out what you think should be done. Many companies insist that key information be given at the beginning of the study so that busy decision makers need not read a report cover to cover. After reading your purpose/recommendation statement, a reader will have a general sense of the scope of your investigation and proposed solution.

Formulate your purpose statement to describe both the project and solution, as the following examples do:

> The report assesses the economic feasibility of using artificial lighting in construction zones in northern Indiana and finds that such lighting can decrease the number of accidents.

> To fulfill Public Law 34283, this study determined that the most advisable location for the new Greene County Hospital was at New Hanover not Morton.

> This report recommends changing Route 34 from a two- to a four-lane road to accommodate increased daytime traffic from the Mayfield plant.

> In accordance with our contract, we have reviewed the cost effectiveness and desirability of continued use of the existing high service steam pumping equipment at the McNeil Street Water Treatment Plant [in Shreveport] and we recommend that the operation of this equipment be discontinued and a new plant be built.

The Problem. The objective is the reason for your writing the feasibility study in the first place. Describe it as clearly as possible, tailoring your description to the needs of your audience. Explain the significance of the proposed project and why it needs to be undertaken. You might begin with an explanation of any problems that led to the study—when and how difficulties (inconveniences, costly repairs, environmental dangers) arose. A brief history is enough for a reader familiar with the project, but a more detailed account is necessary for an audience that knows little or nothing about the technical complexities of the problems. In the case of the McNeil Street Plant study cited above, the readers were city officials who did not have engineering backgrounds, and so the writer correctly supplied information about outdated equipment, shifts in population in Shreveport as they affected the amount of power generated, additional requirements placed on the system, and increasing fuel costs.

The Scope. This section states the options you studied and the criteria used to evaluate them. Tell the reader why you considered the options you did and why you excluded others, especially if there is a chance that an excluded option is one a reader might have thought important to your work. If you investigated several options (more than three or four, say), you might assign each one a number or a shortened name for easy reference. After listing the options investigated, describe the criteria used for evaluation. This information is absolutely essential so the reader can understand the basis for your discussion and subsequent interpretations. Criteria will, of course, vary depending on the topic, the problem, and the needs cf your audience. The following criteria are often used to assess options in feasibility studies: costs, efficiency, durability, safety, difficulty, transportation, location. In the McNeil Street study, the existing system was evaluated according to electrical power distribution, efficiency of the treatment facilities, intake according to electrical power distribution, intake capacity, and, most important, cost of energy.

The Body

The body of a feasibility study supplies two types of information—the facts gathered about each option and the evaluation of these facts. The body of a brief feasibility study might be successfully limited to one or two pages in a memorandum or letter. A longer, more complex feasibility study could include individual chapters, each devoted to a major option or criterion.

The organization of the data will determine how accessible and therefore how useful it will be to readers. In this way, your presentation can very much affect the reader's choice. You will be confronted with a variety of details about measurements, times, costs, site descriptions, and so forth. There are basically two ways to organize your data: by options or by criteria. You can organize the body of the study according to the individual options by discussing each one separately (Plan A, Plan B, Plan C, etc.). Each major section of the body of your study, therefore, would analyze a single option and should be labeled with appropriate headings and subheadings. The criteria used to assess the options would then be treated as subsections. For example,

Plan A	*Plan B*	*Plan C*
1. cost	1. cost	1. cost
2. efficiency	2. efficiency	2. efficiency
3. maintenance	3. maintenance	3. maintenance

The model feasibility study on the Alderson correctional facility included on pages 322–335 of this chapter follows this method of organization. The writers organized their long study according to the six different options available to the Bureau of Prisons to meet the needs of women prisoners.

It is also possible to organize the body of a feasibility study by criteria rather than options. Following this method, you would make the individual options subsections of the criteria, as follows:

Cost	*Efficiency*	*Maintenance*
1. Plan A	1. Plan A	1. Plan A
2. Plan B	2. Plan B	2. Plan B
3. Plan C	3. Plan C	3. Plan C

The Tractor-Rite model on pages 308–311 is organized by the criteria of building space requirements, initial investment costs, and yearly operating costs.

Follow whichever plan is best for your audience. Remember that you cannot include every bit of information you gathered. Your reader will be most interested in how one option compares with another on key points of costs, convenience, availability, and performance. Select those details that will be most useful to your readers. Focus on specific benefits and disadvantages. Zero in on those points that will have the greatest impact on your audience's decision. Tell the reader when a difference is significant and when it is not. Your reader will expect to see major interpretations you make in this section again in the Conclusion.

Conclusions and Recommendations

For a decision maker, these sections are the heart of a feasibility study. The conclusion summarizes your evaluation of each option; the recommendation states which alternative best meets the established criteria. Some feasibility studies combine both sections under one heading as we have done here. In other studies, conclusions and recommendations are listed as separate sections. As we pointed out, many firms insist that a summary of the key recommendation be placed in the purpose statement at the beginning of the study, too.

If listed as a separate section, the findings in a conclusion can be numbered to reflect their order of importance or the order in which they apply to points raised in the report. The following passage contains, in numbered order, the conclusions for the feasibility study on the electrification of the McNeil Street Water Treatment Plant.

Conclusions

1. The existing normal daily high service pumping is 8 to 10.5 mgd. High service hourly pumping rates vary between a minimum of 5 mgd and a maximum of 11 mgd. A discharge pressure of approximately 90 psi is maintained at the treatment plant at all times.
2. The high service pumping requirements are expected to remain essentially the same in the foreseeable future.
3. Three existing steam-driven, high-service pumps are provided at the plant. The pumps have a capacity of 2, 6, and 8 mgd, respectively.
4. Two existing electric-motor-driven, constant-speed, high-service pumps are provided at the plant. The pumps have a capacity of 6 and 10 mgd, respectively.
5. The present annual cost of operating the existing high-service steam and electric pumps has been estimated to be greater than the operating costs plus the amortized capital costs of replacement electric-motor-driven, high-service pumping facilities.
6. Electrical power required at the plant is supplied from two existing electrical substations. The two services are not arranged to allow feeding the required plant power load from one or the other source. The electrical system is obsolete and is in need of replacement.
7. Filters No. 1 through 10 have been out of service for over 30 years and have

deteriorated beyond repair. Piping and valves serving filters No. 11 and 12 are also badly deteriorated.

As these findings show, drawing conclusions requires more than repeating bits of raw data. You need to synthesize the data to reach meaningful conclusions, e.g., annual operating costs are excessively high, the electrical system is obsolete, the piping and valves for two filters have badly deteriorated. Conclusions in some long feasibility reports frequently summarize individual options under the headings of "Advantages" and "Disadvantages."

The recommendation section tells your reader what he or she should do. That is, it says which option is the most feasible. Of course, the decision whether or not to follow your recommendations is up to management. Your recommendations should reflect your professional opinion after careful examination of the evidence. Make your recommendations as precise and direct as possible, giving specific details about timetables, locations, facilities, machinery, personnel, and costs. Notice how precise the recommendations are for the McNeil study.

Recommendations

1. Operation of existing steam-driven, high-service pumping units should be discontinued.
2. The existing 6 mgd electric-motor-driven, high-service pump should be placed on standby service for emergency use only. The existing 10 mgd electric-motor-driven, high-service pump should remain in service.
3. A new high-service pumping station should be constructed and provided with 4 mgd electric-motor-driven pumping units. Three pumping units should be variable-speed and two pumping units should be constant-speed.
4. The high-service, electric-motor-driven pumping units should be designed to be operated locally as well as remotely from the Thomas L. Amiss Water Treatment Plant.
5. Provisions should be made to revise the existing electrical power services to allow feeding of all the electrical load from either power source as desired.
6. Filters No. 1 through 12 should be removed from service.
7. Filter No. 21 should be equipped for operation with media, piping, valves, and control equipment when additional filter capacity is required in the future.
8. A new 36-inch raw-water pipeline should be constructed from the McNeil Street Booster Station to the water treatment plant low-service pumping station.

Sometimes it is relatively easy to make recommendations because there are clearcut distinctions among alternatives. One option may be far superior to the others in costs, efficiency, safety, or productivity, for example. But often, two options are devilishly close. (After all, few problems with obvious solutions and few relatively straightforward projects require a full-fledged feasibility study.) At still other times, different options offer competing advantages. For example, one option might offer environmental benefits while another holds promise of technological advantages. It might be environmentally wise to build a road over one pass yet more feasible from an engineering point of view to take another route. After you decide which option is the most feasible, explain your recommendation to the reader carefully.

When you prepare your recommendations, the following guidelines will help you:

1. *Stick to the subject at hand.* Do not stray from the subject by introducing a new topic. For example, if you are considering ways of increasing warehouse space, do not drift into methods of preventing accidents.

2. *Reach a decision.* If you fail to reach a decision, you are asking the readers to make their own recommendation, thus having them do your work. If you cannot choose between options because more research is needed, tell the readers so. If either alternative is suitable, tell your readers and explain why. If none of the options is acceptable, explain why.

3. *Have conclusions that logically follow the supporting data.* Do not argue against your own conclusions. If everything points to the reader following plan X (costs, capability, improved sales, convenience), do not try to make a case for plan Y.

A CHECKLIST FOR WRITING A FEASIBILITY STUDY

1. Identify the project or problem under consideration.
2. List all realistic options.
3. Determine the criteria for analysis, based on the audience's specific requirements and priorities and the particular project or problem.
4. Organize your report either by option or criterion.
5. Draw conclusions after evaluating each alternative.
6. Recommend which, if any, alternative best meets the established criteria. If none does, explain why.

This model is a feasibility study written in memorandum format to a decision maker in the writer's own company. Evaluating the feasibility of four storage techniques used by a tractor company for finished purchased materials, this study is sent to a vice president whose training and interests are in management and finance. The author has concisely stated the purpose of the study and explained the problem for a reader who has many different operations to oversee. This study is organized according to the criteria of evaluation the writer thought would be most useful to her audience. She also includes a table that graphically summarizes all the statistical data upon which her study is based.

A Feasibility Study of Storage Techniques for Tractor-Rite Inc.

Melinda Drew

TRACTOR-RITE

Fairfield, Iowa 53462

TO: Allen Heard
Vice President, Management

FROM: Melinda Drew M. D.
Head, Operations Task Force

SUBJECT: Selection of the most feasible method to store finished purchased materials

DATE: March 4, 1984

Purpose

This report presents the results and recommendations of the Task Force appointed to study the most economical and efficient method of storing all shipments of finished purchased materials at our factory site. After evaluating four methods, the Task Force recommends the Stacker method as the most feasible. 1

The Background of the Problem

For the past 36 months, Tractor-Rite has rented 32,000 sq. ft. of space at an off-site warehouse to store and handle shipments of finished purchased materials. This warehouse was being used because of its proximity to our factory and its easy accessibility to our loading crews. 2

Two months ago, management decided not to renew the rental agreement for space and equipment and to begin storing all finished purchased material at our own factory site. By doing this, Tractor-Rite will save $154,000 a year, including related transportation, equipment, and personnel costs. 3

Based on the available space and storage facilities, the Triax method was used at the off-site location. This method handled an output rate of 346,000 tons a year with a utilization factor of 85%. This method stored 5680 pallet loads of finished material, with 232 pallets moved in and out daily. 4

Although the Triax method gave satisfactory results at the off-site location, management has considered the possibility 5

Mr. Allen Heard
March 4, 1984
Page 2

of employing other methods of storing the pallets at our location. However, any other method will have to equal or surpass Triax's efficiency, given the annual rate of movement cited.

As a result of these decisions, the current Task Force was appointed to examine the different storage alternatives. 6

Scope

We examined the following four storage methods most frequently used in our industry: (1) Triax, (2) Stacker, (3) Wide Aisle truck with racks, and (4) Floor Storage with fork truck. These methods were evaluated according to the following criteria: Building Space Requirements, Initial Investment Costs, and Yearly Operating Costs. 7

Evaluation by Criteria of Four Storage Methods

The attached table summarizes the data gathered about the four methods based on the Task Force's criteria. The Company comptroller supplied us with estimates of prices. What follows is a brief description of the pros and cons of each option based on the four criteria. 8

Building Space Requirements

Due to limited space available for a storage building at the factory site, this criterion was one of the Task Force's most important considerations. 9

According to this criterion, the Stacker storage system is the most feasible because it will require only 17,120 sq. ft. of space, since the palleted loads can be stored to a height of 42 ft. Also, the Stacker method requires only 3.01 sq. ft. of space per each palleted load for storage. The Triax method, used at the leased storage site, is the next most feasible in this category with 19,800 sq. ft. of space required with 3.49 sq. ft. needed for each load in a 10 ft. high space. 10

Because the Wide Aisle and Floor Storage methods can only stack at 5 ft. and 2 ft. heights, the square footage requirement increased to 67,000 and 98,400 sq. ft., respectively. Because of the limited space available these last two methods are the least feasible. 11

Mr. Allen Heard
March 4, 1984
Page 3

Initial Investment

The second criterion we applied to the four methods was the initial investment the Company would have to make for a storage system. 12

Although the cost per square foot for constructing the Wide Aisle and Floor Storage building is $10, less expensive than the other two methods per sq. ft., the need for a larger total area increases the initial building investment to $670,000 for the Wide Aisle and $984,000 for the Floor Storage method. In comparison, the cost per sq. ft. for the Triax method is $16 and the Stacker is $17, thus making the building investment $317,000 for Triax and only $291,000 for the Stacker method. 13

With each of the four systems we will have additional initial costs for equipment. Our company will have to purchase Triax equipment since we rented the equipment at the off-site location. The equipment costs for the Triax method will be the most expensive at $942,000; when added to the building costs, the total initial investment for this method will be $1,259,000. Although the Floor Storage method has the least expensive initial equipment costs of only $14,000, when this figure is added to the building costs, this method becomes the second most expensive at $998,000. Similarly, the Wide Aisle method has an equipment expenditure of $184,000 with a total investment of $868,000. Finally, the equipment for the Stacker method will initially cost $402,000, but in light of less costly construction expenses, the total initial investment will be only $693,000, making the Stacker method the least expensive system to implement. 14

Yearly Operating Costs

The final criterion we considered was yearly operating costs. These costs include all expenditures necessary to maintain a system once it is established—heat, light, power, janitors, material labor, truck leasing, and equipment maintenance. The Floor Storage system will require the greatest expenditure at $238,000. The Wide Aisle follows with $196,000, the Triax at $151,000. The least expensive is the Stacker method at $146,000 annually. 15

Mr. Allen Heard
March 4, 1984
Page 4

Table 1. COMPARISON OF FOUR STORAGE SYSTEMS FOR FINISHED PURCHASED PRODUCTS, BASED ON STORING 5680 LOADS OF PALLETIZED MATERIAL ALL WITH THE SAME ACTIVITY

	Present Triax 10 Feet High	*Stacker 42 Feet High*	*Wide Aisle Truck w/Racks 5 Feet High*	*Floor Storage w/Fork Truck 2 Feet High*
Building Space Requirements				
Truss Height		50′ 0″	15′ 6″*	15′ 6″*
Space Required—Sq. Ft.	19,800	17,120	67,000	98,400
Sq. Ft. Per Load	3.49	3.01	11.80	17.32
Investment in System				
Building Cost/Sq. Ft. (New)	$16	$17	$10	$10
Building Cost (New Basis)	$317,000	$291,000	$670,000	$984,000
Equipment Cost	$942,000	$402,000	$184,000	——
Trucks, Batteries, Chargers, Etc.	——	——	$14,000	$14,000
TOTAL INVESTMENT (Thousands)	$1,259	$693	$868	$998
Yearly Operating Costs (Thousands)				
Heat, Light, Power, Janitors, Etc.	$17	$15	$58	$85
Material Handling Labor	$102	$130	$124	$135
Truck Leasing Cost	——	——	$8	$10
Equipment Maintenance	$32	$1	$6	$8
TOTAL YEARLY OPERATING COST (Thousands)	$151	$146	$196	$238
INDEX OF PRESENT VALUE OF INVESTMENT & OPERATING COST	2,142	1,547	2,000	2,390
ORDER OF INVESTMENT PREFERENCE	3	1	2	4

* Building height shown at 15′ 6″ only for cost comparison purpose.

Conclusions and Recommendations

Based upon the criteria applied by the Task Force, the loads 16
of purchased finished materials, formerly stored off-site, should be stored using the Stacker system. This system is the most cost-effective of the four we investigated because it provides the most amount of storage space for the least amount of money. Moreover, the yearly operating costs for the Stacker system are the most economical of the four methods.

QUESTIONS FOR DISCUSSION

1. Has Drew successfully adapted the Purpose and Background of the Problem sections for Heard? Does Heard need more or less information about the problem or has Drew given him just the right amount of detail? Explain.
2. What kinds of information does Drew give her audience in the body of the study? Why doesn't she describe each storage method in more detail, focusing on the principles and mechanics of each?
3. Is Drew's method of organization appropriate for her reader's needs? Could she have met those needs just as well by organizing her discussion by option rather than criteria?
4. Why and how does Drew interpret the table for her audience? What information from the table does she exclude from the body of her study? Why?
5. How has Drew organized the table to make it easy for the reader to understand and use it?
6. Is this study readable? Comment on Drew's (a) sentences, (b) paragraphs, (c) forecasting and summary techniques, and (d) the overall layout.

SUGGESTIONS FOR WRITING

1. Write a short, informal feasibility study in memorandum format similar to the Tractor-Rite model. Investigate alternatives a company might have to solve one of the following problems:
   a. schedules (for employees or for deliveries of goods and products)
   b. transportation systems
   c. temporary versus permanent employees (e.g., nurses, geologists, computer programmers, technical writers and editors)
   d. security systems in a restricted area such as a laboratory or test site
   e. renting or buying a car
   f. ways of improving quality control
   g. sales techniques (mail campaign, radio advertisements, television commercials, personal visits by sales force, telephone messages)
2. Based upon information in this feasibility study, write a short proposal in memorandum format from Drew to Heard about the company storing its own goods on the factory site. Assume that the company has not considered this move before and that the memo/proposal is to be written three months before this feasibility study.

This feasibility study on creating a central repair facility at a large power company is written to a decision maker by an individual who is himself a decision maker (the manager of a division) as well as an electrical engineer. The study was undertaken as part of Midcontinent Power Company's plan to streamline operations and reduce costs. The writer's job was to assess existing ways of repairing and overhauling equipment in light of a new plan that would consolidate major repair activities at one central facility. Like all large firms, Midcontinent will be skeptical about changing its operations without receiving ample and convincing evidence that a new alternative is better than old methods. Given that task, the author had to gather, organize, and interpret a great deal of statistical data for an audience that might see the data from another perspective. The organization of this study is more complex than that for the previous model from Tractor-Rite. It first identifies limitations with the current repair methods and then shows how the new method would resolve a number of problems, save the company money, and improve the quality of maintenance.

A Feasibility Study on Establishing a Central Repair Facility at Midcontinent Power Company

Jerry Walker

MIDCONTINENT POWER COMPANY

Hillside, Nebraska 68582

DATE: October 24, 1983

TO: M. T. Bowers
Vice President for Operations

FROM: Jerry Walker J.W.
Manager, Division Electric Operations

SUBJECT: Feasibility of Establishing a Central Repair Facility

Purpose of Study

We have conducted an investigation into the feasibility of establishing a central repair facility to perform many of the repair and overhaul functions now being done at various locations throughout the Company and by outside contractors. These functions include: 1

1. annual line recloser overhaul and maintenance
2. capacitor oil switch maintenance
3. tool repair and salvage
4. transformer repair

Many of these repairs lend themselves to "assembly line" activities while others require specialized equipment. This study concludes that establishing a central repair facility will save the Company money and improve the quality of maintenance. 2

Background and Problem

For many years the Company scrapped much of the distribution-class equipment (transformers, oil switches, reclosers, etc.) requiring more than the simplest repairs because the repair cost often nearly equaled the replacement 3

M. T. Bowers
October 24, 1983
Page 2

cost. During the late 1960's and the 1970's the cost of new equipment grew rapidly, and it became necessary to reevaluate the practice of replacement versus repair of such equipment.

In 1976, the Distribution Department began to compile data on typical repair costs of transformers because of the number of failed or damaged units and the high replacement costs of those units. The large number of failed or damaged transformers caused by spring storms in March and April, 1979, focused attention on replacement costs, and the Company decided to pursue the repair of those units. Beginning in early 1980, most failed or damaged transformers were sent to a commercial repair facility for evaluation, repair (based on preestablished guidelines and limits), or scrapping. Today, over 99% of the Company's failed or damaged transformers are being sent to outside contractors for repair. 4

Other types of repairs and maintenance have been done at six separate locations throughout the Company. District electricians at each of these six operating stations have worked on equipment such as capacitors and line reclosers. 5

The Company is faced with unnecessarily high costs in making these repairs. The cost to repair line equipment, for example, has been increasing since 1978 at an average annual rate of 12%. The cost of repairing transformers in 1982 exceeded $200,000, is projected to exceed $225,000 this year, and will rise to $260,000 in 1984. Repair costs of other equipment have experienced similar increases. Moreover, it is apparent that the Company is losing money by having many of the same job functions duplicated at each of the six operating stations. A more economical way to make (and consolidate) repairs and to use current staff and resources needs to be found. 6

Reasons Existing Methods Are Not Feasible

The Company presently has available the following three methods to deal with damaged equipment: (1) Scrapping/Replacement, (2) Using an Outside Contractor, and (3) Using Six District Repair Shops. As was pointed out in the previous section, ample evidence has been gathered over the last ten years to show that method (1) is not a financially 7

M. T. Bowers
October 24, 1983
Page 3

sound practice. Method (2) is costly; last year the Company spent more than $200,000 on transformer repairs alone. These expenses may be unnecessary since the Company potentially has the resources and personnel to make its own repairs on transformers at a far lower cost.

Method (3), having six separate shops, entails the following 8
disadvantages:

a. It is economically prohibitive to install specialized equipment for every type of repair at six different locations. The Company's expense is six times greater. Moreover, such equipment is not equally in demand at every one of the six shops, and the Company thus pays for idle time.

b. Moving specialized equipment between locations is not feasible because of the added expenses for transportation. Also, transporting such equipment involves the risk of damaging it. Furthermore, there is the problem of accessibility for the other five shops that may need to use the equipment.

c. An uneconomical use of labor results when the Company has six different groups of electricians all performing repairs at six separate locations.

The Feasibility of a Central Repair Facility

Because of the problems with existing methods, the Company 9
will experience several benefits by establishing a central repair facility. Only major repairs and overhauls would be performed at this central facility. Minor repairs will continue to be performed at local shops which were originally designed for that work.

1. Because special tools and test equipment will have to be provided only at the central facility, the Company will save approximately $217,500 in initial costs.

   Here is a breakdown of the average costs necessary 10
   because of duplication of equipment currently required at all six district shops to perform all repairs:

M. T. Bowers
October 24, 1983
Page 4

Recloser Test Set	$ 15,000
Transformer Test Set	12,500
Gantry-type crane	10,000
Laboratory instruments	6,000
	$ 43,500
	×5 district shops
	$217,500

Moreover, minor transformer repair, now being contracted out, could be performed by the Company at a savings of $200,000.

2. <u>Quality of maintenance would be improved through standardization and consistency.</u>

The repairs now performed at each of the six district 11
shops are done by employees who have many duties and responsibilities; repairs are supervised by six employees who also have other duties and responsibilities, and some of these supervisors lack the technical expertise to monitor the repair work.

The establishment of a central repair facility will allow 12
the development of a highly skilled, technically competent work force supervised by a knowledgeable person whose prime responsibility is the accurate repair of equipment. It will be easier to keep pace with technological advances with a smaller work force under one highly trained supervisor.

As a result of this change, the Company's expertise in 13
repair would be enhanced and we would have more efficient in-house repairs and eliminate many "outside" repairs.

3. <u>The facility would be a useful clearing house for all tool retirements. Salvage of reusable parts could be made more easily, and inventory control of retired tools would be more easily accomplished.</u>

Although a precise amount cannot be determined, 14
approximately one-half of our capital budget for tool purchases (1983 Tool Budget is $156,000) is earmarked for replacement of tools which, at this time, cannot be economically repaired. A brief review of these

M. T. Bowers
October 24, 1983
Page 5

replacements was made during the latter part of 1982. We estimate that half of the replacements could be retained for several more years if repair parts were readily available. Moreover, approximately $20,000 per year could be saved on tool purchases through the salvaging of parts and repair of tools, including the labor costs associated with these repairs.

4. Removing repair duties from the district electricians and having such duties performed at a central repair facility would decrease labor requirements.

If repair duties alone were removed from the district electricians and performed at a central facility, the net effect would be as follows: 15

	Labor Requirement Removed WH (work hours)
Clearvale	1134
Jonesville	570
Springfield	482
Harristown	876
Lincoln	804
Freeburg	1076
	4924 (WH saved)

A central repair facility would mathematically reduce the District Electrician requirement by 5.2 employees. This reduction in employees would save the Company about $108,300 a year, including taxes, benefits, and overheads.

5. The central repair facility concept would work especially well with the Central Stores concept now being developed by the Company.

It should be possible to transport much of the equipment from district storerooms to the central repair facility and back to coincide easily with delivery of stored items. 16

M. T. Bowers
October 24, 1983
Page 6

Possible Location for the Central Repair Facility

We studied each of the six district shops to see if any one of them would provide a suitable location for the central repair facility. The Clearvale shop can easily meet the requirements for such a facility. Hence, the Company does not have to spend additional money constructing a new building. The largest of the six shops, Clearvale is also the most conveniently located. Equally important, the Clearvale shop is adjacent to the Central Stores Building which will allow the Company to spend less money for transportation of parts and less time in obtaining those parts. 17

Conclusion

1. Scrapping of equipment is not financially sound.

2. Using outside contractors for 99% of transformer repairs is cost prohibitive.

3. Six separate district repair shops involves unnecessary duplication of machinery and personnel.

4. A central repair facility used for major repairs and overhauls could increase the quality of maintenance and assist the Company in identification and inventory of parts. The district shops could then perform the repair work they were originally designed for.

5. A central repair facility would save the Company $500,000 each year in costs, equipment, and personnel. In fact, this facility would reduce by 5.2 the number of electricians needed at the district shops.

6. Because of its size and proximity to the Central Stores, the Clearvale shop affords the most feasible location for the new central repair facility.

Recommendation

The Division of Electric Operations recommends to Management that a central repair facility be established at the Clearvale Service Center. 18

QUESTIONS FOR DISCUSSION

1. Based upon the language and level of technical detail in this report, characterize M. T. Bowers. What kind of background and job responsibilities might he have? What kinds of questions or reservations might he have regarding Walker's report?
2. What types of information relevant for his reader does Walker include in the Purpose section of the study? Why does he include the types of repairs done by or for Midcontinent?
3. Evaluate the effectiveness of the section labeled "Background and Problem." Why would Bowers need to be reminded of the 1976 Distribution Department study? Does this section include too much or too little information? Explain.
4. By what criteria does Walker evaluate existing methods? Why does he spend more time on Method 3 than the others?
5. Walker gives Bowers five reasons why the central repair facility is the best choice for Midcontinent. Comment on the order, format, and types of details he gives Bowers. Could these five points be listed in another, more effective way? Explain.
6. Why does Walker include a section on the possible location for a central repair facility? Could this section be placed anywhere else in the report? Explain.
7. How effectively do Walker's conclusions summarize and interpret data for Bowers?

SUGGESTIONS FOR WRITING

1. Assume you are M. T. Bowers. Write a memorandum to Walker pointing out some problems in implementing the central repair facility immediately. Stress, though, that the idea is a good one but needs more time to be workable.
2. Write a three- or four-page feasibility study in memo format to a decision maker on one of these topics:
   a. food service on campus (you might consider catering, student participation in the preparation, use of local pizza or fast food operations for some of the meals, or other options)
   b. ways of financing a college education (types of loans, work-study appointments, ROTC programs, etc.)
   c. ways of increasing circulation of your college newspaper
   d. choosing among 3 or 4 sites to relocate a business (e.g., cleaner, garage, movie theater, record store, beauty shop, book store, hardware store, disco, etc.)
   e. renting versus buying furniture for an apartment or an office

f. evaluating three or four different models of a piece of equipment (e.g., pumps, tires, bicycles or motorcycles, computers, tape recorders) or product (pesticides, cereals, soaps)
g. installing more lights on a city street
h. changing a traffic pattern in your city
i. adding a new course in your major (or deleting one)

This long feasibility study on the Alderson Correctional Facility was required by a public law passed after Congressional hearings were held on the problems at the women's prison. "To fulfill the mandate of Congress," the Director of the Bureau of Prisons appointed a task force to provide and evaluate alternative uses for the facility. After months of research, both at Alderson and in Washington, the task force presented a draft report to the Executive Staff of the Bureau, which reviewed the report before sending it to Congress.

Typical of a long report, this feasibility study contains many different, clearly marked sections—executive summary, introduction, background, objectives of the report, discussion of special issues, alternative uses of the prison, summary and recommendations, and appendices. The longest and most detailed section—on alternative uses for the prison—is organized according to the six options identified by the task force rather than by the criteria used to evaluate these options. But the criteria play a central role in the preparation of this study. Based on them, the report systematically weighs the pros and cons of each option in order to recommend the most feasible one. In analyzing each option, the study helps readers to make comparisons easily. Facts about Alderson and other facilities—in use or planned—are logically organized and assessed.

This feasibility study has several audiences. The primary audience is Congress, but it is unlikely that representatives read every word of this study. No doubt key members of their staff did. Executives at the Bureau read an earlier draft as well as the final version. Other prison officials also read the study, especially those involved in the administration of the Alderson, Danbury, and Lexington facilities cited in the study. These individuals have to implement the findings of the study. The findings would also be of interest to individuals involved in civil rights, penal reform, and women's rights groups.

A Feasibility Study of Alternative Uses for the Federal Correctional Institution at Alderson, West Virginia

Bureau of Prisons

EXECUTIVE SUMMARY

The Alderson Task Force was appointed in June 1980 to explore the feasibility of alternative uses for the Federal Correctional Institution (FCI) at Alderson. 1

Analysis of relevant data and a review of related issues concerning Alderson led to the development of six options for alternative use of FCI, Alderson. 2

Option 1: *Convert the Addiction Research Center adjacent to FCI, Lexington to a secure, single-sex facility; change the proposed mission of Danbury Satellite Camp to a camp for women; and change the mission of FCI, Alderson to a co-correctional facility.*

3 Option 1 favorably addresses several of the issues raised concerning FCI, Alderson. Under this plan, improvements are anticipated in visiting, proximity to release destination, and programs and services. The issue of confining female inmates in oversecure environments is largely eliminated. Capital costs are minimal and operating costs will be less than the current level at FCI, Alderson.

Option 2: *Continue to operate FCI, Alderson at its present level and include the establishment of a camp for women in the long-range planning process.*

4 Option 2 proposes no change in the status of FCI, Alderson, but does suggest long-range planning for a camp facility. The establishment of a camp for women would result in favorable changes in accessibility for visitors, proximity to release destination, and a less secure environment. It would also result in excess capacity and, therefore, a more costly operation.

Option 3: *Close FCI, Alderson and build a 500-bed replacement facility.*

5 FCI, Alderson could not be closed without opening another facility for female prisoners. The construction of an appropriate replacement facility in an appropriate location could improve opportunities for visiting. The location of any single institution will be closer for some people and farther for others, but a facility in or near the Northeast Corridor will place the greatest number close to their release destination. Program levels would remain comparable to those at Alderson. Costs for the construction of a new facility would run approximately $26 million, although operating costs would be somewhat less than Alderson's.

Option 4: *Make no major changes at FCI, Alderson but increase the use of Community Treatment Centers for women.*

6 Option 4 would improve visiting opportunities and closeness to home for those women with extended stays in Community Treatment Centers; these women would also be in less secure facilities. There would, of course, be no change for the women remaining at Alderson. This option is not a realistic one because the Community Treatment Center (CTC) placements at Alderson already meet or exceed existing Bureau policy on CTC use.

Option 5: *Convert the Addiction Research Center adjacent to FCI, Lexington to a secure, single-sex facility, acquire a surplus camp facility, and change the mission of FCI, Alderson to a co-correctional facility.*

7 Option 5 is similar to Option 1 except for the acquisition of surplus property for a camp. The unavailability of a surplus camp facility in the Northeast makes this option infeasible at this time.

Option 6: *Make no major changes in FCI, Alderson but increase the number of women placed in State institutions, including the District of Columbia.*

Option 6 would improve visiting and place women closer to home, but 8
programming and services would be reduced in both quantity and quality. In most cases women would be confined in facilities that are even more secure than Alderson. In addition, overcrowding generally exists in those States which do have institutions for women. This option in infeasible because of the small number of appropriate State facilities.

Recommendation

The Task Force recommends implementation of Option 1; it offers the best possi- 9
ble solution to the issues and problems identified at FCI, Alderson.

INTRODUCTION

The Department of Justice Appropriation Authorization Act, Fiscal Year 1981 10
(Public Law*) required the Bureau of Prisons to conduct a feasibility study of possible alternative uses for the Federal Correctional Institution (FCI), Alderson, West Virginia.

To fulfill the mandate of Congress, the Director of the Bureau of Prisons 11
(BoP), Norman A. Carlson, appointed a Task Force to conduct the feasibility study. Members appointed to the Task Force include: Kenneth Neagle, Warden, FCI, Alderson—Chairperson; Ronald Waldron, Chief, Office of Program Development, BoP Central Office, Washington, D.C.; Patrick Kane, Executive Assistant, Northeast Regional Office, Philadelphia, Pa.; Linda Lancaster, Correctional Programs Administrator, BoP Central Office, Washington, D.C.; and Verna Muckle, Budget Analyst, Department of Justice, Washington, D.C.

The Task Force first convened in the Bureau of Prisons Central Office in 12
Washington, D.C., on July 11, 1980, to meet with Director Carlson and the Assistant Directors to discuss the issues raised by the Department of Justice Appropriation Authorization Act, Fiscal Year 1981, and clarify the direction and the purpose of the Task Force.

To conduct their study, two additional meetings were held by the Task 13
Force, on August 7, 1980, at FCI, Alderson and on September 4, 1980, in the Central Office. The Task Force reviewed available documents and conducted special studies on topics related to the specific issues before them. A draft report was prepared and submitted to the October and December 1980 Executive Staff

* Act not yet enacted when study was submitted.

meetings for review and comment. Subsequently, a final report was prepared on the Task Force's findings and recommendations.

BACKGROUND

During the last decade, the United States has experienced increasing concern over the rights and welfare of women in prison. The Congressional Subcommittee on Courts, Civil Liberties, and the Administration of Justice has focused particular attention on the needs and confinement of women at the Alderson, West Virginia facility, in use since 1927. The Committee heard testimony from prison officials on the facility's geographic location, rehabilitative program, health services, and marketable job training assistance. Major concern was voiced about Alderson's remote location which prevented the maintenance of close family ties, especially for the residents of Washington, D.C., who comprise 25 percent of the prison population. Moreover, prison officials were concerned that perhaps many women were placed in overrestricted environments. Forty percent of all offenders at Alderson are classified as Security 1 (or minimum custody) inmates, yet they are housed in an institution that does not meet their needs. 14

OBJECTIVES OF THIS STUDY

Following the Congressional hearings, the Alderson Task Force was appointed in June 1980 to prepare the following report exploring the feasibility of alternate uses for the Federal Correctional Institution (FCI) at Alderson as required by the Department of Justice Appropriation Authorization Act, Fiscal Year 1981. Specific alternatives the Task Force was asked to consider include (1) closing the Alderson facility as a correctional institution and using it instead for another Federal, State, local, or private purpose; (2) exploring alternative placements for women, particularly those sentenced out of the District of Columbia; and (3) seeking placement for all women appropriately designated in community based facilities and if feasible in minimum security camps or other minimum security facilities. 15

DISCUSSION OF SPECIAL ISSUES

The Task Force reviewed the October 1979 hearings on the Female Offenders conducted by the Subcommittee on Courts, Civil Liberties, and the Administration of Justice of the Committee on the Judiciary, House of Representatives. The mandate of Congress in the Appropriation Authorization Act, Fiscal Year 1981, was also examined. Review of these and other documents suggested several issues for consideration by the Task Force. 16

Visiting

It is a generally recognized principle of corrections that the maintenance of family and community ties by the inmate is ultimately helpful to the inmate's morale during incarceration and to the inmate's adjustment when released. The facts that Alderson is located in a rural setting in West Virginia and that many of the inmates incarcerated there are from metropolitan areas in the Northeast make it difficult for inmates who desire to do so to maintain family and community ties. No doubt if a Federal correctional facility were located within the Northeast corridor and closer to the metropolitan areas, the amount of visiting for inmates would increase. 17

Proximity to Home

A review of the legal residence of women confined at Alderson revealed that the inmates are from 41 different states. Obviously, no single institution could be established that would place all inmates close to their legal residence. Several small institutions would place inmates closer to home, but the cost of this approach would be prohibitive. 18

Least Restrictive Environment

Approximately 40 percent of the inmates at Alderson are now designated as level 1 inmates, requiring the least amount of security, and approximately 72 percent have community and out custody status. For those women classified as security level 1 and assigned to Alderson the environment is more restrictive than necessary. This information suggests the need for assignment to less secure facilities or community placement for a significant number of Alderson inmates. 19

Inmate Programs

The Task Force found a wide range of program offerings were available at FCI, Alderson. Among the programs available were office skills, cosmetology, mechanical services, automated data processing, and instructional courses from the elementary school level through college. The quantity and quality of programs at Alderson appears to the Task Force to be more than adequate. 20

Medical Services

A review of the medical services at Alderson revealed that more than adequate services were available. By way of comparison, Alderson has a ratio of 3 medical 21 staff members for every 100 inmates while the Bureau overall has a ratio of 2 medical staff for every 100 inmates.

Because a wide range of medical services are provided, the Task Force 22
concluded that the medical needs of the inmates are being met.

District of Columbia Inmates

The relationship between the Bureau of Prisons and the District of Columbia 23
(D.C.) Superior Court and the D.C. Department of Corrections is unique. The District of Columbia, unlike most of the states, does not maintain a facility for long-term female offenders. Only female D.C. Code violators serving terms of one year or less are confined at the D.C. Detention Center.

A Memorandum of Understanding between the agencies states that the 24
Bureau of Prisons will house all female D.C. Superior Court cases who are committed to terms of more than one year. Generally, these women are designated to FCI, Alderson, and constitute the largest group of female inmates with a single release destination, approximately 25 percent.

The District of Columbia has generally preferred that the Bureau of Prisons 25
continue to assume the responsibility to confine female D.C. Code violators.

Operationally and financially, it is to the advantage of the District of Colum- 26
bia to continue the present arrangement. The District spends approximately $1.9 million each year to contract for the incarceration of their female prisoners in the Federal system as opposed to the $3.7 million it would cost annually to operate their own 200-bed facility. Construction costs for such a facility are estimated to be over $21 million.

The Federal Prison System recognizes the advantages for female offenders 27
should the District of Columbia build its own correctional facility for women. Although the Bureau of Prisons would support the District in an attempt to build such a facility, it is unlikely that this will occur anytime in the foreseeable future. The Bureau of Prisons, therefore, must continue to plan and provide for the care and custody of female D.C. Code violators.

Other Federal, State or Local Use of FCI, Alderson

The Department of Justice Appropriation Authorization Act, Fiscal Year 1981, 28
recommended that the Bureau explore other Federal, State or local uses for FCI, Alderson. If FCI, Alderson were closed, the Bureau would not have sufficient capacity for those female offenders in security levels 2 through 6, requiring incarceration in a secure facility. An existing male institution would have to be converted for female use or a new facility would have to be constructed or acquired. If an existing male prison were converted to female use, Alderson would need to be converted to male use to make up the loss capacity for males. If a new facility were constructed or acquired for females, Alderson should be

converted to a Security Level 1 facility for males, as there is presently a shortage of Level 1 capacity for males in the Northeast. Therefore, if Alderson were closed for females, there exists sufficient need for the facility for other Bureau of Prisons needs and, therefore, the facility should not be made available for other Federal, State or local use.

ALTERNATIVE USES FOR ALDERSON

Review of the issues surrounding FCI, Alderson suggested several options for 29
consideration by the Task Force. To assess the impact of the options proposed by the Task Force, each of the following factors were evaluated:

- *Accessibility for Visitors:* Is the facility more accessible to inmate visitors?
- *Proximity to Release Destination:* Are inmates closer to their release destination?
- *Level of Programs and Services:* Would program offerings increase over the present level?
- *Appropriate Security:* Are offenders placed in facilities of the appropriate security level?
- *Capital Costs:* One time construction and equipment costs.
- *Operating Costs:* Annual salary and expenses necessary for operations.
- *Per Capital Costs:* The operating costs per inmate per day.
- *Physical Capacity:* The number of beds by which the option will increase or decrease the current base.
- *Personnel:* The total number of personnel required to implement the option.
- *Feasibility:* That which prohibits or allows the implementation of the option.

Option 1

Convert the Addiction Research Center adjacent to FCI, Lexington to a secure, single-sex facility; change the proposed mission of Danbury Satellite Camp to a camp for women; and change the mission of FCI, Alderson to a co-correctional facility.

This option consists of a three-part proposal: 30

- Establish a camp for Level 1 inmates within the Northeast Region,
- Convert FCI, Alderson to a co-correctional facility, and,
- Establish a secure, single-sex unit for female offenders who are not appropriate candidates for co-correction.

Camp for Level 1 Female Offenders. The establishment of a 100 bed camp 31
for women in the Northeast Region would allow approximately 100 Level 1 female offenders to be located closer to their intended release destinations. It would also place more offenders in an appropriate security level institution.

Among Bureau of Prisons facilities, possible sites for such a camp in the 32

Northeast Region include the existing Federal Prison Camp (FPC) at Allenwood, Pennsylvania; the Satellite Camp (SC) at Petersburg, Virginia; SC Lewisburg, Pennsylvania; and the planned SC Danbury, Connecticut. For reasons of location, size, and security levels, the Allenwood, Lewisburg, and Petersburg camps were eliminated from further consideration. The Danbury camp, on the other hand, did appear to offer a workable solution. The camp, adjacent to FCI, Danbury, is scheduled for completion in December 1981 with a planned capacity of approximately 100 beds. The Danbury camp will be easily accessible to residents of the New York City area. Although Washington residents would still have to travel some distance to visit women at SC, Danbury, public transportation is much more available.

Because the facility is currently under construction, it would not require transferring an existing population. This would also eliminate the disruptive influence that changing the mission of an existing facility would have on the staff. Program resources could be developed to meet the special needs of women and the FCI, Danbury population (Level 2 males) lends itself to the concept of joint programming between the two facilities. 33

Conversion of FCI, Alderson to a Co-Correctional Facility. The second component of this option entails converting Alderson from an all female institution to a Security Level 1 co-correctional facility. 34

SC, Danbury was originally intended to house 100 Level 1 males and the justification for its construction was based on the need in the Northeast Region for Level 1 beds. If the SC, Danbury mission is converted from a male facility to an all female facility, beds would still be needed for Level 1 males previously intended for Danbury. A change in mission of FCI, Alderson to a co-correctional facility would enable Level 1 males to be housed at Alderson. 35

The physical layout of the FCI, Alderson lends itself to such a concept. The change to a Level 1 co-correctional facility would require minimal, if any, modifications to the existing facility. The grounds at Alderson are already divided into "Upper Campus" and "Lower Campus," and the addition of some high mass lighting coupled with internal adjustments to correctional practices would make co-corrections at Alderson a practical alternative. 36

Secure, Single-Sex Unit. The third component of Option 1 recognizes that there are female offenders who require a secure single-sex facility. 37

Adjacent to FCI, Lexington, Kentucky, is the Addiction Research Center (ARC), operated by the National Institute of Mental Health. The ARC initially had a capacity for 200 beds, fifty of which recently have been converted to a psychiatric unit for Federal female offenders. The one hundred fifty beds remaining in the ARC could be converted to a single-sex secure facility. Because of its proximity to FCI, Lexington, and the existing female population presently at Lexington, programming needs for women could be met with minimal effort. Such a facility would provide appropriate security needs for those women offenders requiring a secure correctional facility. 38

The total proposal, then, encompasses changing the planned mission of 39 the Danbury camp and converting it to an all female camp (100 beds), changing Alderson's mission to a co-correctional facility (250 male), and using the 150 bed facility presently occupied by the National Institute for Mental Health at the Addiction Research Center in Lexington, Kentucky, for a secure single-sex unit for women.

40 Possible deterrents to this proposal include the transition of the ARC facility from NIMH to the Bureau of Prisons. The ARC will be available by January 1982, barring any major construction delays on the new NIMH facility in Baltimore. Secondly, the Federal Prison System would have to provide the necessary staff and other resources in order to activate the ARC as a Bureau of Prisons facility. This change would entail some relocation of staff and some hiring at Lexington.

41 Because of the nature of the population at this proposed secure unit, the inmates themselves could not mix regularly with the FCI, Lexington population. As a result, such services as Education, UNICOR and Food Service generally would have to be provided within the new unit itself. Custodial supervision would need to be increased to meet the security requirements of these offenders. Because comparable programs already exist at FCI, Lexington, staff and program resources could be drawn from the main institution, but not without some additional cost.

Analysis of Option 1:

- *Accessibility for Visitors:* Improved.
- *Proximity to Release Destination:* Overall, an improvement over existing conditions.
- *Level of Programs and Services:* No significant change.
- *Appropriate Security:* The majority of the women will be placed in more appropriate security level facilities.
- *Capital Costs:* $0.2 million. It is assumed that the conversions at Alderson and Danbury would entail no capital costs.
- *Operating Costs:* $7.5 million.
- *Per Capita Costs:* $27.49 per inmate per day.
- *Physical Capacity:* No net increase in female beds and an increase of 150 male Level 1 beds.
- *Personnel:* 274 positions.
- *Feasibility:* This option is a workable one.

Option 2

Continued operation of FCI, Alderson at its current status with a long-range plan for the addition of a female camp.

42 In the evaluation of the future use of FCI, Alderson, one option that must be considered is the continued use of FCI, Alderson as a Federal correctional facility for women. Admittedly, this action by itself would not address many of the issues raised concerning Alderson, but if it were coupled with the establish

ment of a female Federal Prison Camp in the Northeast corridor, it would provide a facility for women of the lower level security somewhat closer to the legal residence of many of the offenders at Alderson.

Three approaches could be used to acquire a camp for women: (1) construct 43
a new camp; (2) acquire military surplus property and renovate it into a camp; or (3) use an existing or planned satellite camp (such as the planned camp at Danbury) for women. Construction of a new camp would be prohibitive ($3.8 million for 100 beds). Moreover, the Bureau has not been able to acquire a suitable surplus property site in the Northeast for a camp. The use of the planned Danbury camp appears to be the most viable option, though it would remove the badly needed male security Level 1 beds from use by the Northeast.

Analysis of Option 2:

- *Accessibility for Visitors:* No improvement for inmates at Alderson, with some improvement for inmates assigned to the camp.
- *Proximity to Release Destination:* Those who could be housed in the camp, for the most part, would be closer to home.
- *Level of Programs and Services:* The activation of a camp would improve programming in some cases and decrease it in others, for a net overall effect of no improvement.
- *Appropriate Security:* Overall, more women would be in the appropriate security level, once the camp is activated.
- *Capital Costs:* No capital costs.
- *Operating Costs:* $6.7 million.
- *Per Capita Costs:* $31.01 per inmate per day.
- *Physical Capacity:* An increase of 100 female beds and a decrease of 100 male beds.
- *Personnel:* 247 positions.
- *Feasibility:* Feasible but not practical considering its effects on physical capacity.

Option 3

Close FCI, Alderson and construct a 500-bed replacement facility.

To construct a new 500-bed facility to replace FCI, Alderson would cost 44
in the neighborhood of $23 million. The facility should be multi-level (i.e., part secure and part camp) and ideally should be located in the Northeast corridor. The current climate of fiscal restraint and the difficulty of locating a prison site close to urban areas where it is accepted by the community make this option highly unlikely. Nonetheless, analysis of this option reveals the following points.

Analysis of Option 3:

- *Accessibility for Visitors:* Assuming that the location of such a facility would be more centrally located for inmate visitors of the Northeast Region, opportunities for visiting could improve.

- *Proximity to Release Destination:* A new facility in the Northeast Corridor would place a large number of inmates closer to their release destination.
- *Level of Programs and Services:* All levels of programming could be at least as good as, if not better than, the programs currently offered at FCI, Alderson.
- *Appropriate Security:* A secure replacement facility with a satellite camp would improve the assignment of inmates to the appropriate security level.
- *Capital Costs:* $23 million. If capital costs were amortized over 30 years at 13 percent it would represent an additional 0.9 million in yearly operating costs.
- *Operating Costs:* $5.9 million, or $6.8 million with capital costs amortized.
- *Per Capita Costs:* $32.43 per inmate per day.
- *Physical Capacity:* No change.
- *Personnel:* 217 positions.
- *Feasibility:* Not feasible given the current fiscal climate and difficulty of locating prison sites in urban areas.

Option 4

Increase the use of Community Treatment Centers.

One way to place offenders closer to home in less restrictive environments 45
would be to increase the use of Community Treatment Centers (CTCs). However, the Bureau's current goal is to have offenders remain an average of 120 days. Presently, those women who have participated in CTC placement stay an average of 138 days, far in excess of the current goal. In addition, 58 percent of the releasees at Alderson participate in CTC programs compared to 47 percent Bureau-wide. It was the opinion of the Task Force, in view of existing goals and the above facts, that further expansion of CTC placement for women offenders is not feasible.

Analysis of Option 4:

- *Accessibility for Visitors:* Would increase visiting opportunities.
- *Proximity to Release Destination:* Would place women closer to their release destination.
- *Level of Programs and Services:* Overall decrease in availability of programs and increase in the availability of community services.
- *Appropriate Security:* Improved assignment by security level.
- *Capital Costs:* No capital costs.
- *Operating Costs:* $7.0 million.
- *Per Capita Costs:* $32.35 per inmate per day.
- *Physical Capacity:* Would result in a surplus of beds for women at current population levels.
- *Personnel:* 230 positions.
- *Feasibility:* Not feasible.

Option 5

Activation of ARC, Lexington. Acquisition of surplus camp and the conversion of FCI, Alderson to a co-correctional facility.

The issues presented in Option 1 are applicable here, except for the acquisition of surplus property for a Federal Prison Camp. Under this proposal, the Bureau of Prisons would attempt to locate an appropriate camp in the Northeast among properties regarded as surplus by other agencies. 46

The Bureau has maintained a continuing check of government surplus property over the last several years and has recently been successful in establishing Federal Prison Camps in Boron, California and Big Spring, Texas. Such an approach has saved the taxpayers millions of dollars in construction costs and has provided the Bureau with badly needed bedspace. Unfortunately, the Bureau has been unable to locate suitable surplus property in the Northeast. Of the four sites reviewed recently by the Bureau, two were as isolated as Alderson, one was objected to by the Navy, and the Bureau was advised by a congressional representative that the other would not have community support and in fact would be strongly opposed. At this time, therefore, the possibility of acquiring surplus property for a female camp in the Northeast does not appear likely. 47

The analysis of this option would be essentially the same as Option 1, except as it pertains to the acquisition of surplus property for a Federal Prison Camp. 48

Analysis of Option 5:

- *Accessibility of Visitors:* Improved visiting opportunities.
- *Proximity to Release Destination:* Women would be housed closer to their release destination.
- *Level of Programs and Services:* Same as Option 1.
- *Appropriate Security:* Same as Option 1.
- *Capital Costs:* $1.0 million.
- *Operating Costs:* $9.5 million.
- *Per Capita Costs:* $29.05 per inmate per day.
- *Physical Capacity:* Increase of 350 Level 1 beds for males.
- *Personnel:* 356 positions.
- *Feasibility:* Not feasible because of the unavailability of surplus camps.

Option 6

Make no major changes at FCI, Alderson but increase the use of State placements for women.

At present, 85 women convicted of Federal crimes, approximately 6% of the female prison population, are housed in State facilities. The development of contracts with the individual States would, in most cases, allow women to be incarcerated in facilities that are closer to their intended release destination. This plan would eliminate the need for any Federal prisons for women and 49 place the issue of incarcerated women in the hands of the States. Although this would place women closer to their intended release destination, the option is not feasible because requests by the Northeast office of the Federal Prison System to transfer Federal female prisoners to State custody are more frequently denied than accepted.

State Request Was Made to	***Number of Requests***	***Accepted or Denied***
Nevada	4	Three were accepted, and one was denied because the female was in need of psychiatric care which the State facility was unable to provide.
Oregon	1	Denied for lack of an adequate State psychiatric program.
Connecticut	1	Denied because the inmate had adjustment problems in the Federal system.
Washington	1	Denied because the inmate had adjustment problems in the Federal system.

All of the States in the Northeast Region, except New Hampshire, have a 50
facility to house female offenders; however, each State has only *one* facility and generally these are overcrowded. In addition, the facilities in Maine, Massachusetts, New Jersey, and Vermont house both male and female offenders and therefore are not suitable for inmates who are unable to be placed in a co-correctional program. Furthermore, the facility in the District of Columbia is only a detention center which houses short-term male and female detainees. As in the Northeast, other States do not have space for Federal female offenders; consequently, this option is not feasible.

Analysis of Option 6:

- *Accessibility of Visitors:* Would increase visiting opportunities.
- *Proximity to Release Destination:* Would place offenders closer to their release destination.
- *Level of Programs and Services:* Given the generally overcrowded conditions in State women's institutions and their reduced level of funding and other resources, it is unlikely that Federal female offenders would experience any improvements in program offerings.
- *Appropriate Security:* The types and security levels of contract facilities would vary from State to State, but in many cases only one secure, single-sex facility is available for women. There would be no improvement over the security provided at FCI, Alderson and it is highly likely that many State facilities would actually be more secure than FCI, Alderson.
- *Capital Costs:* No capital costs.
- *Operating Costs:* $7.0 million.
- *Per Capita Costs:* $32.35 per inmate per day.

- *Physical Capacity:* Would result in an excess of beds for women at current population levels.
- *Personnel:* 230 positions.
- *Feasibility:* Not feasible.

SUMMARY AND RECOMMENDATIONS

Table 1 includes a summary analysis of the options reviewed in this report. 51
The Task Force rank ordered the options according to their viability, and concluded Option 1 was the most viable. It offered the best possible solution to the issues and problems at FCI, Alderson identified in this report.

Table 1. SUMMARY OF OPTIONS

Options	Accessibility for Visitors	Proximity to Release Destination	Level of Programs & Services	Appropriate Security	Capital Costs	Operating costs†	Per Capita Costs	Inc./Dec.* Physical Capacity M	Inc./Dec.* Physical Capacity F	Personnel**	Feasibility
1. ARC/BOP Camp/Co-Corr. Alderson	improved	closer for more	better for most, worse for some	yes	$0.2M	$7.5M (0.8M)	$27.99	+150	0	274 (30)	yes
2. Continue present status Alderson with Camp Dan.	same	same	same	no	$0	$6.7M (0.1M)	$31.01	−100	+100	247 (3)	yes
3. Close Alderson build 500 bed facility	improved for most inmates	closer for most inmates	same or better	yes	$26M	$5.9M (−0.3M)	$32.43	0	0	217 (−13)	no
4. No change at Alderson but expand CTC utilization	improved	closer for a few	worse, minimal programs other than work	yes	$0	$7.0M (0.8M)	$32.35	0	0	230 (0)	no
5. ARC/Surplus Camp/Co-Corr. Alderson	improved	closer for more	better for most, worse for some	yes	$1.0M	$9.5M (3.1M)	$29.05	+350	+50	356 (112)	no
6. No change at Alderson but increase state placements	improved	closer for a few	worse	probably not	$0	$7.0M (0.8M)	$32.35	0	0	230 (0)	no

* Represents an increase or decrease over existing physical capacity, which includes SC, Danbury.
** Figures in brackets ( ) represent increase or decrease over current base.
† Figures in brackets ( ) represent increase or decrease over current base.

QUESTIONS FOR DISCUSSION

1. All audiences for this study will read the executive summary. How does the summary assist a reader in grasping the main points of the study quickly and accurately. What kinds of information included in the study are omitted from the summary? Why?
2. What is the purpose of the following sections of this feasibility study: (a) introduction, (b) background, (c) objectives of the study? What specific questions do these parts answer and for what readers?
3. Why are the special issues discussed before the options are evaluated? What types of information does the issues section give the reader? Why is there so much emphasis on the D.C. inmates? What audience(s) would require this information?
4. What criteria does the study use to judge the six options? Where are these criteria listed for the reader?.
5. Option one "consists of a three-part proposal." What are the parts of this option? How does the discussion of this option resemble a proposal?
6. How does the study use information about one option to evaluate others?
7. Could the study have been organized in any other way? If no, why not? If yes, how?
8. What are the advantages (or disadvantages) in using a table for the conclusion (summary) section? Will the reader feel cheated by not having this information listed in paragraphs?

SUGGESTIONS FOR WRITING

1. Assume that you work for the Bureau of Prisons. Write a letter (see Chapter 12 on correspondence) to three of the following audiences summarizing the research and recommendations of this study. Make sure that you include appropriate details, explanations, and emphases for these audiences.
   a. the director of education at Alderson
   b. the executive director of a penal reform group
   c. the Mayor of Washington, D.C.
   d. the Mayor of Danbury, Connecticut
   e. the warden of the Lexington, Kentucky, facility
   f. a member of the staff of the Congressperson representing the district in which Alderson is located
   g. the Chair of the House Ways and Means Committee
   h. the editor of a Washington, D.C., newspaper
2. Write a feasibility study on a subject your instructor chooses or one of your own choice. Include the following sections in your report: an abstract, an introduction with objectives, analysis of at least three options similar in technique and format to the Alderson study, a conclusion, and a recommendation. Specify your intended audience.

chapter 10

Progress and Performance Reports

Progress and performance reports are documents essential to most operations in business and industry. Both types of reports supply the reader with information about accomplishments as well as problems that have occurred during a specific period of time in the life of a project. Progress reports are usually written before a job is completed; performance reports are submitted after things are in operation. Such reports can be organized in a variety of ways—by task, by function, or by chronology. Whatever method of organization you use, information needs to be organized logically so the reader can see what exactly has occurred, why, and how. The techniques discussed in Chapter 2 (Description) and Chapter 5 (Process Explanation) are especially useful in writing progress and performance reports.

PROGRESS REPORTS

A progress report gives an account of the type and amount of work completed during one phase of a job. Since many jobs in business and industry are not completed in one phase, preparing progress reports may be an important part of your written work. For example, major work done for the federal government must, according to contractual agreement, include regularly scheduled progress reports. A progress report can be submitted on almost any kind of ongoing activity—laboratory research on flame retardant materials, construction of a bridge, installation of a solar heater, or even a new employees' training program. Depending on the size of the project, progress reports can be submitted weekly, monthly, quarterly, or biannually. A company embarking on a long project will prepare a chart, similar to the one below, showing milestones, or points where major parts of the job are expected to be completed.

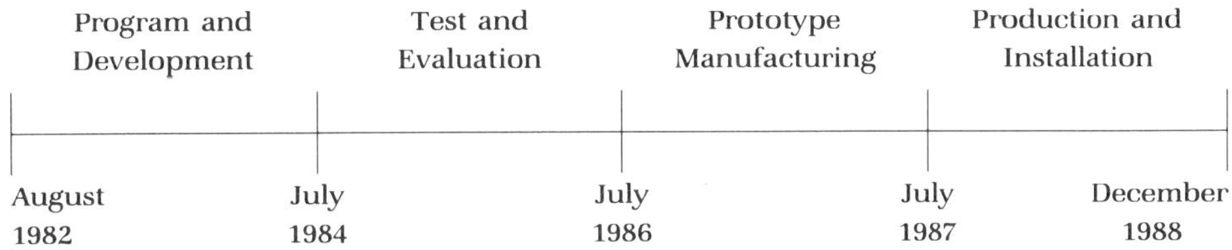

Progress reports will be due as each milestone is reached and at many points in between. For example, in developing the prototype machine upon which every assembly-line model will be based, a company will expect progress reports on materials, testing, and safety features. Depending on the progress the company makes, these milestones can be compressed, when a phase is completed ahead of schedule, or expanded, when a phase takes longer than planned.

When a project is completed, you may have to write a formal, final progress report for management, summing up the work done on the job, evaluating its successes and failings. The progress reports you have already written will often provide the information needed to write that final report. Finally, progress reports might be considered "hindsight documents" when you plan a new project, for they can point to the problems an earlier job presented and suggest ways you might avoid future difficulties.

Audience

Progress reports are generally written for decision makers at your own company or at the client's. In some instances, the decision makers may also be experts or, in the case of a complex job, may ask experts for comments on a progress report. One large electronics plant we know of requires that a brief progress memo be written for executives each month on every major project at the company. Since executives and clients are rarely present at the actual work site (be it laboratory, plant, oil field, or training center), they need details about the work in order to evaluate it. A written summary lets them know if things are progressing satisfactorily or, if not, helps them decide what changes need to be made. They may need to budget more funds for materials or personnel, slow a project down, accelerate it, temporarily discontinue it, or even abandon the job altogether. Progress reports may also be checked by coworkers to give them a clear sense of what has been accomplished and what remains to be done.

You can expect readers to raise the following questions as they examine your progress reports:

1. What is the job?
2. What is the time period?

3. What has been accomplished so far?
4. What specific work was completed in this time period?
5. Were there any problems?
6. Are we on schedule?
7. Are we within budget?
8. What will be done next?
9. What remains to be done?

Content

In order to answer the reader's questions posed above, be sure your report contains the following information:

1. *The type of job you are reporting on.* Since managers may be supervising many ongoing projects, identify yours clearly by name, contract number, location, authority for the work, or even code name.

2. *The period of time your report covers.* Provide beginning and ending dates as well as the specific sequence of time your report covers—e.g., second of five reports. Do not just indicate "third quarter."

3. *A quick rundown of what's happened so far with this project.* Supply the reader with a brief summary of what's been done since the project started, making clear the connection with your current report. By doing this you will spare the reader the inconvenience of having to go to the files to determine what activities immediately preceded your work.

4. *The work done during this reporting period.* Define the exact tasks accomplished. Do not dwell on descriptions of routine procedures or include unnecessary technical documentation. Keep in mind that many of your readers—business executives, clients—may not have your expertise and will not need every technical detail to get a clear picture of what's been done. Rather than describing each test in detail, point out how many tests were conducted and what conclusions were reached. Of course, if you are testing a new procedure, and *its* progress is important to the reader, then by all means describe the major parts of that test.

5. *Any problems encountered.* Not every progress report brings good news. You may have encountered trouble in one or more phases of a project. What do you say without getting fired? First and most important, inform your audience about the problem so that they can make necessary modifications in scheduling, budgeting, or personnel. In informing them, classify the problem as major or minor. Give the readers an explanation of what you did about the problem; show them that you took responsible, decisive action. Where possible, offer a

solution to the problem, even if it is temporary. Bad news is easier for a reader to bear if you indicate that the likelihood of the same problem happening again will be eliminated or reduced. Ideally, try to be guardedly optimistic without misrepresenting the facts. Also, if you describe a problem in one progress report, refer to it in a subsequent report, even if it was solved in the interim. Do not leave the reader guessing about what happened.

6. *Schedules.* Again, do not leave the reader wondering. State exactly why you are or are not on schedule. If there was a delay, give the reason and specify how long it was. The reader needs this information to know if more time will have to be scheduled during the next stage to correct the situation or to meet a contract deadline.

7. *The budget.* The budget is among the most important considerations for the decision maker. Give the source of any overruns—more equipment was needed; weather delayed construction and idled workers; estimates were too low. Tables or graphs are useful when listing amounts in dollars and cents.

8. *Work planned for the next reporting period.* Many progress reports call this section "Work Forecast," thereby suggesting the tentative schedule to be followed. In it, spell out what work is planned next.

9. *Work planned for the rest of the project.* Again, the readers will expect a general overview of major tasks to be completed. An unnecessarily detailed account of procedures, or equipment functions, will not help readers. In planning for the future they will be more concerned with deadlines, budgets, and major accomplishments.

Format and Organization

The format and organization of your progress report will, of course, depend on the size of the job you are doing and on the facts you must emphasize. A report on a small job can be in memo format and will require fewer parts and less commentary than a formal report on a complex project. Any reader of a progress report, however, will want to have the most important data first. Looking at information in the order of its importance, the reader will first want to know *what's been done;* second, *why and how;* and third, *what's next.* A long progress report, resembling a formal report in both scope and format, tells the reader about the most important details in abstract and summary sections. A shorter report in memo format will focus on the *what* in the first paragraph or two. Whether long or short, your progress report should begin by informing the reader if you have accomplished a major task, if you are on schedule, if you have stayed within the budget, and if you have encountered any problems.

Two Ways of Organizing a Progress Report

Following are two ways of organizing a progress report. The first outline is suitable for a brief report and the second for a more detailed one.

APPROPRIATE OUTLINE FOR A SHORT PROGRESS REPORT

I. Overall achievement of project objectives so far
II. Work completed to date
III. Work done this period
IV. Work left to do

APPROPRIATE OUTLINE FOR A LONG, MORE COMPLEX PROGRESS REPORT

I. Abstract
II. Objectives and scope of the project
III. Summary of progress to date
IV. Detailed description of progress this reporting period
V. Work planned for the next reporting period
VI. Work planned for the rest of the project
VII. Conclusions
VIII. Literature cited

Number IV in the outline for a complex report might be organized by task or by topic. If by task, you should number each task, even if it occurs simultaneously. Subcategories under each task can be devoted to "Work accomplished" and "Work forecasted," as the following example shows:

IV. DETAILED DESCRIPTION OF TECHNICAL PROGRESS THIS REPORTING PERIOD

If you organize the body of your report by topic, you might do it this way:

IV. DETAILED DESCRIPTION OF TECHNICAL PROGRESS THIS REPORTING PERIOD
   A. Research
   B. Equipment/Material
   C. Personnel
   D. Problems
   E. Schedule
   F. Budget

Each of the topics here would receive comment under appropriate tasks in the task-organized report.

A CHECKLIST FOR WRITING A PROGRESS REPORT

1. Find out whether your reader has any special requirements (time, format, etc.) or concerns (schedule, cost, personnel, etc.).
2. Decide upon an appropriate principle of organization.
3. Identify the job clearly.
4. Specify the precise time period covered in the report.
5. Summarize the total work done so far on the job.
6. Give the detailed account of the work done in the specified period.
7. Mention any problems (schedule, personnel, equipment, budget, procedures) and what was done to solve them or prevent them from recurring.
8. Alert the reader to what's scheduled next on the job.

PERFORMANCE REPORTS

Like progress reports, performance reports (also called *evaluation reports* or *activities reports*) cover a specific period of time and give information about accomplishments as well as problems. The difference is that performance reports either evaluate projects that are already complete and in operation, or report on activities (sales, demographic changes, ecological changes) that may continue indefinitely and are not always easily controlled or predicted. Progress reports will end when a job is finished; then performance reports take over to assess operations or the effectiveness of a project. Conclusions offered in a progress report may sometimes be tentative, for exact outcomes (completion dates, results of tests, compatibility of assembly-line models to prototype) in many instances may still be unknown. Performance reports, however, provide final data about each period they cover.

Readers of a performance report are most likely to ask the following questions:

1. What's being reported on?
2. What period does the report cover?

3. Did equipment, systems, processes, personnel, or markets operate as planned?
4. Were there any problems? What caused them? Have they been corrected?
5. How have any modifications made after the last report affected performance this period?
6. What are the most significant changes during this period as compared with last period?
7. What changes, if any, should be made to maintain or improve performance during the next report period?

Audience

Performance reports can be written for all kinds of readers, both within your company and for clients. Engineers may examine them to evaluate the effectiveness of a system or a piece of equipment. Experts look at performance reports to see whether a model conceived in the laboratory is working out as planned. Performance reports let them know if it is necessary to introduce new parts, modify old ones, or phase out a particular feature or operation. Operators may read performance reports to learn more about how a piece of equipment is assembled and to be aware of malfunctions that might recur on the assembly line or in the shop. A well written performance report will alert them to precautions they must take. Decision makers study performance reports closely to determine if equipment, personnel, or projects are being managed economically and productively. These readers will want to know if productivity has been sustained from one report period to another, if it has dropped, or if it has increased. If there were problems, a performance report should tell decision makers how serious and costly they were. Cost overruns and increases in personnel need to be documented and explained.

Content

Performance reports may include the following information:

1. A description of a system, piece of equipment, market, or product in that market, and, possibly, an overview of performance history.
2. An account of the actual performance of a system, piece of equipment, product, or market during the specific report period. When the performance of machinery is being assessed, the reader will want facts, for example, about scheduled and nonscheduled maintenance.
3. An analysis of the performance of the system, market, or product based upon a comparison with the figures and results from the last reporting period.
4. Conclusions offering detailed recommendations to continue or improve performance over the next reporting period.

Format and Organization

As with progress reports, the format and organization of performance reports vary widely. A memo or letter format is suitable for brief reports, whereas the structure and parts of a long report are necessary for complex reports.

Performance reports can be organized according to types of activity, chronology, or topics. Performance reports should always begin with an overall assessment of the operation for the specific period of time covered. At the beginning of the report, specifying what topics the report will discuss not only summarizes events for the reader but also helps him or her follow the report. For longer projects, you might organize your report according to the major types of systems or activities you are reviewing. Wherever helpful to your reader, compare and contrast performance among systems or events. Another way to organize a performance report is by chronology. Following this plan, you provide an account of activities in sequence. To avoid turning your performance reports into a diary or log book, however, do not list every event that occurred each day or week. Some days will be uneventful or the jobs being done will be routine, and providing information about such details is unnecessary and time consuming. Instead, carefully summarize the major occurrences—achievements and problems—and interpret their significance for the reader.

A CHECKLIST FOR WRITING A PERFORMANCE REPORT

1. Determine the requirements and needs of your audience.
2. Select a principle of organization appropriate to your topic.
3. Identify the topic of the report.
4. Describe its performance in detail (personnel, equipment, procedures, sales or labor market, program).
5. Assess performance relative to past and future performance.
6. Identify problems and how they affected the performance of a system, piece of equipment, program, or policy.
7. Conclude with any appropriate recommendations.

PROGRESS REPORTS

Big Harbor Memorial Hospital, a medium-size health care facility, has decided to establish Phase III of a cardiac rehabilitation program for post-MI (myocardial infarction, or heart attack) patients. The model below is the initial progress report sent to the hospital administrator informing her what an appointed committee has done during the first four months of their work. The report begins with a short, useful summary telling the administrator why the report is being sent and alerting her to some potential problems. A background section refreshes the reader's memory about the history of the program. The committee's major accomplishments are then listed in chronological sequence by numbers for easy reference. The report concludes with a section on work planned for the next period and a date by which another report will be sent to the administrator.

Progress Report #1 of Phase III of a Cardiac Rehabilitation Program

Lee Young

BIG HARBOR MEMORIAL HOSPITAL

BIG HARBOR, CALIFORNIA 95738

TO: T. R. Dubrowski
Administrator

FROM: Lee Young, L.Y.
Chair, Phase III Committee

RE: Progress Report #1 on Phase III
Cardiac Rehab Program

DATE: April 2, 1983

Objective and Scope

This is the first of three progress reports you will receive on the establishment of Phase III (outpatient exercise program) of the Cardiac Rehab Program. To date, we are on schedule, but have two potential budget problems to report, discussed in numbers (3) and (5) of the "Work Completed" section. 1

Background

Two years ago, Big Harbor Memorial Hospital (BHMH) instituted Phase I (acute) and Phase II (convalescent) of the Cardiac Rehabilitation Program for all post-MI patients. In the last 19 months, the Cardiac Rehab Unit has received more than 125 inquiries about a monitored exercise program available to MI patients after discharge. At the rehab meeting on November 22, 1982, we decided, with BHMH's approval, to establish Phase III of an outpatient exercise program with an October 15 completion date. The Phase III Committee was appointed with representatives from Medicine, Nursing, Physical Therapy, and Social Work. The Committee started work on December 2, 1982. 2

Work Completed During This First Report Period (December 2, 1982–March 30, 1983)

The following tasks have been completed during this period: 3

T. R. Dubrowski
April 2, 1983
Page 2

1. The committee formulated a set of guidelines for patient admission to Phase III which were approved by the House Physicians Staff on December 17.

2. During the week of January 7-13, the committee conducted a preliminary investigation to find an appropriate off-site location for the program by studying the seven public gyms in the Big Harbor area. The directors of each facility were interviewed by telephone to find out about space, schedules, and, most important, availability for Phase III patients. After these interviews, the Committee narrowed the choices to Taylor University, Crosby County Junior College, Hastings High School, and the Big Harbor YWCA. After an inspection of these facilities, we chose the YWCA based upon its proximity to BHMH (.5 miles), its availability (7-9 a.m., M-W-F), and its layout. The facility's exercise room and track are excellent. In addition, it was the only gym with storage area available for our equipment.

3. On January 28, the committee received bids on the two EKG machines and the two treadmills to be used in the program from Medico, Inc. and Big Harbor Hospital Supply Corporation. Members of the committee thought both bids were high—$8,250 from Medico and $9,500 from Big Harbor Supply. The only other local firm, Armco, refused to bid on our job because it was too small. Consequently, we will have to ask for bids from other firms outside the Big Harbor area which may delay awarding a contract. Moreover, if these out-of-town bids are too high, we might be forced to accept the Medico or Big Harbor Supply offer and would then have to adjust our budget to cover the price of this equipment.

4. On February 13-15, Dr. Mary Phillips, Dan Jones, R.N., and I attended a seminar at State University Hospital on "Goals and Objectives for Establishing an Outpatient Cardiac Rehabilitation Program." Dr. Granville Richs, the Director of the University Hospital's Rehab Program, gave us a helpful history of their facility and outlined some of the problems we might encounter as we enter Phase III --particularly patient recidivism.

T. R. Dubrowski
April 2, 1983
Page 3

5. On March 25, the Committee submitted a grant proposal to the National Community Health Bureau in Washington applying for $47,500 in federal funds to match money put in escrow by BHMH to establish the Phase III program. We met the Bureau's March 30 deadline and expect to hear from them by June 1. Approximately 60 percent of the committee's work for this period was devoted to researching and drafting this proposal. Problems in starting the program by October 1 may arise if our proposal is not funded, or if the Bureau wants us to make extensive revisions and resubmit the proposal. BHMH currently lacks the resources to sponsor the Cardiac Rehab Program alone.

Work Planned for the Next Report Period (April 1 to July 1, 1983)

In the next three months, the Committee plans to accomplish 4
the following tasks:

1. Outline the exercise program to be offered.

2. Prepare a brochure explaining the program to patients.

3. Begin an in-hospital seminar for interested staff.

4. Begin writing the final proposal, assuming our preliminary grant proposal is accepted. If it is not, the committee will apply to a private funding agency or another government one.

5. Award the contract to an equipment vendor.

Our next report will be sent to you by July 5. 5

QUESTIONS FOR DISCUSSION

1. What specific types of information about the Cardiac Rehab Program would Dubrowski need? Why? Identify details in the report that are appropriate for this reader. Who else at Big Harbor Memorial Hospital would read Young's report? Has Young prepared his report with those readers in mind?
2. How does the report describe progress on the program in language appropriate for the reader? Give examples from the report where the writer has adapted technical information for the reader. How might this report be written if the primary audience were a nurse or physician?
3. How does Young explain the problems for the reader without minimizing or exaggerating them?
4. How does the report provide continuity for the reader between each section?
5. What kinds of information contained in this report might not be included in subsequent reports?

SUGGESTIONS FOR WRITING

1. Assume you are a member of Young's committee asked to write the evaluation of the locations cited in point (2) of the progress report. Submit a one-page memo to Young assessing these locations so that the committee could include the findings in this progress report.
2. Write an initial progress report to a manager or supervisor on one of the following topics:
   a. preparing a proposal in response to an RFP
   b. landscaping an office or apartment complex
   c. remodeling a store or area of a factory or laboratory
   d. building a road
   e. changing from conventional to solar heating in an office
   f. installing new safety equipment in a plant or store
   g. designing or testing a new model "X"

Like the previous model, the next progress report is written in memo format. But this model will show you how to write a progress report to an instructor on a project that is nearing completion. It was submitted by a physics major to an instructor after the student had completed 80 percent of his work on a junior-senior research project. The student's purpose in writing the report was to document the work he had done during the last two and one-half months and to assure the instructor that the project would be completed on time. Not everything the student did on the project merits inclusion in the report. He has been selective and has included only important details about equipment, research trips, tests, and problems. From this information, the instructor has a clear idea of the progress the student has made and is in a better position to offer assistance and advice.

A Memo to a Physics Instructor on the Progress of a Student Research Project

Gordon Hall

TO: Dr. Tim Franklin
Department of Physics

FROM: Gordon Hall G.H.
Senior Physics Major

DATE: November 20, 1984

RE: Progress Report #3 on Rotation of Density-Stratified Fluids Project

This report will bring you up to date about my research project for 1983–1984. It is the third report you have received and covers the period of September 10 to November 20, 1983. My work is progressing satisfactorily and the few problems I encountered this period were easily solved. 1

Background and Summary of Work Completed June 28 to September 9, 1984

On June 28, 1983 you tentatively approved my plan to develop an experimental test of the model for the motion of a linearly density-stratified fluid when rotating in a cylindrical container. After I completed my literature review and submitted a preliminary proposal on July 30, you approved my project. I then made preliminary designs for the apparatus and submitted questions concerning the problems with the experimental tank by August 10. By August 28, I had built a device out of plexiglass that rotates a cylindrical container. I also modified a Westphal balance in order to measure the density of the fluids. The dimensions of this balance were recorded in the log book. 2

Work Completed September 10 to November 20, 1984

My objective for this period was to set up the experimental equipment. I received a used Graham variable transmission, model number BD4MW60, from Central State University on September 13. However, the transmission needed repairs, and the service manual and parts list requested from Graham Transmission did not arrive until September 28, delaying repair of the transmission until October 2. The dimensions of the transmission and the rotation apparatus were recorded so that the equipment could be assembled according to specifications. I drilled a gear to drive the belt leading to the rotation apparatus so that the gear would fit the axle of the transmission. Also, I constructed a wooden stand 3

Dr. Tim Franklin
November 20, 1984
Page 2

(10 x 30cm) for the rotation device so that the two gears would be at the same level.

From October 5 to 7, I attended the Fluid Dynamics Research Workshop of the American Physical Society at Technical State University. While there I observed a number of experiments which helped clarify some methods I could use to obtain my objectives. On October 20, I conferred with Dr. M. L. Bathus, Research Advisor for the American Institute of Physics, about my research activities, especially using the "slip-under" technique in a rotating container. 4

This "slip-under" technique--employing two lab stands, a wooden plank, 2 three-liter beakers, and tubing--was used to create linearly density-stratified fluids. A variable speed motor, number NLS-101453, drove the stirrer. A five-molar NaCl solution was mixed as the original fluid to be combined with distilled water. I used the collection container from the actual experiment. A diagram of the physical set up is recorded on page 63 of the log book. 5

Preparations were begun on November 13 to use the Westphal balance for measuring the density at each level in the stratified fluid. The riders for the balance were labeled and recorded in the log book. The number and position of riders on the Westphal balance were determined both with and without the aluminum rod. 6

The first attempt to construct the fluid successfully was hindered by my placing the original fluids in the wrong container on November 15. On November 16, my repeated attempt was successful. (A more suitable fluid for experimentation will be constructed.) The density was measured as a function of depth by noting that the amount of downward force exerted by the riders needed to offset the upward force exerted by the fluid. This force is proportional to the density by a factor equal to the volume of the aluminum rod. The mass of each rider was later determined, using a Mettler H30 analytical balance, number NLS-11916. 7

My preliminary report on the theory of the motion of a linearly density-stratified fluid is now approximately 80 percent complete. 8

Dr. Tim Franklin
November 20, 1984
Page 3

Projected Activities for the Next Period, November 21, 1984 to February 14, 1985

During the next report period I intend to complete the following: 9

1. A computer program calculating the density of the fluid with respect to depth.

2. Photography techniques evaluating the significance of the data and aiding presentation of observations to colleagues.

3. My report on the theory, carefully adjusted to fit the data.

QUESTIONS FOR DISCUSSION

1. Evaluate Hall's first paragraph. What information does it give the reader? Could it have been deleted?
2. Why does Hall include a "Background and Summary" section? Wouldn't Dr. Franklin already know these details? How would Dr. Franklin use the information in this paragraph?
3. How has Hall organized his information so that Dr. Franklin will be able to judge the experiment's progress?
4. Why does the student cite specific pieces of equipment by number? Why is the reader referred to a log book?
5. Why does Dr. Franklin need to know about what happened on Nov. 15 when the student corrected the problem the next day?
6. What functions does the last section of the report serve? Why are the parts of this section numbered?

SUGGESTIONS FOR WRITING

1. Write the second or third progress report on a project you are doing, or have done, for an instructor. Model your report after Gordon Hall's.
2. Write a middle or final progress report to a chairperson or dean at your school about one of the following topics: (a) a team's progress during the season; (b) funding for purchasing new equipment through alumni contributions; (c) recruiting new majors; (d) having a tract of land rezoned for experiments being conducted.

Larger projects require longer and more detailed progress reports than the two preceding models. The following model is a long progress report that assesses the many different activities occurring during the construction phase of a pulp mill owned by an international corporation. Based upon shorter reports submitted by section chiefs in construction, transportation, engineering, etc., this report was written by the head of the Leaf River Forest Products, Inc. to his superior at the parent company, Great Northern Nekoosa Corporation. The firm Brown & Root referred to in the report is an independent contractor whom Leaf River hired for this construction job. The report contains detailed information about weather conditions, equipment, personnel, contracts, site inspections, as well as problems and delays. In addition to illustrating the organization of a long progress report, this model provides a marvelous insight into how a business is run and interacts with the government and other businesses.

The Progress Report for December 1982 on the Construction of a Pulp Mill for the Leaf River Corporation

V. Mattson

Leaf River Forest Products, Inc.
a company of
Great Northern Nekoosa Corporation

Inter-Office Communication

TO: R. Hellendale

FROM: V. Mattson

RE: Leaf River Progress Report/December 1982

DATE: January 11, 1983

1 In spite of unprecedented amounts of rain and threats of flooding in December, most phases of the pulp mill construction progressed on schedule. Brown & Root has worked out many of the site drainage problems. In addition, most of the construction is now above ground and less affected by weather conditions. Work on the underground piping, the effluent lagoon, and the administration building was, however, severely hampered. Final work on the railroad was also delayed.

Construction

2 At the end of the year Brown & Root had approximately 1,034 employees and the subcontractors had approximately 135 employees on the site.

3 Although December was exceptionally wet, a number of significant construction projects were started or completed.

1. The reactor-clarifier concrete work was completed, and the mechanical components are being erected. The raw water intake caisson was sunk to a depth of about 30 feet. Several pours were made on the filter plant walls.

2. The wall pours have been completed on both primary clarifiers, and interior slabs are being poured.

3. The underground piping was completed in both secondary clarifiers, and the center well was poured in No. 1 clarifier. Interior slabs are being placed in No. 2 clarifier.

4. The footing and first section of wall were poured for the effluent lift station.

Leaf River Progress Report/December 1982
Page 2

5. Underground piping progresses in all areas with continued emphasis on fire protection lines. Testing of the underground fire lines has started.

6. Work continued on the short wood flume, and foundations were completed for the jackladder and barking drums. Foundations for the chip screen building were completed, and foundation work continues on the chip pileout and reclaim system. Chip conveyors are being erected in the chip screen building, and construction of the barking drums has begun.

7. Structural steel erection has started on the ships and stores building and is continuing on the recovery and power boiler buildings.

8. Foundation work in the turbine generator area is nearing completion. Foundations in the pulp mill area progressed well considering the wet conditions, and backfilling is underway.

9. Work continued on permanent mill roads and railroads. Track laying is expected to be completed at the south end of the site in early January to receive rail shipments.

The construction monitoring staff continues to inspect all 4
areas of construction activity.

Engineering/Organization

Personnel: Sixteen of our staff of 261, or a total of 6% of 5
the production crew, have been hired. An average of seven people are devoted to Brown & Root Engineering.

Engineering: During December the following drawings were 6
reviewed:

0	Flow Sheets	100.0% complete
3	Equipment Layouts	98.5% complete
14	Engineering Flow Sheets	93.0% complete
40	Electrical Drawings	27.0% complete
80	Electrical 11 × 17	5.0% complete
164	Instrument 11 × 17	11.0% complete
35	Piping	12.0% complete
180	Others	64.0% complete

Total engineering work is about 48% complete. 7

Leaf River Progress Report/December 1982
Page 3

Cost: During the month 11 major cost changes were approved. 8
The Brown & Root contract is now $521,598,829, compared to
the original cost of $522,000,000.

Training: Our training program has started. We are 9
collecting material and have prepared rough drafts of eight
of the proposed 310 modules. The major emphasis in training
during December was to coordinate our efforts with those of
the State of Mississippi. Eighty percent of our supervisory
training manual is now complete. Preliminary personnel and
cost figures have been sent to Arnie Nemirow.

Spare Parts and Rolling Stock: Our philosophy of keeping a 10
minimum of spare parts entails some risk. Since most of our
equipment is foreign-made, we must carry more spare parts
than a typical minimum. Our spare parts budget is $5 million
of which over $2 million is required for pulp dryer rolls
and kamyr digester replacement parts. We now know that we
cannot supply the rest of the spare parts required with only
$3 million, but we expect savings on the $4.2 million
alloted for mobile equipment and machine tools. With this
advantage we will try to keep the spare parts below the $5
million budgeted plus this savings.

Contracts

Reichhold Chemical Company is developing a firm price for a 11
tall oil plant and an ROI. Leaf River can pay for the plant
in installments or on a per ton of tall oil delivered basis.

Patchen Mingledorff has quoted a price of $814,900 for the 12
steam condensate line to the sawmill. Because this is below
Brown & Root's final quote and includes pumps and tanks not
included in Brown & Root's quote, we plan to give this
system to Patchen Mingledorff in January.

Black & Clawson quoted a laser optical scanner for their 13
merchandizer. They found that the system quoted was not
practical and changed to a less expensive system without our
approval. Since they have not agreed to charge a lower price
for the less sophisticated system, we have cancelled the
optical scanner option completely. It is possible that we
may buy the Black & Clawson equipment if they come back with
a fair price as we expect they will.

Leaf River Progress Report/December 1982
Page 4

Transportation/Railroad

We received the shocking news that an ICC review board denied our application for an operating certificate for our railroad. On January 5, 1983, we filed an appeal and have already secured affidavits or letters of support from David Smith, State Senator; Perry County Board of Supervisors; the State Economic Development Board; Reichhold Chemical Company, and Olin Corporation. The appeal emphasizes the validity of our request based on similar commission decisions and points out errors in factual judgment by the review board. Because of this documentation, we are optimistic that the review board's decision will be overturned and the certificate will be issued. 14

Railroad Contract 1-C (Queen City) construction has been delayed during December by excessive rains. All road crossings, however, have been installed, and track connections can be completed on two days' notice, assuring prompt delivery to the mill site upon receipt of cars. Completion is expected by the end of January. 15

During the month our locomotive was delivered to New Augusta and has been used to spread ballast on a portion of the completed track. An engine heater has been purchased and will be installed within the next week, thus permitting operation of the engine in cold weather. 16

We were given the opportunity to negotiate rates on woodpulp to be exported from Ashdown, Arkansas to European ports in 1983. Contract rates via the KCS from Ashdown to Baton Rouge were reduced from $19.00 to $12.00 per short ton. Unloading costs at Baton Rouge were reduced by 35¢ per ton. Ocean rates of $45.00 per metric ton to Rotterdam/Antwerp and $46.00 to other ports were obtained from Forest Lines. These rates represent a substantial reduction from those quoted during 1982. In fact, they are the lowest berth term rates published by any known carrier handling woodpulp from Gulf and South Atlantic ports to North Europe. In addition, our contract with Forest Lines includes service to all major ports in North Europe, with no minimum quantities specified. Such an arrangement gives us greater delivery flexibility than any other carrier being used by other exporters. 17

General

Our first Manitowoc 3900 crane was put into service by Brown & Root at the end of December. The second crane is expected 18

Leaf River Progress Report/December 1982
Page 5

later in January and should be in service with Brown & Root by February 1. Although these cranes have been purchased for our woodyard, they are being leased to Brown & Root during the construction period.

All of the bids on the construction of the propane system are now in and are being evaluated by Ellis Stevens. We have requested quotes for the dirt work for the road and area around the propane tanks. 19

The contract proposal from Dixie Electric, the electric power co-op serving the mill, is being analyzed by our utilities and power group here at Leaf River as well as by our legal staff in Stamford. A major problem will be to determine how much and what type of standby protection we need and wish to pay for. 20

Woodlands

Major emphasis during December has been placed on evaluation studies of the Masonite lands within a 30- and 35-mile radius of New Augusta. We have retained Zeb White to help us with this work. The 30-mile radius includes approximately 100,000 acres of high quality timberland about half of which is already in plantations. Control of this land would, of course, be an extremely valuable asset. 21

Area superintendents have visited all wood dealers who requested dealerships with Leaf River Forest Products. If 60% of the volumes can be produced by the dealer production, it would amount to 450,000 cords. By adding the contractor production and longwood supplier volumes to the above figures plus chips, we will have met the required wood needs of the new pulp mill. We will continue to monitor wood dealers, suppliers, and prices monthly for any changes. 22

Present weather conditions have increased woodyard activities. Pine sawlog prices have increased over last month. Latest USFS sales near Richton sold for $231 stumpage to Scotch Plywood at Waynesboro. Scotch Plywood Company is anticipating a market surge with a need for stumpage for second shift start up. 23

V. Mattson

V. Mattson

QUESTIONS FOR DISCUSSION

1. How does this report satisfy the audience's need for information about costs, completion dates, and the implementation of company policy?
2. Identify sections of the report where Mattson gives R. Hellendale only summary statements and where he provides more detailed explanations. When does he use summary statements as opposed to explanations? What different purposes do the two kinds of descriptions serve?
3. Where and how often does Mattson offer predictions about activities at the pulp mill?
4. This report contains good and bad news. How does Mattson report the bad news (e.g., spare parts shortages and budget, ICC review, Black & Clawson equipment supply) so that the reader knows appropriate decisions were made?
5. Explain how the format of this report helps the reader to understand and evaluate the amount and type of progress at the site.

SUGGESTIONS FOR WRITING

1. Assume you are coordinating efforts to build a community recreational facility on a 25-acre tract of land and that you must submit a progress report each month. Write a progress report midway through the project in which you supply relevant information under the following headings:
    1. Overall progress this month
    2. Site Work (including tree removal, grading work, drainage of one swamp, clearing two fields)
    3. Construction (a small office for a staff of 3; a tennis court; five pedestrian bridges; six rest stations; four pavilions; and 17 picnic set-ups—tables, benches, outdoor grills)
    4. Planting (trees, shrubs, grass, flowers)
    5. Lighting (roadway lights, tennis court lighting, building lights)
    6. Roadwork (three main roads within the facility plus four parking lots, ten gravel paths, and 30 dirt paths)

    Identify your audience. Supply appropriate visuals to help your reader understand what progress you have made. The only major problems have been with material delays and inclement weather. Explain these problems so that the reader knows how you handled the difficulties successfully.
2. Write a progress report suitable for the following project:

    A number of local businesses as well as private individuals have asked the college you work for to offer an informal, non-credit course on how to buy and use a word processor. As a member of the Data Processing Department, you were asked to develop such a course and must submit five progress reports spread over 15 weeks. Write either report number 1 or 5 for your department head who will show it to the academic vice president, the director of public relations for the college, and a number of local business executives. In your report specify the progress you have made in the following areas: (1) scheduling, (2) teaching materials, (3) equipment, (4) publicity.

PERFORMANCE REPORTS

The performance report below comes from the Crew Surgeon's report on the health of the Skylab 3 Crew. It evaluates the health status of the three-person crew from 30 days before takeoff to their postflight recovery. The writer supplies medical details about the crew's activities as they relate to their physical fitness and describes the causes, symptoms, and treatments of the crew's health problems. The report concludes with a brief comparison with earlier Skylab flights. Information contained in this report is important to a variety of readers—the medical personnel assigned to Skylab missions, NASA administrators, and the pilots themselves. The report identifies conditions that might cause trouble on future human-powered flights and thus alerts NASA officials so they can reduce, eliminate, or take precautions about them.

A Performance Report on the Health Status of the Skylab 3 Crew

No infectious diseases or other medical problems were experienced by the crew during the 30-day preflight period, the last 21 days of which included the Flight Crew Health Stabilization Program. 1

Launch and orbital insertion were nominal. Shortly after orbital insertion, the Pilot began to experience nausea; this was aggravated by head movement. One hour after insertion, the Pilot took an antimotion sickness capsule, scopolamine/dextroamphetamine sulfate, with good relief. The crew entered the Orbital Workshop 9 hours and 45 minutes after lift-off. Following strenuous work to activate the Orbital Workshop, the Pilot vomited once. During the second mission day, the Commander and Scientist Pilot also experienced some motion sickness during continued Orbital Workshop activation; they took scopolamine/dextroamphetamine sulfate, as required, for alleviation of symptoms. This indisposition caused a loss of work time during the first 3 days of flight. Two additional days elapsed before all symptoms had dissipated. Since medical experiments were not run until mission day 5, subjective voice reports by the crew were the only means of health assessment during this time. On mission day 5, after the first medical experiments were conducted, objective clinical data were available to aid in evaluating the crew's health. In general, the crew members remained in excellent health except for a few minor clinical problems and rare sporadic early terminations of the Lower Body Negative Pressure experiment by the Commander and the Scientist Pilot. 2

The Pilot reported a painless sty on the left upper eyelid on mission day 29, which responded to an ophthalmic antibiotic ointment and cleared by mission 3 day 32. On mission day 33, the Commander reported the beginning of a boil under his right arm. Instructions from the ground to the Commander were to avoid using stick-type deodorant, and wearing garments which fitted tightly under the arms. No medications were recommended and the condition cleared in about 48 hours. A recurrence of the boil in approximately the same area on mission day 50 again lasted only 48 hours, and did not require any medication.

4 The crew maintained high levels of daily exercise during the mission. Extravehicular activities were successfully completed on mission days 10, 28, and 57 without medical problems.

5 The crew slept 6 hours on the night prior to entry and were awake approximately 15 hours prior to splashdown on mission day 60. The Scientist Pilot took an antimotion sickness capsule approximately 40 minutes prior to the entry burn, while the Commander and Pilot took their antimotion sickness medication approximately 5 to 10 minutes after the burn. Prior to the burn, all three crewmen inflated their orthostatic counter measure garments. The entry was nominal. At about 20 to 30 minutes after splashdown, while still in the Command Module, the Scientist Pilot checked the pulse rate of each crewman and obtained the following values: Commander, 88; Scientist Pilot, 70; and Pilot 62 beats per minute. Pulse checks by the Crew Surgeon immediately after the Command Module was aboard the recovery ship were similar. Blood pressures were within acceptable ranges for these crewmen. All three crewmen egressed the Command Module on their own power.

6 Postflight, the cardiovascular deconditioning observed was carefully documented, but no clinically serious events occurred. As in Skylab 2, vertigo, postural instability, hyperreflexia, dry skin, and slight fissuring of the hands were noted. On recovery, a previous back strain suffered by the Commander recurred from a situation combining "lifting" and loss of balance. On recovery day the Commander developed presyncope during the stand test. The Pilot had a vagal response, also associated with presyncope, during the recovery phase of the Metabolic Activity experiment M171. The overall rate of recovery postflight was more rapid than that observed in the first manned Skylab mission.

QUESTIONS FOR DISCUSSION

1. How does this report differ from a progress report? To answer this question, identify the specific rhetorical aim of the Crew Surgeon's Report—its function, relationship to other reports, and its usefulness to specific audiences.
2. How is this performance report organized? Comment on the function of the opening and concluding paragraphs? Could the report be organized in any other way? Explain.
3. Discuss the style and tone of this report. What types of vocabulary and sentence patterns does it use? Why?

4. Identify summary sentences that help the reader to recall specific sections and conclusions of the report.

SUGGESTIONS FOR WRITING

1. Write a brief abstract (50–100 words) of the crew's health status during the flight for a consumer audience. See pages 219–220 on writing summaries.
2. Write a 1- or 2-page performance report on one of the following topics or one chosen by your instructor:
   a. an athelete's workouts for one month
   b. a salesperson's activities for a two-week period
   c. the success of a new machine used in your shop, laboratory, or office
   d. the effectiveness of a new policy at a business—safety and security, split shifts, parking, recruiting, corporate relations with a given segment of the community, company-sponsored seminars on writing, or travel procedures
   e. a new software package for a company's computer
   f. an employee's yearly evaluation
   g. your work during a semester in a technical writing course

The last model in this chapter shows how a long performance report can be organized and written. It is a monthly performance report on a solar heating system used at a Bell Telephone building in Pennsylvania. Written in compliance with the National Solar Heating and Cooling Demonstration Program, the report interprets data on one particular use of an alternate form of energy to officials in various governmental agencies, including the Department of Energy, executives of Bell Telephone, and individuals responsible for the construction and maintenance of the system. Companies like Bell Telephone receive tax advantages and save energy costs by participating in the Demonstration Program. The performance report is based upon data gathered from a computer-generated monthly report, which records only a series of numbers without commentary. Your job responsibilities may require you to write reports based on such computer-generated information. The following model will show you how one group of writers organized, explained, and highlighted numerical data found in such statistical reports.

A Monthly Performance Report on the Solar Heating System at Bell Telephone of Pennsylvania

I. SYSTEM DESCRIPTION

The Bell Telephone of Pennsylvania site is a one-story office and warehouse building in West Chester County, Pennsylvania. The solar energy system is designed to supply approximately 62 percent of the annual space heating requirements for the building. The site has an array of Heliotherm flat-plate collectors with a gross area of 2,112 square feet that faces south at an angle of 55 degrees from the horizontal. Solar heated water is stored in a 6,000-gallon capacity tank, but contains only 4,600 gallons. The tank is located above ground in the Mechanical Room. Thermal energy is distributed to the loads by circulating water from the storage tank through forced-air heat exchangers and through 20 individual heat pump evaporators. When solar energy is insufficient to meet the needs for space heating, auxiliary energy is supplied to the circulating water by electric resistance heating in the boiler. 1

The system, shown schematically in Figure 1, has three instrumented modes of solar operation. 2

Mode 1—Collector-to-Storage. This mode is entered when the collector-absorber plate temperature exceeds the temperature in the middle of storage by at least 18°F. Water from the tank is then circulated through the collectors 3

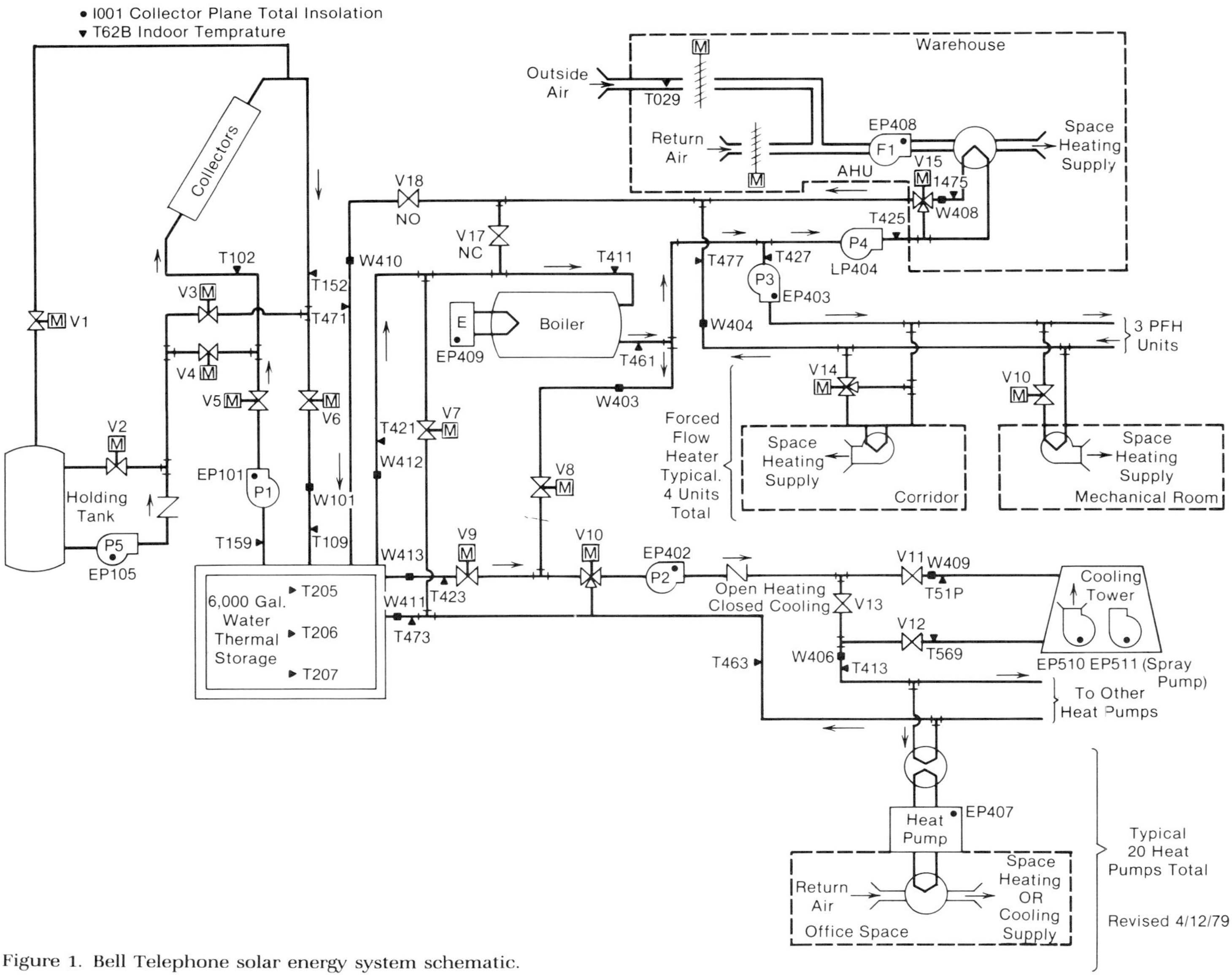

Figure 1. Bell Telephone solar energy system schematic.

until this temperature differential is less than 3°F. The collector subsystem has a drain-down feature to prevent freezing. Overheat protection is provided by automatic draining of the collector array whenever the collector output temperature exceeds 190°F.

Mode 2—Storage-to-Space Heating. This mode is entered with a demand 4
for heating from manually preset thermostats in either the office corridors or the warehouse area. Water is circulated by pumps P3 and P4 between the storage tank and the air-handling unit (AHU) heat exchangers and the forced-flow heater (FFH) units. The AHU and the FFH units are activated to provide heat to the warehouse area and the corridors, respectively. Pumps P3 and P4 are deactivated whenever the space heating demands are satisfied at the respective areas.

Mode 3—Storage-to-Space Heating (Heat Pumps). This mode is entered 5
when there is a demand for space heating from individual offices, and the sensed temperature of the water to the heat pumps is below a nominal 70°F. Valve V10 blends water from the storage tank with the heat pump recirculation flow as required. Flow from the storage tank ceases when the heat pump fluid operating temperature again reaches the 70°F set point. When the sensed storage tank temperature is less than 70°F, the boiler then becomes the thermal energy source for the heat pumps by the opening of valve V8 and the closing of valve V9.

II. PERFORMANCE EVALUATION

The system performance evaluations discussed in this section are based primarily 6
on the analysis of the data presented in the attached computer-generated monthly report [not included here]. The attached report consists of daily site thermal and energy values for each subsystem, plus environmental data. The performance factors discussed in this report are based upon the definitions contained in NBSIR–76–1137, *Thermal Data Requirements and Performance Evaluation Procedures for the National Solar Heating and Cooling Demonstration Program.*

A. Introduction

The solar energy system at the Bell Telephone of Pennsylvania site satisfied 69 7
percent of the measured space heating load (14.88 million Btu) during March. The solar energy system saved 7.23 million Btu of electrical energy. The system operation was interrupted by a cooling tower rupture on March 31, which subsequently allowed pump P2 to empty the storage tank.

B. Weather

The March outside temperatures averaged 9°F warmer than the long-term aver- 8
age temperature of 42°F as reported in the *National Oceanic and Atmospheric*

Administration Bulletin for the Philadelphia International Airport. The measured isolation in the plane of the collector array averaged 1,412 Btu/ft²-day, which is slightly higher than the long-term average of 1,322 Btu/ft²-day as estimated for the area from the *Climatic Atlas of the United States.*

C. System Thermal Performance

Collector. The solar energy system collected 21.90 million Btu, or 24 percent 9
of the 92.47 million Btu of the incident insolation on the collector array. The collector array efficiency is 25 percent lower than the value computed for February. The primary factors that adversely affected the collector array performance for March included: reduced collector-pump operation from March 20 through March 23 to avoid overheating the storage tank, average storage tank temperatures were 36°F higher for March than for February, and reduced collector pump operation due to drain and fill control problems. A total of 55.72 million Btu of insolation were incident when the collector pump P1 was operating. The operation of collector pump P1 and refill pump P5 required 0.57 million Btu of electrical operating energy.

Storage. A total of 23.52 million Btu of thermal energy was added to storage 10
from two sources during the month. The solar energy system provided 21.90 million Btu. The electric boiler added 1.62 million Btu, due to the circulation of storage water between the boiler and storage. A total of 19.60 million Btu of thermal energy was extracted from storage during the month. Of this amount, 13.10 million Btu were supplied to the source water for heating the warehouse and the office building corridors. The remaining 6.50 million Btu were supplied to the continuous recirculation loop for the 20 heat-pump units. The storage energy level increased by 0.28 million Btu. The energy loss from storage was 3.64 million Btu, which was computed by subtracting the sum of the change in storage energy and the energy removed from storage from the total thermal energy delivered to storage. The storage tank losses increased by 2.36 million Btu over February due to the significantly higher storage temperatures and partially due to the uncontrolled removal of storage energy when the cooling tower heat exchanger ruptured on March 31.

Space Heating Load. The measured space heating load was 14.88 million 11
Btu. The solar energy system provided 19.26 million Btu of energy to the space heating subsystem. Of this amount, only 10.27 million Btu of solar energy were used to satisfy system loads. The solar fraction of the space heating load was 69 percent. The remaining 8.99 million Btu of solar energy removed from storage was apparently added to the interior of the building through pipe losses. The electric boiler and the heat pump compressors provided an additional 5.32 million Btu of thermal energy to the space heating subsystem. The operation of circulating pumps P2, P3, and P4, the AHU fan F1, and the FFH unit fans required 3.27 million Btu of electrical energy.

D. Observations

The warmer March outside temperatures necessitated the manual switching of the system operation from the winter to the summer mode on March 21. In the summer mode, valve V10 is held closed to the storage and boiler heat source lines and fully open to the heat pump return line. In addition, valve V13 is closed and valves V11 and V12 are opened to allow heat pump loop water to flow through the cooling tower located in the building's Mechanical Room. The system remained in the summer mode through March 31. 12

The solar energy collection system was operated manually on March 28 due to a control failure that prevented the collectors from draining whenever outside temperatures fell below the nominal 45°F control setting. In addition, manual intervention was required to transfer heat, via valve V10, from the storage tank to the heat pump loop as required to satisfy space heating demands on the heat pumps. 13

On March 31 at 6:31 P.M., water apparently began to be expelled from the heat pump loop at a rate of 32 gallons per minute as calculated by differencing the flow measurements W409 and W406. The water temperature measured by sensor T519 jumped from 80°F to 143°F. At 9:06 P.M. the flow, measured by both W409 and W406, became zero. In an apparently related event, the collector refill pump P5 was activated at 6:36 P.M. and attempted unsuccessfully to fill the collector arrays. The cause of these events is under investigation. The site contractor indicated that the heat pump loop circulating pump P2 operated to empty the 4,600-gallon water storage tank onto the Mechanical Room floor. The energy content of the expelled storage water was estimated to be approximately 4.0 million Btu. 14

The conventional method for the calculation of the thermal energy removed from storage and delivered to the heat pump loop was unsatisfactory this month because of an incompatibility between the measurement range of flow sensor W413 and the range of the actual flow to be measured. Consequently, an alternate procedure was used to estimate the amount of energy removed from storage and delivered to the heat pump loop. By establishing a thermal energy balance at the storage tank on an hourly basis, we calculated the amount of energy delivered to the heat pump loop. To establish this balance, the storage heat loss rate was determined at a time when no energy was being added or removed from storage. Using measurement data from the September, 1978 Bell Telephone of Pennsylvania Performance Report, we determined the storage heat loss rate to be 50.33 Btu/°F-hour. Using this heat loss rate in conjunction with other storage measurements, storage energy balances were obtained. 15

Occasionally, negative energy tabulations of solar energy used will be obtained on a daily basis using this computed energy from storage parameter. The negative tabulations of solar energy used on March 6 and 7 illustrated this problem. These results are inherent in the calculations. However, because of the cancelling effects of the positive and negative values for this parameter, the monthly errors are expected to be small. A total of 6.59 million Btu of storage energy was contributed to the heat pump loop. 16

E. Energy Savings

The Bell Telephone of Pennsylvania solar energy system saved 7.23 million Btu 17
of electrical energy during March. The reported savings were based on the amount of solar energy used to satisfy calculated heating loads.

III. ACTION STATUS

A site visit was made by IBM personnel on March 15 to investigate the flow 18
incompatibility problem of flow sensor W413 and also to implement the latest measurement improvements. The problem with the measurement of flow at W413 could not be resolved without replacing the flow sensor with one of lower range. This action would require a smaller pipe size which might slow the system operation. Consequently, the current method of calculating the amount of thermal energy delivered to the heat pump loop will be continued.

Site personnel indexed the heat pump system to summer operation on 19
March 21. The solar energy system was not fully operational on March 28 due to testing of the automatic collector drain and fill control system.

QUESTIONS FOR DISCUSSION

1. How is this report organized? How will the reader use each part of the report? Could the authors have employed a memo format?
2. What kinds of language are used to describe the operation of the system? For what specific primary audience is this description intended?
3. This report not only includes the raw data obtained from computer printouts but also assesses it. How effectively do the writers notify the reader that the report contains explanations as well as documentation? How carefully do the writers combine a discussion of facts with an explanation of what those facts mean? Cite examples from the report.
4. Why do the authors include information on electrical energy from the boiler in their report? Where is information about electrical energy mentioned? Why?
5. Major problems occurred at the Bell Telephone facility during March. How objective have the authors been in describing and evaluating these problems? Describe the authors' language in the section labeled "Observations."

SUGGESTIONS FOR WRITING

1. Overall, how successful was the solar energy system during March at the Bell Telephone site? Write a 2- or 3-paragraph summary of events as described

in this performance report for consumer-readers, perhaps a summary that might be included in the monthly news pieces a telephone company sends with bills to customers, e.g., "What's Happening at Penn Bell."

2. Write a performance report modeled after the Bell example. Include a description of the facility and equipment as well as a precise account of major events and the action that was taken on them during the month. You can use the health facility proposed by the L.E.R.C. (pages 272–279) or the Cardiac Rehab Program described on pages 347–349, discussing performance one month after it has been implemented.

chapter **11**

Technical Publicity Announcements

This chapter provides models of common types of technical publicity announcements: news releases, product briefs, specifications, catalogs, and journal advertisements. A public relations or information department may not always relieve you of the responsibility of writing these materials. You may work for a small company where the employer expects you to prepare copy for these documents. Even at a large company, you may be asked to assist a public relations staff explain technical subjects to a reader. Government agencies at all levels sometimes have no professional writing staff adequate to handle all publicity chores. Moreover, you may write such publicity documents for a professional society to which you belong. Hence, it is important to know about these materials.

NEWS RELEASES

A news release is a short announcement sent to different media to get publicity for your company or agency. An effective news release can assure free and favorable exposure. News releases are useful for introducing a product, announcing a policy decision, or informing the public about internal news at a firm.

News releases are sent to editors of newspapers, wire services, trade and technical publications, and radio and television stations. Your first and most significant reader, the editor, welcomes news releases that help gather news or build a file for future stories. Having a limited amount of space for free publicity, an editor uses only a fraction of the releases received. Those that are printed are usually cut, sometimes drastically. Whether your release is used or not will depend on an editor's answers to these two questions:

1. Will the information in this release be of interest to my readers?
2. How much rewriting will I have to do to prepare the release for publication?

Choosing a Newsworthy Topic

The editor's first and most essential question concerns the newsworthiness of your release. If it does not offer relevant, current news, it is useless to him or her. Remember that not everything that happens at your firm deserves news coverage. Routine matters should not be elevated to seem impressive. News about established products, services, or policies are as valuable as last week's paper to an editor of a trade publication. A release about a controversial subject at your company should always be approved by your employer. In fact, it is a recommended policy to have your releases cleared by both management and research.

What, then, constitutes newsworthy material? The following four topics generally interest editors and their readers:

1. New products or services—construction, availability, appearance, differences among models, improvements.
2. New procedures or policies—significance for the audience, date they go into effect, differences from previous procedures, benefits, and precautions, if any.
3. Personnel changes—appointments, promotions, honors, retirements, and even resignations when they reflect positively on your company.
4. Financial and corporate news—quarterly earnings, stock reports, reorganizations, mergers, acquisitions.

Be careful to match the news to the publication. Identify specific publications whose readers are interested in one or more of these topics. For example, local newspapers and newsletters are eager to print consumer-related subjects and information about local residents. Trade publications serving various professional groups are most concerned about technical products and procedures. These publications generally have a tabloid format, are printed on glossy paper and often look like catalogues intended for specialists. Publications like *Forbes, Barron's,* or *Financial World* closely follow changes in corporate policies and personnel, taxes, and profits.

Study the newspapers and magazines to which you intend sending your release, especially regular departments or sections—"New Products," "Products and Services in Review," "Companies in the News," "Literature," or "People." However, not all journals in a single field will be equally interested in your story. Not every journal in health care will publish a release on new hospital equipment. Material relevant for the reader of *Nursing Management* may not apply to a reader of *Occupational Health Nursing.* Likewise, items suitable for the reader of *Concrete Construction* may not automatically find a place in *Modern Packaging.*

Should you send the same release to both the general and the trade press? Probably not. An item of interest to a local newspaper very likely will not be suitable for a technical journal. These publications reach two different audiences. Releases about industrial chemical products would not receive much coverage in the general press whereas the topic would be appropriate to trade publications aimed at industrial chemists. Conversely, a release on discounts given to cash customers of an oil company would primarily interest a reader of a daily newspaper.

When you do write about the same topic for different readers, adjust the technical level of your release to meet these readers' different needs. In a release for nonspecialists, it is wise to omit or clearly define any technical terms. A release intended for experts, however, should contain sufficient technical details to answer these readers' questions.

Writing an Effective News Release

The second question that determines whether your release will be used by an editor is how effectively it is written. Conciseness is essential. Compare and contrast the two releases in Figures 11–1 and 11–2. They contain annotated releases—a good and a bad example—taken from the *Postmaster's Guide to Communications,* which assists postmasters to develop a good working relationship with the media. The good example is only 124 words long, yet it gets the job done much better than the longer, wordier version.

Most releases are short—one or two pages. By making your release compact and easy to read, you also make it easy to edit and print. Follow the example of good newspaper writing. Get to the point quickly, arrange material in a descending order of importance, and exclude your own opinions. As in a news story, the most important details in a release should come first. In fact, editors sometimes cut beginning with the last paragraph, moving toward the first, thus making the first few paragraphs of your release extremely important.

The first and most important paragraph is called the *lead.* In addition to arousing reader interest, the lead contains the main facts of the story by answering most, if not all, of the following questions: why, what, when, who, where, and how. Study the following first sentence from a news release about sodium labeling that answers these questions:

> Responding to consumers' needs for nutritional information [why], Campbell Soup Company [who] is voluntarily [how] including sodium labeling on its soups and broths [what], Campbell President R. Gordon McGovern announced today [when].

Unless the subject of your release is a person, avoid starting off with a name. Of course, not all five questions have to be answered in every lead you write. Nevertheless, the more questions you do answer, the easier it will be for an editor to use your release.

In the second and subsequent paragraphs, do not overwhelm the reader,

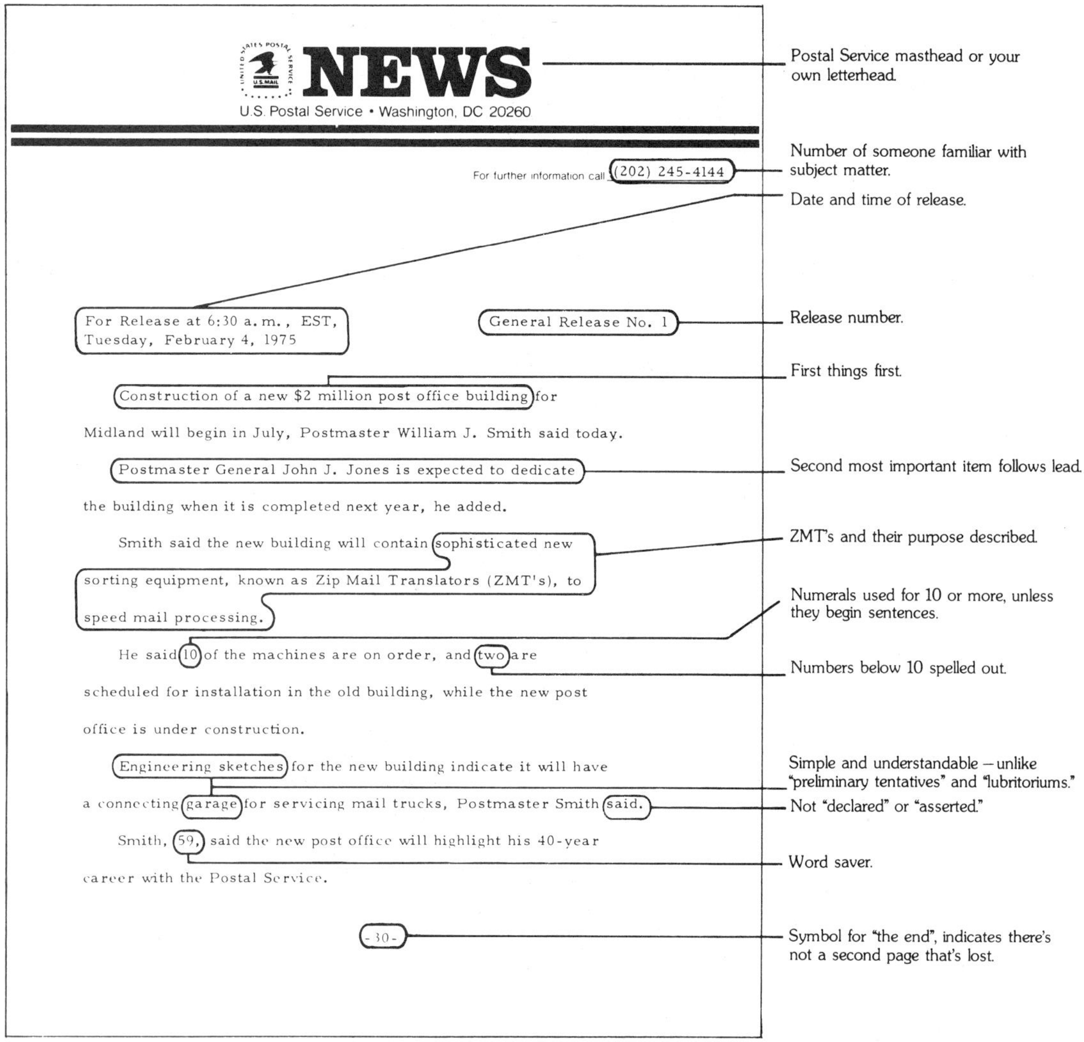

NEWS
U.S. Postal Service • Washington, DC 20260

For further information call (202) 245-4144

For Release at 6:30 a.m., EST,
Tuesday, February 4, 1975

General Release No. 1

Construction of a new $2 million post office building for Midland will begin in July, Postmaster William J. Smith said today.

Postmaster General John J. Jones is expected to dedicate the building when it is completed next year, he added.

Smith said the new building will contain sophisticated new sorting equipment, known as Zip Mail Translators (ZMT's), to speed mail processing.

He said 10 of the machines are on order, and two are scheduled for installation in the old building, while the new post office is under construction.

Engineering sketches for the new building indicate it will have a connecting garage for servicing mail trucks, Postmaster Smith said.

Smith, 59, said the new post office will highlight his 40-year career with the Postal Service.

-30-

Figure 11–1. An example of a good news release.

even an editor of a technical publication, with specifications, formulas, or performance characteristics. Save that kind of documentation for the product briefs and catalogs discussed later in this chapter. Highlight the major points of research or product development as they relate to the potential users of the product or service. Quotations from executives of your company are an effective way to summarize. However, do not turn your release into an interview with these executives. Another way to summarize is to write only about the one or two best, most innovative features your product or service offers.

To keep your release short and serviceable, direct the reader to other sources

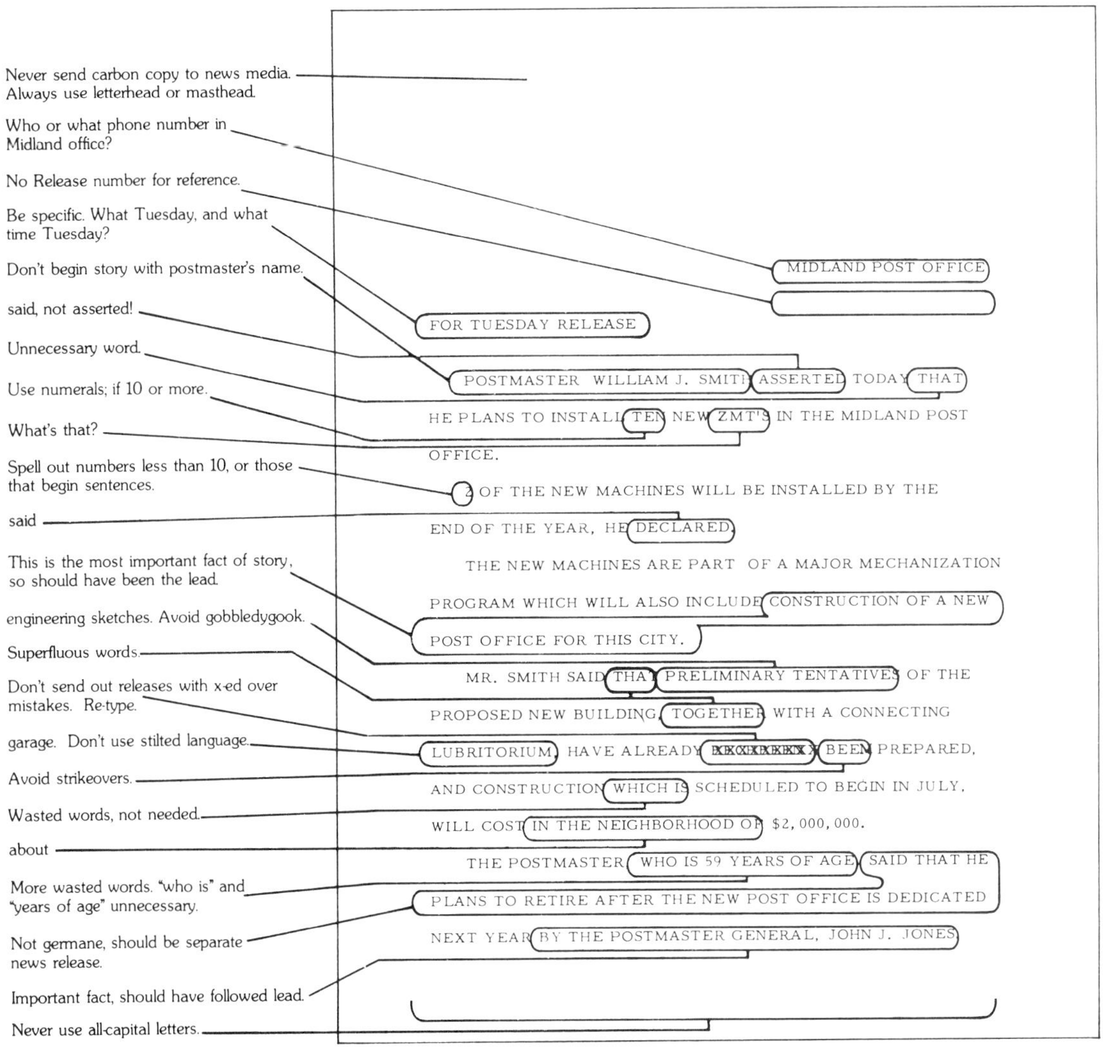
MIDLAND POST OFFICE

FOR TUESDAY RELEASE

POSTMASTER WILLIAM J. SMITH ASSERTED TODAY THAT HE PLANS TO INSTALL TEN NEW ZMT'S IN THE MIDLAND POST OFFICE.

2 OF THE NEW MACHINES WILL BE INSTALLED BY THE END OF THE YEAR, HE DECLARED.

THE NEW MACHINES ARE PART OF A MAJOR MECHANIZATION PROGRAM WHICH WILL ALSO INCLUDE CONSTRUCTION OF A NEW POST OFFICE FOR THIS CITY.

MR. SMITH SAID THAT PRELIMINARY TENTATIVES OF THE PROPOSED NEW BUILDING, TOGETHER WITH A CONNECTING LUBRITORIUM, HAVE ALREADY XXXXXXXX BEEN PREPARED, AND CONSTRUCTION WHICH IS SCHEDULED TO BEGIN IN JULY, WILL COST IN THE NEIGHBORHOOD OF $2,000,000.

THE POSTMASTER WHO IS 59 YEARS OF AGE SAID THAT HE PLANS TO RETIRE AFTER THE NEW POST OFFICE IS DEDICATED NEXT YEAR BY THE POSTMASTER GENERAL, JOHN J. JONES.

Figure 11–2. An example of a bad news release.

for further information. Some releases provide attachments to help editors—fact sheets, financial statements, backgrounders (which give brief histories), copies of policies, or necessary visuals.

Format

To be sure your news releases follow an accepted format, remember these six points:

1. Type your release on *company letterhead* using only one side of the page. Leave margins of at least one inch all the way around. Send an original or an original-looking photocopy to every editor. Never give one editor the original and the rest of the editors copies.

2. Include a *release date,* or the time when the news can be made public—"AT ONCE"; "FOR IMMEDIATE RELEASE"; "Thursday, February 9, 1984"; or "For Use After Monday, April 16, 1984." List a calendar date as well as the day of the week. Adjust your release date to the publication schedule of the journal or paper you want to reach. A monthly publication will not have the same release date as a daily newspaper, as it requires more than just a few days' advance time.

3. Specify a *contact person* whose name most often appears at the top of the release. Some firms, however, prefer only the company name to follow the story. A telephone number for this contact person will assist the editor who has questions or who wants to verify the details of a story.

4. Supply a few words at the top of the release that an editor might use as a headline. Although not mandatory, this information is helpful for the editor.

5. If your news release exceeds one page, type the word *MORE* at the bottom of each page except the last one. On the last page, type and center THE END, ########, or "30" at least four or five lines after the story, thus alerting editors that no more copy follows.

6. If you send a photograph with your release, make sure that editors can easily reproduce it. Photographs should be glossy, clear, and large enough so that corners might be cropped (clipped). Identify the photo by including the name of your product in pencil on the back of the photograph. Be sure to tell editors, at the bottom or end of your release, that a photograph or other visual is included.

A CHECKLIST FOR WRITING A NEWS RELEASE

1. Choose a newsworthy topic.
2. Follow the acceptable format editors expect.
3. Arrange material in descending order of importance.
4. Adjust technical detail to the audience.
5. Clear any news release with management and, if appropriate, research.
6. Determine carefully which publications to send the news release to.

TECHNICAL SALES LITERATURE

Technical sales literature comprises a wide range of materials: catalogues, brochures, and specification sheets (specs). These materials come in many sizes and shapes. Specs are almost always printed on glossy 8½ x 11" paper for easy filing, sometimes with perforated holes so that they may be inserted in looseleaf notebooks. Sales literature is mailed to customers as part of a regular or special sales program; it is also part of a salesperson's portfolio, and is often left behind for customers to study.

Sales materials serve a variety of functions. *First,* they are scientific and legal documents. Customers will judge a product's design and performance by them, and if there is a discrepancy, the manufacturer is liable, unless a disclaimer is printed on the literature. Clear, descriptive writing and accurate details are essential. *Second,* these materials can be used as reference tools. Technical catalogues are more than buying guides; they must provide necessary information on capabilities and exact specifications—of precision tools, for instance. *Third,* and most important for the purpose of this chapter, such literature must convince prospective customers that a specific product or service is the best one available—and is right for them.

Technical sales materials differ in the amount of expertise they expect from the reader. The audience for a product brief may be the salesperson in the showroom or the customer interested in that particular model. It should be detailed but not highly technical. As its name and function imply, a *specification sheet,* however, assumes that the reader does require technical information on such facts as dimensions, materials, performance, characteristics and ranges. Before writing a piece of sales literature, find out for whom it is being written and why. A brochure sent to the company buyer, who is a business executive, should be far less technical than a booklet intended for the chief engineer.

A piece of technical sales literature needs to answer the following kinds of questions readers are most likely to ask:

1. Do I need this?
2. Why and how is it useful?
3. How much does it cost?
4. How much will it save me and my firm?
5. What does it look like?
6. How does it work?
7. What is its capacity?

Headlines, Copy, and Visuals

Technical sales literature, like that written for the consumer market, uses headlines, copy, and visuals. A successful product brief or booklet merges these three elements into an effective message for the reader.

What makes a good headline or convincing copy? The answer is simple—emphasize the benefits for the reader. Study the kinds of problems a reader faces, anticipate any questions, and then offer a persuasive solution. A successful headline lets the reader know that you are aware of his or her needs. It can stress a useful feature or equipment part, ask a leading question, use statistics, or supply an endorsement. Let your headline pull your reader into the first line of your copy. Once there, the reader will judge the suitability of your product by the relevance of the information you supply. Give details about operating principles, parts, specifications, ranges, comparisons with other models, and availability. Simply supplying technical facts, however logically arranged, will not sell your product or service. Each line of copy must say, "Buy this product, use this service; it will make your job easier as well as save you time and money." Focus on such reader benefits as cost, convenience, easy handling, safety, productivity, flexibility, and availability. In emphasizing your product's advantages for the reader, remember that you are writing for a skeptical audience, not for the already convinced company executive for whom you work.

Visuals also help to sell products and services. An effective product brief or catalogue depends upon them as much as on your writing, perhaps more so. How a piece of sales literature looks—its layout—may determine if it is read and how carefully. Visuals are eye catching, dramatic, convincing, and provide a valuable means of summarizing information. In fact, visuals can bring more product features and benefits before a buyer in less space than your words ever could. In all likelihood, you will not be expected to prepare the visuals for your company's technical sales literature. That is the province of graphic designers. But the more you know about the importance of visuals, the more informed will be your opinion when a graphics department asks for it.

ADVERTISEMENTS IN TECHNICAL PUBLICATIONS

Advertisements in trade journals are yet another form of sales literature. In writing ads, you also have to know your audience and the product or service you want to sell. Impart technical information clearly and quickly, avoiding exaggeration and error. Visuals and copy need to support each other persuasively. But writing an advertisement presents additional requirements. Ad writers have a smaller amount of space and time to catch a busy reader's attention and hold it. Unlike a piece of sales literature, an ad does not constitute a separate sales package. One ad has to compete with all other ads—and news—in the same issue of the publication in which it appears. Moreover, ads cannot contain the detailed information that specification sheets or brochures offer. Ads are not generally kept on file for reference. Simply put, ads are not designed in the same way as sales literature.

Preparing advertising copy for the technical reader is a complex and challenging job. Teaching you how to flourish on Madison Avenue is beyond the scope and intention of this book. For that kind of information consult one of the many detailed and richly illustrated textbooks on advertising, marketing, or public relations. We have included model ads intended for technical readers to help you sharpen your overall writing skills. By studying these ads, you can learn to (1) enhance your ability to analyze the needs of a selected audience,

(2) summarize specialized information persuasively in a few sentences or paragraphs, (3) determine the amount of documentation to use for the specialized reader, and (4) attract the attention of the technical reader with appropriate touches of human interest and humor.

All the components of a good ad—visuals, headlines, descriptions, applications—are audience-specific. Target your audience to catch interest and win customers. A great deal of attention should be given to visuals. They can destroy or enhance an ad. An appropriate visual can speak hundreds of words that you could never include in the space the ad takes up. A well-chosen photograph, for example, can arouse the reader's interest and reinforce and expand the message presented in your copy. A poor visual may prompt the reader to turn the page without even glancing at your copy.

After visuals, the best attention-getter in an ad is the headline. Whether set in very large type (18 point) or not, headlines need to entice the reader to study the entire ad. Whenever possible, key your headline to the type of visual you have chosen. As with brochures or product briefs, a variety of techniques are used to create effective headlines. Some headlines use humorous puns. Some ask a question. Some cite revealing statistics. Look at the headlines for ads that appear in publications in your field to see the variety of techniques ad writers use.

Given the limited amount of space copywriters have, how much information should the reader be given? You must arouse enough interest to publicize the significance of your product, but don't weigh the reader down with so many facts that you have to resort to using the small print of specification sheets. Keep your copy lively and informative; lure the reader into asking for more facts through a telephone call or letter. Make sure you give numbers and addresses. Use an easy-to-read format within the ad to disseminate information. For example, use bullets when listing key facts about operations or services.

A CHECKLIST FOR WRITING TECHNICAL SALES LITERATURE, INCLUDING ADVERTISEMENTS

1. Be careful with the accuracy of your claims—this literature can be regarded as legal or scientific documents.
2. Determine the technical expertise and needs of your reader and tailor details accordingly.
3. Focus on specific benefits your product or service offers to the reader.
4. Provide details about operation, parts, specifications, service, and comparisons with other models, whenever appropriate.
5. Use attractive and illustrative visuals whenever possible.

NEWS RELEASES

This news release introduced Abbokinase, an anticoagulant or blood thinner, at a crucial time in its marketing history—just after the drug had been approved by the U.S. Food and Drug Administration (FDA). The release was directed to financial analysts, science writers, and the general public. The author clearly explains the action, effectiveness, and distribution of the drug for a nonspecialist audience while also providing the essential scientific facts required by the FDA.

A News Release Announcing a New Product from Abbott Laboratories

ABBOTT
Health Care World Wide

PUBLIC AND PROFESSIONAL RELATIONS
ABBOTT LABORATORIES
ABBOTT PARK, NORTH CHICAGO, ILLINOIS 60064
AREA CODE 312/688-3930

FOR IMMEDIATE RELEASE
Contact: T. A. Craig

NORTH CHICAGO, Ill., January 17, 1978—Abbott Laboratories announced today that its new drug application for Abbokinase, brand of urokinase, a blood-clot dissolving agent, has been approved by the U.S. Food and Drug Administration. 1

A product of more than a decade of Abbott research and development, Abbokinase has been approved for treating massive pulmonary emboli (blood clots in the lungs). Various estimates indicate that pulmonary embolism causes about 200,000 deaths each year in the United States, third only to heart disease and cancer. 2

Effectiveness of urokinase in this disease has been evaluated in clinical studies in the U.S. that began in 1968. Approval of Abbokinase was conditional on broader clinical evaluation ("Phase IV") of the agent. The company also will continue clinical studies in the U.S. and Europe to determine the effectiveness of Abbokinase in other diseases such as deep-vein thrombosis. 3

To make it readily available throughout the U.S., Abbokinase initially will be distributed from North Chicago and 43 strategically located hospitals. 4

The drug is recommended for use only by physicians with wide experience in the management of thrombotic disease and in hospitals where recommended clinical and laboratory monitoring can be performed. 5

Urokinase is a naturally occurring human protein substance produced by the kidneys and found in minute concentration in the urine. Originally, it could be produced only from vast quantities of human urine. Abbokinase is produced more efficiently and in larger quantities from kidney cells grown in tissue culture. 6

Urine-derived urokinase has been marketed in Japan and several European nations but not in the U.S. 7

-more-

Abbokinase
Abbott

Abbokinase activates the body's system for breaking up 8
fibrin, the protein that is the essential part of a blood
clot. The drug converts plasminogen, an inactive substance
in the blood, to the enzyme, plasmin, which degrades fibrin
clots. The soluble degradation products then are carried
away by the blood.

The action of Abbokinase is in contrast to that of the 9
anticoagulant, heparin, marketed by Abbott under the brand
name, Panheprin, which prevents the formation of clots but
does not dissolve those already formed.

Because of the profound effect of Abbokinase on the 10
patient's clotting mechanism, special precautions, including
laboratory monitoring, must be taken in the care of the
patient to prevent complications from bleeding.

Abbokinase is administered by vein. It is supplied as a 11
sterile powder to be reconstituted with sterile water and
diluted with normal saline immediately before using. Dose
depends on body weight.

###

QUESTIONS FOR DISCUSSION

1. Identify the different parts of a news release in this model, and briefly explain their function.
2. What specific questions does the lead answer for the reader? How effectively does the lead accomplish its job?
3. How does this release meet the needs of its diverse readers? Comment on the author's use of language and examples. How has he made technical details easy to understand?
4. What are the most important facts about the drug? Where are they found in the news release? If you were an editor who could use only portions of this release, what sentences would you cut and why?
5. What function do paragraphs 5, 10, and 11 serve?

SUGGESTIONS FOR WRITING

1. Assume you are an editor of a health magazine read by the general public and that you have only enough space to use half of this news release. Prepare a shortened and edited version.
2. Write a news release for the general public about a new technical product or service in your major.

The following release also announces a new product but it is written primarily for a technical audience. This release uses the specialized language (PAR flood system, parabolic reflector, neoprene gaskets) familiar to lighting engineers and technicians. The release calls attention to the new features available, compares the system to existing ones, and gives an overview of the testing done to verify the performance of the system.

A News Release Announcing a New Sylvania Outdoor Lighting System

News

GTE

GTE Lighting Products
Public Affairs Department
Sylvania Lighting Center
Danvers, Massachusetts 01923

From: Richard H. Dowhan 617/777-1900 (After 5 p.m. 603/964-5756)

Release Date:
FOR IMMEDIATE RELEASE (11-19-80)

SUMMARY: New Sylvania outdoor lighting system provides energy savings and precision beam control. 1

DANVERS, Mass., Nov. 19—A family of Sylvania precision control lighting fixtures which come equipped with either a 50-, 70-, or 100-watt high pressure sodium lamp has been introduced by GTE. The new outdoor lighting system provides significant energy savings over comparable incandescent, mercury vapor and quartz light sources. 2

Called the BeamKat lighting system, it is available in three beam patterns for security, safety, and other lighting applications. Long distance control with a powerful center beam is achieved with the narrow pattern refractor version while the medium beam refractor model provides general purpose lighting with uniformity. For close-up broad area landscape lighting or wall wash applications, a widespread beam refractor version is available. 3

ENERGY-SAVING FEATURES

The features of high lumens per watt output of the high pressure sodium lamps and precision optics designed into the fixture combine to provide significant energy savings for the new Sylvania lighting system. A 70-watt narrow pattern BeamKat system is designed to replace a 300-watt incandescent PAR flood system, with only a 10 percent loss of illumination. Based on 4,000 hours per year operation and five cents per kilowatt hour electricity cost, it will save the user $52.40 annually. Other BeamKat fixture-lamp combinations result in comparable savings. 4

GTE - 2

The 24,000-hour life of the 50-, 70-, and 100-watt medium- 5
base, high-pressure sodium lamps translates to five to six years of dusk to dawn operation. This long lamp life means less frequent lamp replacement and added economy.

The high efficiency of the Sylvania BeamKat lighting system 6
was achieved by use of a computerized design. The parabolic reflector, tempered glass refractor, and high-pressure sodium lamp were computer-matched for optimum performance, including high efficiency and beam control. This results in full utilization of the lumens produced by the lamp, good cut-off angles, and glare reduction.

TESTING VERIFIES PERFORMANCE

Testing has verified the precision optics and efficiency 7
achieved with the new family of computer-designed lamp systems. At 100 feet, for example, the 100-watt, narrow-pattern refractor BeamKat lamp will illuminate a 17-by-17-foot area to four footcandles; at 200 feet, the fixture-lamp combination will produce nearly one footcandle over a 36-by-36-foot area.

The BeamKat unit is modern in appearance with a bronze 8
finish that blends with any architectural background. It comes prewired for easy installation with the lamp compartment hinged for convenient service and sealed with neoprene gaskets for weather protection. The unit comes complete with hardware for either wall or pole mounting.

The heavy-duty adjustable bracket allows 180 degree 9
horizontal and 110 degree vertical swings at three-inch increments for precise aiming of the light beam once the unit is mounted. An optional energy-saving photocell control is available to provide automatic dusk to dawn operation.

With ballasts for 120, 208, 240 and 277 input voltage, the 10
BeamKat fixture will mount either on a standard electrical box or a one-inch to one-and-one-quarter-inch arm.

For information on the Sylvania BeamKat lighting system, 11
write Sylvania Lighting Equipment, GTE Lighting Products, Fall River, Mass. 02724.

###

QUESTIONS FOR DISCUSSION

1. Evaluate the lead. Is it effective? What function does the second sentence in the first paragraph serve? How does the information in this sentence help the writer to organize the release?
2. Why does the section on energy savings precede the section on testing?
3. As was pointed out in the introduction to this chapter, the use of summaries is very important in writing news releases. Identify effective summary sentences from this release and explain why they are helpful to the reader.
4. Provide a list of at least three technical journals/magazines to which this release might be sent and explain why these publications might be interested in printing news of the BeamKat.
5. Which details about the BeamKat might be omitted if the writer were sending the release to a local newspaper?
6. How does the format of this release differ from that used for the news release from Abbott Laboratories? Which one might be easier to read for a busy editor who has to write a story?

SUGGESTIONS FOR WRITING

1. Find an advertisement for a new product in a technical journal. Using information from that advertisement, write a news release for similar journals. Attach a copy of the advertisement to the news release that you submit to your instructor.
2. Write a news release announcing the Sylvania lighting system for distribution to a consumer audience that has no technical background in electronics.

The following release is about an individual, Elizabeth E. Bailey, a government official who is leaving her post with the Civil Aeronautics Board (CAB) to assume new duties in higher education. The release emphasizes her professional accomplishments as well as her specific duties—past and future. Executive changes are always regarded as highly newsworthy events, and this release brings favorable publicity to both the CAB and the Carnegie-Mellon Institute, the university Bailey is joining. The audience for the release included individuals in the CAB and related federal agencies and most of the wire services.

A News Release about a CAB Official

CAB NEWS

CIVIL AERONAUTICS BOARD, WASHINGTON, D.C. 20428

Contact: Alan M. Pollock
(202) 673-5990

FOR RELEASE
Thursday
April 22, 1982
CAB 82-66

VICE CHAIRMAN ELIZABETH E. BAILEY TO LEAVE CAB IN 1983 TO BECOME DEAN OF CARNEGIE-MELLON SCHOOL OF INDUSTRIAL ADMINISTRATION

WASHINGTON, D.C. (April 22)—Vice Chairman Elizabeth E. Bailey of the Civil Aeronautics Board has accepted the position of Dean of the Graduate School of Industrial Administration at Carnegie-Mellon Institute in Pittsburgh, Pa., starting in the summer of 1983. She will be the first woman dean of a leading graduate business school of this country. 1

Mrs. Bailey became a member of the CAB on August 8, 1977. President Reagan designated her Vice Chairman of the Board in October 1981. 2

Mrs. Bailey is the first woman member of the CAB. At the time of her appointment, she headed the Economic Research Department of Bell Telephone Laboratories in Holmdel, N.J. In that capacity, she was responsible for research in areas of economics related to the telecommunications business and regulatory processes affecting it. Also at the time of her appointment to the CAB, Mrs. Bailey had been an adjunct Professor of Economics at New York University for four years. 3

Author of the book, Economic Theory of Regulatory Constraint (1973) and editor of Selected Economic Papers of William J. Baumol (1976), Mrs. Bailey has written numerous articles on economics for such technical journals as the American 4

MORE

- 2 -

Economic Review, Journal of Political Economy, Bell Journal of Economics and Journal of Public Economics.

A native of New York City, she received her undergraduate 5
degree with high honors in economics from Radcliffe College in 1960, a master's degree in mathematics and computer science from Stevens Institute of Technology in 1966, and her doctorate in economics from Princeton University in 1972.

Mrs. Bailey is a member of the Board of Trustees of 6
Princeton University and a member of the Executive Committee of the American Economic Association. She is head of the standing Committee on the Status of Women in the Economics Profession and is a member of the Executive Council of the Federation of Organizations of Professional Women.

Mrs. Bailey and her two sons reside in Chevy Chase, Md. 7

In announcing the acceptance of her new position, Mrs. 8
Bailey said:

"I am proud of the Board's accomplishments in bringing 9
deregulation to the domestic air travel industry. It has been an exciting experience and, over the next year, I will continue vigorously to support further moves toward a fully competitive airline environment.

"The position of Dean of a business school that is as 10
innovative and respected as Carnegie-Mellon offers a marvelous opportunity. I look forward to contributing to the tradition of excellence in research and teaching of the school and to assisting the faculty and administration in finding creative solutions to the compelling intellectual challenges confronting business in the 1980's and 1990's."

- 30 -

QUESTIONS FOR DISCUSSION

1. Why would the CAB want to send out a news release about the Vice Chairman quitting her job?
2. What kinds of information about Bailey does the release give the reader?
3. Why does the author of the release include the quotations last?

SUGGESTIONS FOR WRITING

1. Write a news release based on the following information. You will have to add a few facts as well as delete some information below that you think is unnecessary.

   A. T. Boyd, 43 years old, native Alaskan, with a B.S. in finance and an M.B.A.; in personnel management work for the last three years; appointed administrative assistant to the mayor of Juneau; has served six years in the U.S. Navy; honorably discharged; worked for five different accounting firms in Los Angeles, Houston, and Seattle; married with three children, one of whom will be a freshman at the University of Alaska, Fairbanks, next year; hobbies are skiing and cooking; member of the First Congregational Church of Seattle.

2. Write a letter to Elizabeth E. Bailey congratulating her on her new appointment. Assume you have worked with her at the CAB for the last 8 months. See Chapter 12 for information on the format, style, and function of letters.

Prepared by the Nuclear Regulatory Commission (NRC), the following news release announces a change in governmental policy regarding cloisonné jewelry that contains radioactive uranium. This NRC news release will give you a model to follow when you have to announce a change in policy for an employer or organization. First it carefully identifies the policy to be changed, then it explains reasons for the change, and finally it concludes with specific instructions the readers are to follow regarding the new policy. The audience for this release included officials in various regulatory government agencies, executives in the wholesale and import jewelry businesses, and news agency wire services.

A News Release from the NRC about Importing Cloisonné Jewelry

UNITED STATES
NUCLEAR REGULATORY COMMISSION
Office of Public Affairs
Washington, D.C. 20555

No. 83-112 FOR IMMEDIATE RELEASE

Tel. 301/492-7715 (Monday, July 25, 1983)

NRC SUSPENDS EXEMPTION FOR RADIOACTIVE CLOISONNÉ JEWELRY

The Nuclear Regulatory Commission is suspending immediately an exemption from its regulations that permitted cloisonné-type jewelry containing uranium to be imported into the United States without a specific NRC license. Consumers, retailers, and others who already have the jewelry will continue to be allowed to possess, sell and transfer it without a license. 1

The suspension will maintain the status quo and prevent the amount of radioactive cloisonné jewelry in circulation from increasing while the NRC reevaluates the exempt use of glass enamel containing uranium in consumer products such as jewelry. 2

On January 25, 1983, the New York State Department of Health announced that it had found that some pieces of cloisonné jewelry were radioactive. The radiation was a result of uranium used to produce golden-yellow and beige colors in the brightly colored glass enamel of the jewelry. 3

After the announcement by the State of New York, other states and the NRC examined thousands of pieces of cloisonné jewelry. The Department of Energy's Radiological and Environmental Sciences Laboratory in Idaho Falls, Idaho, also performed a radiological analysis of the jewelry. This information indicated that about 10 percent of the jewelry is radioactive as a result of the uranium content in the glass enamel. The uranium content of the enamel of the radioactive pieces was in the range of 3 to 7 percent by weight. The radiation dose rate at contact with the surface of the enamel ranged from 3 to 7 millirems per hour, but the 4

- 2 -

exposure rate decreased rapidly with distance from the surface of the jewelry. No removable radioactivity was detected on any sample.

Based on these findings, the NRC has determined that a piece 5
of cloisonné jewelry worn with the radioactive enameled
surface in direct contact with the skin (such as a ring or
pendant) for a period of 10 hours per week for a year could
deliver a dose of about 2 to 4 rems to a small area of the
skin. This corresponds to a skin cancer incidence risk of
about 2 to 4 in a million. The risk of death from these skin
cancers is considerably less.

When the radioactive enameled surface of the jewelry is not 6
in direct contact with the skin, the radiation dose and
corresponding cancer risk are much lower. . . .

Although the use of the radioactive cloisonné jewelry does 7
not constitute an immediate or significant health hazard,
the NRC believes the use of the jewelry does constitute an
unnecessary exposure to radiation. This view is based on the
principle that there should not be any artificial radiation
exposure without the expectation of benefit resulting from
such exposure.

The NRC believes that all radioactive cloisonné jewelry is 8
imported. If such jewelry was imported or ordered for import
into the United States before July 25, 1983, it is not
subject to the suspension of the exemption.

If there is a manufacturer of the jewelry in this country, 9
the suspension of the exemption has the effect of
prohibiting him or her from any commercial distribution of
the jewelry not made before July 25, 1983.

Further details on the NRC's suspension of the exemption are 10
contained in a Federal Register notice published on July 25,
1983.

#

QUESTIONS FOR DISCUSSION

1. How does the release meet the needs of its diverse audience?
2. Evaluate the first paragraph in terms of presenting the most important information. Does it accomplish that goal effectively?
3. What kind of information is contained in the second paragraph of the release? Why doesn't the release begin with this information?
4. Why does the release writer wait until the third paragraph to define cloisonné jewelry? Are there any other places in the release where the writer uses definition?
5. Why do the third and fourth paragraphs provide details about the studies conducted by the New York State Department of Health and the Department of Energy?
6. Obviously, this release reports on a potential health hazard. How does the writer report the facts objectively yet without alarming the readers?
7. Why does the release end by referring the reader to the *Federal Register?* If the new regulations are listed in the *Register,* why has the release been written?

SUGGESTION FOR WRITING

1. Compose a news release based upon the following facts from the Mining Safety and Health Administration (MSHA) for MSHA district managers, mine operators, labor organizations, and mining associations:
   - Title 30, Code of *Federal Regulations,* Parts 48.8 and 48.28 require that each miner receive a minimum of eight hours annual refresher training and that such training include a specified course of instruction.
   - The MSHA has changed its policy to make it more flexible in response to requests from the mining community.
   - If courses listed in the Code of *Federal Regulations* are not applicable to a specific mine, they may be omitted from its training plan. Such courses must be cited in the mine's training plan. In the exceptional situation where a course is applicable to the mine but not to a particular miner, the miner is not required to take that course.
   - Time spent on individual courses may now vary from year to year or from one area of the mine to another, depending on specific safety or health problems. The mine's accident experience will significantly determine the amount and type of training miners receive.
   - Two or more courses may now be covered in one training session or safety meeting. Separate blocks of instruction are no longer required for such topics as safety standards, ground control, or prevention of accidents.
   - Safety meetings of at least 30 minutes duration and con-

ducted by an approved instructor are acceptable training sessions that satisfy the annual refresher requirements.

- Training plans may be revised to reflect the training needs of individual plans.
- Requests for revisions in training should be directed to the appropriate MSHA District Manager.
- Questions should be directed to MSHA, Educational Policy and Development, Arlington, VA.

2. Write a news release about a change in a policy at your school or place of employment. Some policies you might consider include registration procedures, progress or travel reports, vacation, filing for graduation, use of company vehicles, changes in computer use and security.

PRODUCT DESCRIPTIONS IN SPECIFICATION SHEETS AND CATALOGS

The following product spec (specification) is sent to individuals responsible for maintaining heavy-duty, off-highway vehicles. The spec sheet gives these readers the relevant facts about the product—what it is, how it works, when and where to use it, what it's made of, and where to obtain it. By providing this information, Shell hopes to satisfy current customers and attract new ones.

A Product Specification for Shell Rimula Oil

QUESTIONS FOR DISCUSSION

1. A product spec sheet cannot be vague or ambiguous, for these writing errors can result in malfunctioning equipment, user injuries, and lawsuits. How has the author of this spec sheet explained the product so that the reader will have no trouble understanding it? Comment on the clarity of the writer's words, the types of sentences and paragraphs used, and the specialized abbreviations employed.
2. How does the layout of this spec sheet make it easy for the reader to find information? Like other spec sheets, this one may be filed away by a reader for future reference. What specific layout features would assist the reader to file this sheet?
3. How is the table organized? Why are the categories in the table of interest to the reader? Give a practical example.
4. How does the writer make technical information about the product a part of the sales approach?

SUGGESTIONS FOR WRITING

1. Research a product of your choice and write a spec sheet about it.
2. Write a news release about Shell Rimula Oil for an appropriate technical publication read by maintenance personnel.

The next example of a piece of technical sales literature appears in the catalog of a well-known tool manufacturer. A great deal of information about the Kwik-Chek hole gage is clearly listed on one uncluttered, carefully designed page. In fact the layout helps to explain the usefulness of the tool for the readers and thus to convince them to order one.

A Catalog Description for Kwik-Chek Hole Gages

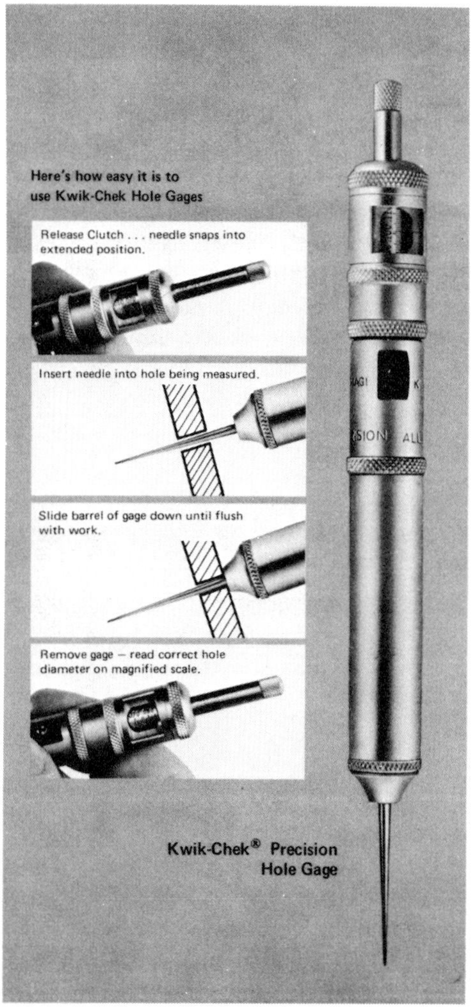

KWIK-CHEK® HOLE GAGES

Kwik-Chek Hole Gages are designed for the rapid measurement of small plain or counter-sunk hole diameters. The four models available cover a range of hole diameters from 0.010" to 0.380" or from .25mm to 9.50mm.

These gages work on the principle of directly transferring diameter measurements from a sliding, tapered needle to a linear scale readout graduated in 0.001" or 0.02 mm.

These highly versatile gages are especially suitable for measuring small holes in sheet metal, plastics, printed circuit boards, fuel injection nozzles and thin-wall tubing.

OUTSTANDING FEATURES

- No skill required – untrained personnel can make fast and accurate measurements.
- Rapid and simple calibration – setting master for plain and countersunk holes provided with each gage. Permits rapid periodic check on calibration and field readjustment, if necessary, by shop personnel.
- Magnified, direct reading scale – permits easy reading of hole diameters to nearest 0.001" or 0.02mm.
- Hardened and plated needle – precision ground tapered needle is made of hardened 52-100 steel and chrome plated for maximum wear resistance.
- Positive Clutch – locks needle and scale at exact dimension of hole being measured.
- Completely portable – convenient pocket size and weight makes these gages ideal for roving inspectors and mechanics.

Individual Gages (Specify inch or metric)

Hole Size Ranges	Plain Hole	Countersunk Hole	Plain & Countersunk Holes
0.010" – 0.028" 0.25mm – 0.70mm			Model 10
0.025" – 0.130" 0.60mm – 3.30mm	Model 20A	Model 20C	Model 20B
0.130" – 0.255" 3.30mm – 6.50mm	Model 30A	Model 30C	Model 30B
0.255" – 0.380" 6.50mm – 9.50mm	Model 40A	Model 40C	Model 40B
0.010" – 0.028" 0.25mm – 0.70mm	Kwik-Chek Orifice Gage		Model 10-0

All prices include gage and setting standard in case as shown. (Extra head with Model B only).

Two interchangeable heads provided with each Model B gage.

QUESTIONS FOR DISCUSSION

1. Who is the audience for this catalog entry—operators, executives, experts, all of these groups? On what are you basing your answer? How does the writer demonstrate the relevance and convenience of the product to the reader?
2. What kinds of information about the Kwik-Chek do the first three paragraphs contain? Provide a subhead for each of these paragraphs.
3. The catalog description uses four different kinds of visuals. What are they, and what function does each serve? Where is each visual placed and why? Explain how these visuals are keyed with the text of the description.

SUGGESTIONS FOR WRITING

1. Research a product you're familiar with or interested in and write and design a catalog description of it.
2. Based upon information in the model, write a physical description of the Kwik-Chek that does not include any advertising information.

TECHNICAL ADVERTISEMENTS

This and the model advertisement following it, both from AT&T, will show you how one company's communication service can be marketed to two different audiences. The first advertisement appeared in Traffic Management, *a trade magazine read by executives in the transportation business, who are frequently confronted with communication problems involving individuals in many different locations. The AT&T advertisement describes the benefits of its service to meet the specific needs of transportation managers. The second AT&T advertisement is directed to a variety of business readers and communication specialists. In fact, the advertisement ran in such diverse trade magazines as* Telecommunications *(devoted to the total spectrum of communications worldwide) and* Textile World. *AT&T knows that readers of these magazines will be concerned primarily with how well a communication system will work in their office for customers across the country and across the world.*

Two Ads from a Telephone Company

AT&T IS IN ORANGES.

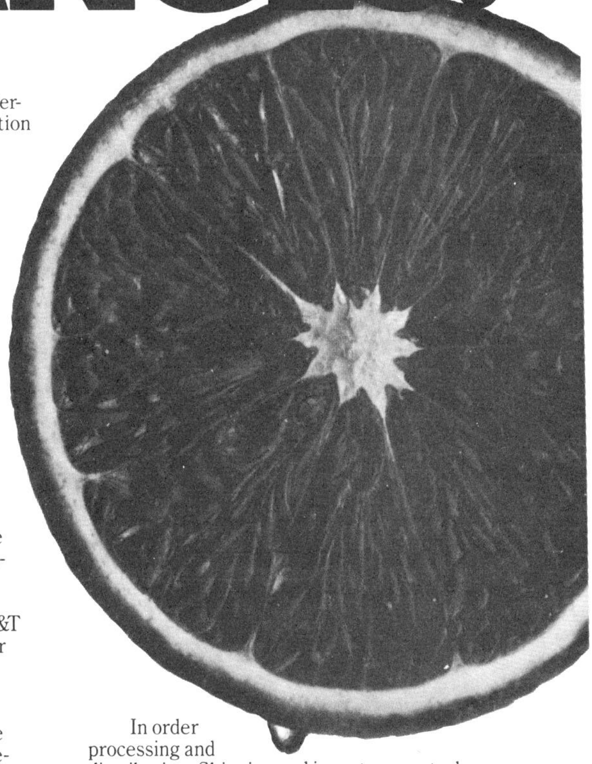

And grapefruit and tomatoes and all kinds of produce. How?

The people of AT&T Communications understand the ins and outs of the freight transportation business in ways you never thought of.

We can show you how to expand your business, offer new services and run your operations more efficiently.

Consider the case of the Orange Blossom Special, a unique transport system, conceived by the CSX Corporation. It's an efficient means of combining their high-speed Seaboard System Railroad with their trucking unit, Chessie Motor Express.

The Orange Blossom Special carries perishables from the growers in the Southeast to markets in the Northeast and Canada. Because of the nature of perishables and the competition among freight carriers, timely communication between grower and carrier is vital.

That's where the people of AT&T come in. We showed them how an AT&T 800 Service number can allow growers to arrange for same-day pickup of their produce, shortening the delivery time.

And how to take advantage of another AT&T 800 Service number so their people can answer questions and negotiate rates prior to actual ordering.

Results: $11 million in revenue in the first year. And the Golden Freight Car Award for the CSX Corporation. In this, the second year, revenues are nearly $20 million.

The people of AT&T Communications are thinking about your business in ways you never thought of.

In order processing and distribution. Shipping and inventory control.

You get specially trained people with AT&T, to help your business work better.

To find out more about how we're thinking about your business in ways you never thought of, call your Account Executive at **AT&T Communications. Or 1 800 222-0400.**

AT&T

THE AT&T INTERNATIONAL BUSINESS NETWORK:

IT COSTS LESS TO SEND YOUR FACTS BY FAX.

SAN FRANCISCO

TOKYO

An exact copy of the original.

That's fax. And fax is the only system that can transmit blueprints, legal documents, and signed contracts almost anywhere in the world.

This unique ability makes fax one of the most exciting electronic mail technologies available today.

All you need is a telephone and a facsimile machine.

And in as little as 25 seconds, an exact duplicate arrives at its destination.

With fax, faster is cheaper.

And different machines have different capabilities, different speeds.

With the AT&T Network, you can transmit *up to* 400 words on a 1-minute call.

Most telex machines transmit at a standard 66 words per minute.

Any fax machine can match that rate. And most beat it easily.

In fact, you can save as much as $8.00 per page with AT&T's lowest rates compared with sending the same page via telex. With some fax machines, you don't even have to be there.

They can be programmed to transmit during Discount and Economy periods. Telex has no daily reduced rate periods.

Since fax is a copy of the original, you don't need a typist. No special operators.

And since there's no retyping, no redoing of any kind, fax is free of transcription errors.

So if your international customer or business associate needs a document, a blueprint, or a signed contract in a hurry, you can fax it on the spot.

Transmitting facts with fax.

Another cost-effective service for your business from the AT&T International Business Network.

International service to and from continental US

FOR A FREE BROCHURE, CALL TOLL FREE

1 800 874-8000

AT&T

QUESTIONS FOR DISCUSSION

1. How effective are the headlines in these two model advertisements? How do they arouse the reader's interest?
2. Comment on the layout of these advertisements. Are the visuals carefully coordinated with the text of the advertisements? Do they appear in the most effective places? Explain your answer by providing specific details from the advertisements.
3. How does the first advertisement try to convince readers that the AT&T service is tailor-made for executives in "the freight transportation business"?
4. The second advertisement uses short and, many times, incomplete sentences. Is this an effective style for the writer's purpose and the reader's needs?
5. The last line of the second advertisement on the Fax service is a good summary of the entire advertisement. Explain how this line summarizes the copy and urges the reader to find out more about it?
6. Does the Fax advertisement include too much, too little, or just the right amount of description of the Fax service for readers?

SUGGESTIONS FOR WRITING

1. Write and design an advertisement for some telephone equipment or services—the phone service used at your office or dormitory, an answering service, etc. Specify some publications in which your ad might run.
2. Write a news release about the Fax service for a consumer audience.
3. Write an audience-specific advertisement, modeled after the two AT&T ads, with an accompanying visual on the same product to at least two of the following groups:

   a. accountants
   b. nurses
   c. safety engineers
   d. paralegals
   e. mechanics
   f. school administrators
   g. computer programmers
   h. electricians
   j. dietitians
   k. real estate agents

The PQ Corporation advertisement below appeals to potential customers by describing the way its product works. It appeared in Chemical and Engineering News, *a publication read by experts in these fields. The advertisement amusingly yet with scientific accuracy explains the behavior of silicates in the production of beer. First the copywriter identifies the problem a brewer can have with proteins and then shows how the PQ Corporation can solve it. You may want to contrast the process explanation here with that found in "The Brewing Process" on pages 164–165.*

An Advertisement on Beer Production from the PQ Corporation

Look what happens when you combine silicate chemistry with an active imagination.

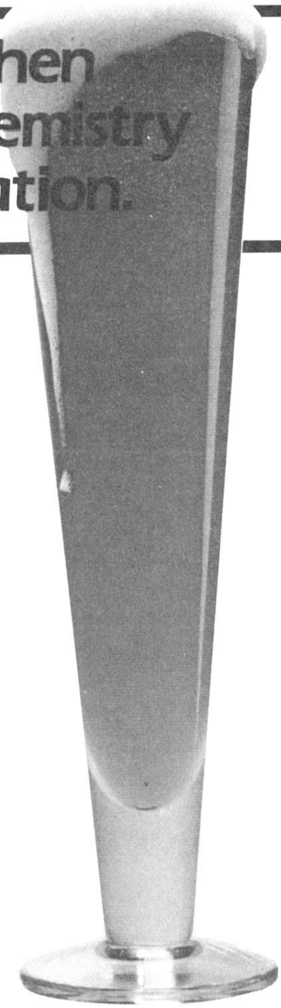

How to keep your brilliance without losing your head.

That's not just a problem for your resident genius. It's an abiding concern of your friendly local brewmaster.

Because giving you what you want—a crisp, cold brew, brilliantly clear, with a foamy head—is not easy.

The problem is that in the chemistry of beer, cold and clear don't go together. Proteins get in the way.

Polypeptide proteins make the head, which is good. But when beer is chilled, other proteins make it hazy, which is bad.

Just to complicate matters, there are a few tannins, carbohydrates and esters floating around which, in their complexity, add up to taste and aroma.

So how do you get the "haze" proteins out, without removing all the useful molecules?

Ever since 1900, brewers have tried to solve the problem by introducing certain protein-eating enzymes into their brews.

But that's a trade-off, because the enzymes aren't selective enough about what they hydrolyze. They consume "head" proteins just about as readily as they dine on "haze" proteins.

And once you put enzymes into beer, they stay there, with the natural products of fermentation in your glass.

The natural answer.

Enter silica hydrogel made by PQ and marketed in food grade to the brewing industry. The hydrogel is a highly adsorptive silica sponge with holes of a tightly controlled angstrom size, which can easily be added to, and then removed from, your beer.

On its enormous internal surface area, the silica hydrogel adsorbs the proteins that will fit within its pores—and the *only* proteins that will fit are haze proteins.

So the "head" proteins, the tannins, carbohydrates, and esters, keep up their good work.

Brewers add only a small amount of our hydrogel to their brews. And then filter out every adsorptive particle, satisfying the master brewer and consumer alike with a brilliant, full-headed natural beer.

The silica solution.

Silica hydrogel is just one of the ways that we turn sand, air, and water, into things that are useful and interesting.

If you'd like an adsorptive something with an affinity for something else, well, then, we'd certainly like to talk.

Over a beer, maybe?

Imagine: after 150 years, we're continually finding new uses for silicates. And, after 150 years, we're still delivering silicates of consistent quality, at competitive prices.

That's why you can rely on PQ.

The PQ Corporation
P.O. Box 840, Valley Forge, PA 19482

QUESTIONS FOR DISCUSSION

1. Who is the audience for this advertisement? What details in the advertisement (including the journal in which it appeared) help you to answer this question? How does the explanation given here differ from that in "The Brewing Process" on pages 164–165.
2. What does the PQ Corporation want the reader to do immediately after seeing their advertisement?
3. How does the advertisement use humor to get its message across? Is the use of humor appropriate? Would the advertisement be less effective if the humor were omitted?
4. What image of the PQ Corporation does the advertisement project?
5. The advertisement uses short sentences in the active voice. The tone is friendly, almost informal. Are these stylistic features appropriate to the copywriter's purpose?
6. Discuss the effectiveness of the layout of this advertisement, especially the placement and use of visuals, type faces, and the logo.

SUGGESTIONS FOR WRITING

1. Rewrite the advertisement making it suitable for the consumer.
2. Write an ad for a technical service you or your company can supply. Include an appropriate visual.

chapter 12

Correspondence

Everyone knows that letters and memos play a crucial role in business. Used to communicate with the reader outside a company, letters request and furnish information; elicit prompt action; and influence the reader to buy a product, to approve a policy, or to hire you. Companies pay a premium for their letters. The National Office Management Association estimates that it costs between seven and twelve dollars to process a single letter—from composing and formatting to storing and mailing a copy. Memos, especially longer ones, can be equally expensive. Primarily sent to individuals within a company, memos explain and monitor the internal workings of a company; they play an important role in helping a company run smoothly. Organizations could not exist without correspondence.

This chapter begins with an overview of the fundamental principles of effective correspondence and then explains how to write the types of letters and memos you will be most likely asked to prepare.

Your letter and memo writing ability can advance or hinder your career. The higher up the organizational ladder you wish to climb, the more skillful you will need to become at writing letters and memos. Be on guard against some frequently expressed misperceptions about correspondence. After each misleading idea below appears a reason why you should reject that erroneous statement.

1. *Most communication in business and industry is done over the telephone or in person, thus relieving me of much of the burden of writing.* Although there is a great deal of oral communication in the world of work, customers and employers require frequent written notification about business transactions and may often need a written report as a follow-up to a telephone call or conference. Maintaining written records is therefore important for business, professional, and legal reasons. As a general rule, if something has not been written, it has not been planned, agreed to, or accomplished.

2. *A secretary will write my letters for me, or at least save me from mistakes of fact and errors in spelling and grammar.* The fact is that you are often on your own. Companies do not have the money to hire trained word watchers for employees, nor is there a readily available corps of secretaries whose function is to be accomplished writers.

3. *Most of my correspondence will be dictated and hence save me from the agony of writing.* This statement contains a half truth. Because of the pressures of time much of your routine correspondence will be dictated. But for nonroutine or important correspondence, you will do plenty of research and editing. You will need more than two minutes in front of a dictaphone to collect, organize, and present your ideas. You might have to write several drafts of an important letter. Experts estimate that as much as two-thirds of an individuals's writing time is spent planning what to say.

4. *Sophisticated data processing equipment (word processors, electronic mail) will decrease the time I have to spend writing and thus make the writing I do close to effortless.* Rather than decreasing your writing, the availability of computer hardware may actually increase your writing time. More sophisticated hardware means that companies will expect greater efficiency and productivity. Moreover, while the transmission and storage of ideas is enhanced by such equipment, using it will not automatically free you from the time and effort that writing requires.

THE "YOU ATTITUDE"

Through letters you establish a relationship with a reader. Depending on what you say and how you say it, that relationship can be cordial, bland, or downright hostile. To obtain the cooperation or approval of your reader, be sensitive to various needs. To demonstrate a high regard for the reader, put yourself in that person's place. What will the reader find of interest or profit in your letter? Adopt the "You Attitude" by emphasizing the reader's problems or opportunities rather than your own. A letter in which each sentence begins with "I" crowds the reader off the page. For example, the following reply to a potential customer's request for information lacks a "You Attitude":

> Our model #332211-J has the specifications we think are most useful—36" length, 24" width, and 30" height. We supply this model in three colors—black, white, and aqua—and have been very pleased by the customer's acceptance of our choices. Please let us know if we have answered the query that has been directed to us. We have a reputation for being thorough and prompt.

Rewritten with the "You Attitude" in mind, the reply makes the reader—not the company—the center of the letter, and thus increases the chance for a sale.

> Thank you for your request for information. You will find that model #332211-J has the specifications that will meet almost all of your needs; it is 36" long, 24" wide, and 30" high. You can choose from three colors—black, white, and aqua.

> For your convenience I am enclosing a brochure describing the model in detail. If you would like further information, please feel free to call or write.

As you will see, the "You Attitude" is particularly essential to the success of a letter of application. An employer will not hire you just because you want the job and believe you have the qualifications for it. You have to demonstrate that your qualifications will help the employer's company.

STYLE

The style of your letter can affect whether the reader responds favorably or not. Letters should be personable and readable, not stiff and awkward. This does not mean that you should be obtrusively informal or use slang or incorrect grammar. But you should avoid *letterese,* a disease afflicting writers who think that they must use pompous words instead of simple ones. You fall victim to letterese when you write sentences such as the following:

> It has come to our attention that recent employees have failed to peruse relevant sections of the company policy manual. Accordingly, herewith please find enclosed the marked sections, which present themselves for immediate attention.

With these words you place the reader in a cold, unfriendly, unnatural light. Your reader will not feel that you perceive him or her as a human being. Using direct, sincere words shows the reader that you are friendly and human.

> As a new employee, you should take time to review the company policy manual; I have marked some relevant sections for you to consider. The manual should help you to become familiar with your duties as well as with company benefits.

All readers appreciate brevity. Don't turn your letters into telegrams, but be concise. Prune deadwood from your letters by replacing windy phrases like "at this point in time" with "now" and "whether the possibility suggests itself in the foreseeable future" with "if." Cut wordiness, too, by avoiding such redundant phrases as "close proximity," "brown in color," "midway between," "cancel out," and "true facts" (are there false facts?).

As you saw in Chapter 2, you increase the chances of giving the reader exactly what is needed when you use concrete words ("13 metric tons") rather than abstract ones ("a large volume"). Tell the reader how much, how long, how many, who, when, and why in specific terms. A letter with vague words raises questions that may require additional and costly correspondence. Moreover, a letter filled with abstractions is hard to get through.

Consider also the impact of your paragraphs and sentences on the reader. A careful letter follows a structure—an introduction, a body, and a conclusion. Provide forecasting statements to let the reader know how you have organized your work. For example, "This and the next two paragraphs discuss the advantages of the gasification system." Use subheadings when they are helpful, indentations, underscorings, and adverbs of time or cause (e.g., *first, second, third,* and *finally*).

Many readers skim letters, paying attention mainly to the first sentence in each paragraph. Consequently, it is especially important to begin each paragraph with a clear topic sentence that informs the reader about the contents of the paragraph. Within each paragraph arrange information logically, keeping important points near the start of the paragraph so the reader can locate them easily. Do not bury key facts or ideas in the middle of the paragraph. Between paragraphs provide helpful transitions using the type of connective words found in Table 5.1 (page 161).

Also consider the visual impression your paragraphs make. Thanks to word processors, which show writers their sentences and paragraphs on screens, readers are more conscious of how a piece of writing looks. Big chunks of uninterrupted prose look formidable and make for difficult reading. The trend in business letters is toward brevity and directness. But do not go to the extreme of writing shoestring paragraphs (for example, a string of paragraphs one or two lines long). Very short paragraphs make your letter look like a telegram and send a clipped, abbreviated message to the reader. These short paragraphs may also be in need of factual supporting details to clarify and explain a message for readers.

FORMAT

The appearance of your letter can encourage the reader to respond positively to your message. Letters should look neat and follow accepted conventions, i.e., placing the correct information in the proper places. A variety of formats exist, but the two most common types are the *block* and the *semiblock*. Figure 12–1 illustrates the block format and 12–2 the semiblock one; Figure 12–1 also identifies the parts of a letter.

The letter in *block format* is easier to prepare. Everything is typed flush against the left-hand margin, and to indicate a paragraph break you simply double space. In the semiblock format, the date and complimentary close are aligned, both typed a fixed number of spaces away from the left-hand margin. A variation of the semiblock format is the *indented letter* illustrated in the second model in this chapter—Robert Eddy's letter of application on page 429.

The rest of this chapter is devoted to a discussion of the most frequent types of letters and memos you will write in your professional career.

APPLICATION LETTERS AND RÉSUMÉS

A letter of application may be the most important letter you will ever write. Through it you sell yourself to a prospective employer who will receive letters from many other applicants who want the same job as you do. Your letter must display confidence but avoid boastfulness. Send an original letter—never a copy—that is error free. A spelling mistake or a typographical error is more than embarrassing; it could be fatal to your chances.

letterhead	**The PQ Corporation** Valley Forge Executive Mall P.O. Box 840 Valley Forge, PA 19482 (215) 293-7200
date	October 19, 1983
inside address	Philip C. Kolin Janeen L. Kolin Southern Station Box 8395 Hattiesburg, Mississippi 39406
salutation	Dear Mr. Kolin and Ms. Kolin:
opening	Thank you for your recent letter and the generous praise for our corporate ad on the use of silica hydrogel in the brewing process. You are welcome to reprint the ad in your text on technical writing.
body of the letter	Credit for the copywriting and creative direction properly belongs to Joseph Meehan at Schaefer Advertising Inc., our advertising agency. This is one of a series of ads in a corporate ad campaign that the agency proposed to build recognition for our company in terms of its basic strength in silicate chemistry. The enclosed brochure of reprints shows other ads in the series. I have asked Schaefer Advertising to send you the necessary artwork and to arrange for its safe return. If you want to reach Mr. Meehan directly, the address is Schaefer Advertising Inc., Leighton Building, P.O. Box 847, Valley Forge, PA 19482.
closing	We would be pleased if you would send us a copy of the textbook when it is published.
complimentary close	Sincerely yours, *John M. Reed*
typed signature	John M. Reed Manager, Corporate Communications
stenographic notation	JMR/kp
enclosure	Encl. 1
carbon copy	cc. Joe Meehan, Schaefer Advertising

Figure 12–1. A letter in block format.

DATA SURVEYS, INC.
2345 West Highway 98A
Milwaukee, WI 56743

May 1, 1984

Mayor Warren P. Green
Office of the Mayor
Town Hall
750 11th Avenue
Porterville, WI 58439

Subject: Public Parks Report

Dear Mayor Green:

I am submitting the accompanying report "The Accessibility of Public Parks in Porterville, Wisconsin" for your review and approval. This report was commissioned by the City Council last December to determine if parents in Porterville think a new public park is necessary and feasible. We sent a questionnaire to over one hundred parents in the Porterville area to survey their opinion.

This report discusses the respondents' backgrounds and ideas on the importance of a new public park. It includes the respondents' opinions on park facilities, location, and use. The parents do not believe that Porterville's existing parks offer ample, accessible recreation opportunities and feel a new public park should be constructed within the next year.

In preparing my report I was assisted by two members of the Board--Mary Ellen Arp and David Whitworth. Patricia Hess of the Porterville Park System granted me three helpful interviews.

If you have any questions or comments about the report, I would be happy to discuss them with you. Please write or call me at 756-9800, ext. 345.

Sincerely yours,

Janis Monk

Janis Monk

Encl. "The Accessibility of Public Parks in Porterville, Wisconsin"

Figure 12–2. A letter in semiblock format.

Before writing the letter, do some homework about yourself. Assess your strengths and weaknesses to have a clear idea about the type of position you want and are qualified for. Take a close look at your course work, career goals, and job experiences. Establish a file at your school's placement office and ask at least four individuals (teachers, former employers, civic leaders) to write positive letters of recommendation for that file.

Do some homework about your prospective employer, too. Find out something about the company or agency (equipment used, current research projects, location of plants, major product innovations) you can include in your letter to convince the reader of your sincere and informed interest. Obtain the name and title of the appropriate contact person. Never address your letter "Dear Sir," which is sexist and nonspecific, or "Dear General Dynamics," which shows a lack of enthusiasm. Call the switchboard to find out the personnel officer's correct name and, if necessary, ask the operator to spell it.

Your letter should persuade the reader to interview you. In your first paragraph state why you are writing and mention the specific job you seek. (If your letter is unsolicited, inquire if such a position is, or will be, available.) Indicate where you learned of the job, especially if it was from someone (a professor, an employee of the company) the reader knows. If possible, mention any appropriate interests or achievements that might distinguish you from other applicants.

The body of your letter should include specific evidence of your qualifications for the position. The "You Attitude" is essential here. Emphasize what you can do for the reader's company; relate your achievements to the company's needs. A vague or boastful statement can hurt your chances for employment. Do not say "I excel in getting a job done." Rather, show what you have done in the past so that your potential employers will know what you can do for them and how you can fit it into their company: "While working at Dearborne Products, I was responsible for making out employee schedules, ordering materials, and performing basic accounting functions. These are the duties I can perform well as your new assistant manager."

End your letter by politely requesting an interview and specifying the times you are available.

Accompanying your letter should be a one- or two-page résumé summarizing in outline form your educational and professional experience. Résumés that are attractive and easy to read win points for job seekers. To make your résumé look professional, take it to a printer to be designed and reproduced. Put your name, address, and phone number at the top of the résumé. Then include separate sections for the following information:

CAREER OBJECTIVE: the job you seek and perhaps your future career plans

EDUCATION (most recent first): degrees (expected or completed); types of, and credits in, major courses; special training (reports or labs); honors and G.P.A. (only if over 3.0)

EXPERIENCE (most recent first): titles, and responsibilities, even if unrelated to your current job search; include military experience

PROFESSIONAL ORGANIZATIONS: include offices held

PERSONAL (optional): hobbies, community service, travel

REFERENCES: state "available on request" or "available from placement office" or list names if reader is likely to know them

The order of "education" and "experience" will vary for different job candidates. Experience would precede education if the person's work experience is a more impressive selling feature than education might be. Individuals with years of experience who return to school should list their experience first. Federal laws prohibit employers from discriminating against candidates because of age, race, sex, marital status, national origin, religion, or handicaps. However, your résumé should include anything about yourself that will reflect favorably on your qualifications and help you to benefit from the employer's hiring goals.

INQUIRY AND REPLY LETTERS

Inquiry letters seek information about a wide range of subjects, from the simple (the availability of a product) to the complex (the financing of a major research project). When you write to a company in response to an advertisement or a catalog, your letter is eagerly solicited because it might bring the firm business. When a student writes to a company for information to complete a report, the letter is unsolicited; it asks for a favor.

Make your inquiry letter easy for the writer to answer by asking reasonable and precise questions. Never ask for "whatever information you have" on a topic. Restrict and outline your specific requests before you put them in writing. A courteous tone is also important. A letter that brusquely demands "Give me the detailed descriptions of your plans for solar converters and send them to my home address within the next ten days" is less likely to receive a prompt and complete answer than one that politely asks, "Would you please send a copy of your plans for a solar energy converter to me so that I might be able to study them and possibly construct one for my home? I would appreciate your sending me the plans as soon as possible." In order to ensure a prompt and full reply your letter of inquiry should answer the following questions for the reader:

1. who you are
2. why are you writing to this particular person or office
3. what specific information you need
4. why you need it
5. when you need it

Formulate precise questions and, if there are several, number and indent them. Be realistic, though. Don't expect the reader to answer ten or more questions listed separately in your letter. If you have a large number of questions, chances are that you have not sufficiently restricted your research. If you want the reader to complete a survey, send a separate questionnaire (see pages 252–255) with your letter.

Close your letter politely, but do not say, "Thank you in advance for your help." Then you are presuming that the reader will and can do exactly what you want and when you want it accomplished. Instead, close by saying you appreciate the reader's considering your request.

A reply to an inquiry letter provides the necessary information the reader seeks so that a second letter is not required. Try to answer questions in the order in which they were asked without repeating the author's original questions verbatim. You can further answer an inquiry by including copies of relevant brochures, catalogs, or nonconfidential reports. If the writer requests information you do not have, citing other relevant sources will be appreciated. Of course, if a writer requests information that is confidential or incomplete, explain why you cannot provide it. You might ask if a nonconfidential, related study you could send would be a useful alternative.

CLAIM AND ADJUSTMENT LETTERS

A claim letter voices your dissatisfaction with a product or service; an adjustment letter responds to a claim letter. Both require tact. If you write a discourteous or angry letter, you are not likely to receive what you want. A diplomatic letter, however, may help you to obtain one of the three things you are most likely to request—a replacement, a repair, or a refund. Companies want to hear if they are doing something wrong so that they can correct the situation and not lose a customer. But before you write, double-check your warranty to make sure of the company's obligations and your own responsibilities.

Guidelines for Writing Claim Letters

If you have to write a claim letter, follow these time-tested guidelines:

1. *State why you are writing.* Alert readers to the problem you want to bring to their attention.

2. *Identify the product or service accurately.* For a product, include model and serial numbers, size, cost, and the place and date it was purchased. For a service, include vendors' names, duties, and dates.

3. *Explain specifically what went wrong.* Tell how you were using the product (so the company knows you were following the provisions in a warranty),

the number of times you used it, and describe what the problems looked or sounded like.

4. *State exactly what you want the company to do.* Do not mutter a weak and vague complaint: "I expect you to do something about this problem." Specify if you want a new part shipped, a defective one repaired, the use of a loaner for a week, a refund in cash, or a discount on future items or services.

Strategies for Writing Adjustment Letters

A letter of adjustment honoring the reader's request is easier to write than one refusing it. Both should be courteous. A letter begrudging the reader a request further threatens the chances for maintaining mutual goodwill; an angry letter telling the customer that he or she made a mistake worsens an already tense situation.

When you grant the request, begin with an apology and a thank-you. Then, tell the reader right away what the adjustment is—don't delay good news. Next, explain why there was a problem and what you did to correct or stop it. End on a friendly note by inviting the customer to write again if you can help.

When you refuse a request, do not start off with the bad news. Ease the reader into it. Identify something you and the customer can agree on, e.g., your company, like the customer, does not appreciate inconvenience. Then give the reason for your refusal. Readers will be less angry if they know that you have taken the time to investigate their complaint and provide more than just a curt one- or two-sentence refusal. End on a positive note by extending to the customer some courtesy—perhaps a discount on a future purchase of your product.

SALES LETTERS

Sales letters are an important part of a company's advertising program. As a consumer you have doubtless received many of these letters. The same techniques used in sales letters for consumers can be used successfully in promoting a technical product or service. Writing a letter that will be read and acted on requires faithful adherence to the "You Attitude." Since self-interest is the motivating force guiding the reader of your sales letter, stress the benefits of your product or service for the reader.

The opening of a sales letter is crucial. If you lose the reader here the rest of your letter is likely to go unread. Begin with something that attracts the reader's attention—a question, an offer to solve a problem, an interesting or little known fact, an impressive statistic, or even a photograph. Make a potential customer want to continue reading your letter so that you can sell your product or service. After your opening, pinpoint a problem that affects the reader and provide a benefit. Then, in the next part of your letter, show how your company can solve that problem or deliver that benefit. Use persuasive language that emphasizes advantages, for the reader not your company. Describe your product or service in graphic language that makes the reader want to see it or use it. When writing for a group of specialists, you may have to use technical terms or symbols. But do not turn your letter into a specification sheet (discussed on page 379). Supply only enough information to get the reader to call or write to your company. That is the last stage of your letter. Make it convenient for the reader to express interest. If you can, provide a toll-free telephone number, ask potential clients to call collect, or include postage-paid envelopes or response cards. Other tactics include offering the reader a free demonstration or a catalog if he or she calls or writes.

MEMOS

Memos (*memorandum,* singular; *memoranda* or *memorandums,* plural) are used primarily to communicate with individuals within your own company or organization, thus explaining the adjective "interoffice" often used before "memo." Memos circulate information that requires the reader's immediate attention and action. For example, a memo may announce a meeting or change in policy or personnel, provide explanations or instructions, outline main points of a telephone call or meeting, or describe a sales strategy. As we saw in earlier chapters, memos can also serve as progress reports (pages 347–361), internal proposals (pages 268–270), and feasibility studies (pages 308–319). Memos travel up and down the organizational ladder (from decision makers to employees and vice versa) as well as laterally (from one employee to another).

To provide information quickly, memos are more streamlined than letters. Notice how the sample memo included in Figure 12–3 differs from a letter. Unlike a letter, a memo does not contain an inside address, salutation, or complimentary close, formalities unnecessary for coworkers. Memos use the following four headings (often preprinted):

TO:	(Names and titles of reader)
FROM:	(Names and titles of sender)
SUBJECT or RE:	(concerning—your message)
DATE:	(always include a calendar date, not just day of the week)

Some memos include a file or reply number to make answering and filing them easier. Memos can be sent on half (5 x 8½") or full sheets of paper with or without letterheads. Memos are often short (one page), but when they require more than one page, they identify second and subsequent pages just as letters do.

The principles of effective letter writing (using the "You Attitude" and having a concise, personable style) also apply to memo writing. Any reader deserves courtesy and clarity, especially your employers who will judge your performance by your written work.

Unlike the audience for your letters, the reader of your memo is more

WOODRITE PRODUCTS

Danbury, VT 02133

INTEROFFICE MEMO

TO: All Regional Managers

FROM: Julie R. Morris, Production Supervisor J.R.M.

SUBJECT: Liquid Slip Sheet Problem

DATE: October 10, 1984

Our office has received a number of complaints from your area salespeople about the film that adheres to our stained panel products because of the liquid slip sheet we use. This condition is neither serious nor permanent.

All stained panel siding products are protected from sticking and transit damage by the application of a liquid slip sheet. This milky white aqueous liquid solidifies on the back of the panel, leaving a thin white film for protection.

Since our panel sidings are stacked back to face, the film may occasionally transfer to the high levels of embossing on the panel faces. If this happens, ordinary tap water will wash off the transferred film.

Please inform your sales force about this condition. If you have any questions, please call me.

Figure 12–3. A sample memo.

likely to know the background of a situation or problem and is frequently (though not always) able to use the technical vocabulary and symbols of your profession.

Begin each memo with a forecasting statement or a statement of purpose. Come to the point at once by giving the reader the most important information first. If you have a number of points to cover, number your paragraphs, use sub-paragraphs as the memo on the configuration of a C-130 does on pages 461–463, and make these items easy to follow using plenty of white space. Memos can end in a variety of ways, depending on your purpose. In a conclusion, you might do one of the following:

1. Ask the reader to call or write for further information or clarification.
2. Provide a list of recommendations.
3. Remind the reader of the implications of a change of procedure.
4. Ask the reader to communicate with others within or outside your organization.

A CHECKLIST FOR WRITING LETTERS AND MEMOS

1. Adopt a "You Attitude" to relate your correspondence to your reader.
2. Be courteous.
3. Pay special attention to style—avoid letterese, strive for brevity, be specific.
4. Prepare each letter and memo carefully to look neat and professional.
5. For application letters, focus on what you can offer rather than on what you want.
6. For inquiry letters, make all requests as precise as possible.
7. For claim letters, identify the product, explain what went wrong, and state specifically what you'd like done about it. For adjustment letters honoring the reader's request, do it early and cheerfully; when you refuse the request, explain why.
8. For interoffice memos, be conscious of the "You Attitude"; begin with a statement of the topic and don't leave the reader in doubt about what you want him or her to do.

APPLICATION LETTERS WITH RÉSUMÉS

The first model in this chapter is a solicited letter of application (with résumé) written by a computer science major looking for a position as a systems analyst. Her letter is one-page long, the standard length for an application letter from an undergraduate student. Although it is brief (5 paragraphs) and to the point, Tina Bond's letter is highly persuasive, personable, and displays the "You Attitude" throughout. The style and organization of her letter are appropriate for her objective—to convince the reader that she is the best qualified individual for the job.

A Solicited Application Letter and Résumé

3500 Taylor Street
Jacksonville, FL 32830

May 1, 1984

Mr. Henry West
Director of Systems Analysis
Consolidated Industries
7900 South Morgan Drive
Ft. Pierce, FL 34507

Dear Mr. West:

I am writing to apply for the position of Systems Analyst that you advertised 1
in the Jacksonville Times. On May 25, I will receive a B.S. degree with high honors in computer science and a minor in general business from Center University. I would be available to start work any time after June 1.

Unlike most computer science graduates from Center, I have completed 18 2
more credit hours than were required for my major, including courses that trained me in FORTRAN, COBOL, PASCAL, SNOBOL, and RPG II. All of my assignments were completed on the VAX/11/780 model computer, which, I believe, is very similar to the VAX/11/750 model your company uses.

My greatest interest is in operations research, and I completed an internship 3
in this area under the direction of Dr. Frank Natoile, a professor of computer science. For my final research report, I prepared a one-year feasibility study on maximization of profits at a small electronics firm, similar in capability to Consolidated's. This project gave me experience in developing programs for manufacturing, production, and finished goods inventory systems, areas in which a systems analyst needs to excel.

My work experience is equally appropriate. The two most significant positions 4
I held were at Jacksonville banks, where I received commendations for my thorough work. My responsibilities included evaluating checking account problems, organizing customer checking files, and operating check microfilming machines.

I would appreciate an opportunity to discuss my academic and job experiences 5
with you and to provide you with any details not listed on the enclosed resume. I will be available for an interview any weekday except Tuesday after 1:00 until May 25. After that date, I am available any time at your convenience.

Sincerely yours,

Tina Bond

Tina Bond

Enc.

TINA BOND
3500 Taylor Street
Jacksonville, FL 32830
Phone: (904) 268-7804

CAREER OBJECTIVE:	To become a computer systems analyst with a large manufacturing company and gain eventual responsibilities in equipment design and application.
EDUCATION:	
1982–1984	CENTER UNIVERSITY, JACKSONVILLE, FL. B.S. in computer science expected May 1984. Top 25% of class with GPA of 3.3. Proficient in FORTRAN, COBOL, PASCAL, SNOBALL, and RPG II. Produced more than 25 technical lab reports on specific breadboarding techniques in Hardware I and II. As part of an internship, wrote a feasibility study on the year-long operations of a small electronics firm.
1980–1982	DAVIS COMMUNITY COLLEGE, JACKSONVILLE, FL. Received A.A. degree in computer science in May 1982. Dean's List all four semesters. Member of Phi Theta Kappa, a scholastic honor society. Earned a scholarship for high achievement in final two semesters.
EXPERIENCE:	
1981–1982	NATIONAL BANK, NORTH SHORE, FL. As a full-time customer service clerk, evaluated customer checking account problems by phone and in person, filed debits and credits, operated microfilming machine, and supervised overdraft charges.
1979–1980	CITIZENS BANK, JACKSONVILLE, FL. As a part-time score phone clerk, conducted in-bank on-line debits via computer; filed closed account sheets, insurance cards, and checks; mailed checking and saving statements; and assisted customers in person and over the telephone.
PROFESSIONAL MEMBERSHIPS:	Association for Computer Machinery Electronics and Electrical Engineering
REFERENCES:	Dr. Frank Natoile Department of Computer Science Center University Jacksonville, FL 32073 (904) 345-8720 Ms. Alice Newcomb Manager Citizens Bank 10 S. Main Jacksonville, FL 32110 (904) 421-3200

QUESTIONS FOR DISCUSSION

1. Describe the style, tone, and format of Bond's letter. What kind of verbs has she selected to emphasize her qualifications?
2. Bond had to be selective in describing her educational experiences. What specific kinds of accomplishments did she choose to include? Why?
3. How has Bond related her accomplishments to her potential employer's needs? In doing so, how has she demonstrated an interest in and knowledge of Consolidated Industries?
4. How would the description of Bond's previous job experiences convince Henry West that she is qualified for the position?
5. How does her letter help to interpret the résumé? What information appears in both the letter and the résumé? In just the résumé? Why?
6. Discuss the format and parts of Bond's résumé. How has she helped the reader to locate and evaluate information about her accomplishments quickly?

SUGGESTIONS FOR WRITING

1. Find an advertisement for a position (in a newspaper, professional journal, placement office notice) that you would like to have upon graduation. Write a solicited letter of application for that position. Bring the announcement of the position to class along with your letter.
2. Write a résumé to accompany the letter in question 1 above.

Unlike Bond's application, the next letter and résumé are unsolicited. The applicant, who has a background in forestry and a graduate degree in business, saw an opportunity to apply for a position and proceeded with confidence. The organization of Eddy's résumé differs slightly from Bond's. Since he has a great deal of work experience, he has placed this category before education to emphasize this fact for the reader. The Eddy letter and résumé illustrate one way individuals with years of job experience might organize these important documents.

An Unsolicited Application Letter and Résumé

620 South 28th Avenue, Apt. 422
Hattiesburg, MS 39401
April 6, 1983

Mr. Stuart Carlson
Manager, Industrial Relations
Leaf River Forest Products, Inc.
7 Professional Parkway
Hattiesburg, MS 39401

Dear Mr. Carlson:

While touring the Leaf River facilities to gather information for a mar- 1
keting class, I had the opportunity to meet with Mr. Don Hawthorne and his
staff. I appreciated the chance to speak with him about Leaf River Forest
Products, Inc. and to learn of the diverse opportunities your company offers
its employees. During our conversation Mr. Hawthorne suggested that I apply
for a management trainee position with leaf River. Would you please consider
this letter as my application for the position?

My studies at Louisiana State University in industrial forest management 2
familiarized me with harvesting procedures and equipment used by large forest
products companies like Leaf River. My work as a procurement forester with
Georgia-Pacific after my graduation taught me to write timber appraisals,
secure procurements, and coordinate landowner assistance programs, activities
highly applicable to work as a management trainee at Leaf River. Most recent-
ly, I was the Southern Farm Manager for Go-Pher Evergreen of Minnesota, with
responsibilities in land management and cultural practices.

In addition to my academic and work experiences in forestry, I have done 3
much related business research. Next month, I will receive an M.B.A. from
the University of Southern Mississippi, where as a part of my studies I am
completing a seminar paper on Leaf River's international approach to the
marketing of pulp. The enclosed resume provides further details about my
achievements.

I would appreciate an opportunity to speak with you further about the 4
management trainee program and my qualifications for it. And I would, of
course, welcome the chance to revisit the Leaf River facility, Mr. Hawthorne,
and his staff. I could come for an interview any Monday, Wednesday, or Fri-
day afternoon.

Sincerely yours,

Robert N. Eddy, III

Enc.

RÉSUMÉ

ROBERT NEWTON EDDY, III
Registered Forester

620 South 28th Avenue
Apartment #422
Hattiesburg, Mississippi 39401
Telephone: (601) 268-6209

CAREER OBJECTIVE:	To work as a professional forester, with responsibility for managing timberlands and procuring raw material for a progressive forest products company.
EXPERIENCE:	
1981–Present	While working full-time on MBA, served as a consultant to Christmas tree growers in Mississippi, and worked on timber marking crew for John G. Guthrie and Sons, a regional forestry consulting group in Picayune, Mississippi.
1981 (Jan.–Aug.)	As southern farm manager for Go-Pher Evergreen, Inc., a Minnesota-based Christmas tree production firm, was responsible for hiring, training, and supervising all full- and part-time employees. Implemented Christmas tree cultural practices and pest control on the 100,000-tree plantation.
1979–1980	Field instrument analyst for Scientific Gas Processing, Inc., New Orleans. Responsible for marketing, installing, and monthly servicing of gas measurement instruments for territory with 25 off-shore production platforms and 5 on-shore production facilities.
1977–1979	Procurement forester, Georgia-Pacific Corporation. Located, appraised, and procured standing timber for the firm's chip-n-saw mill in Marion County, Mississippi. Supervised one technician.
1977 (May–Sept.)	Forest Soils Assistant at Gila National Forest at Silver City, New Mexico.

Responsible for locating, gathering, and performing lab analysis on soil samples for the Soil Science group. Developed map reading and aerial photography skills.

EDUCATION:

1981–Present	M.B.A., the University of Southern Mississippi, May, 1983; have completed 9 semester hours in marketing (including one course in international marketing) and one course in personnel management. Was awarded a graduate assistantship by the Finance Department of the USM School of Business. Earned 100% of educational expenses.
1973–1977	B.S. in Forestry, Louisiana State University, Baton Rouge, Louisiana. Completed 48 semester hours of study in forestry, with an emphasis in industrial forest management. Paid for 70% of undergraduate education.
PUBLICATIONS:	"Forestry Outlook for Louisiana for 1977," 1977 Outlook for Louisiana Agriculture, 3 (December 1976), 25–30.
	"Plantation-Grown Christmas Trees for the South Mississippi Investor," Business Insights, 2 (Fall, 1982), 6–8.
HONORS:	Elected treasurer of Student Chapter of the Society of American Foresters at LSU, 1976–1977.
	Awarded the State President's Recruiting Award while member of Marion County Jaycees, Columbia, Mississippi, 1981.
PROFESSIONAL ORGANIZATIONS:	Mississippi Registered Forester (License No. 561); Society of American Foresters; Student Chapter of Society of American Foresters–LSU; National Christmas Tree Association; Louisiana-Mississippi Christmas Tree Association (Charter Member); Graduate Business Association (Charter Member)–USM.
REFERENCES:	Available upon request.

QUESTIONS FOR DISCUSSION

1. How does the first paragraph of Eddy's letter capture the reader's attention and encourage him to begin with a favorable impression of Eddy?
2. How many times does Eddy refer to the Leaf River Company in his letter? Is this good or bad? Explain.
3. What work and educational experiences does Eddy stress in his letter? How does he show that they are relevant for and of immediate use to Leaf River?
4. Why does Eddy mention Mr. Hawthorne's name again in the closing paragraph? What effect is this paragraph intended to have on the reader?
5. What kinds of information about work and education does Eddy include in his résumé? What specific details in the résumé are omitted from the letter? Why?
6. How does the style of Eddy's letter differ from his résumé? How has he managed to say a great deal about himself in two résumé pages? Discuss the specific ways he has highlighted and summarized his accomplishments.
7. Explain the relevance for his job search of the following details in Eddy's résumé:
   a. the title "Registered Forester" under his name
   b. the career objective
   c. payment for his education
   d. publications
   e. honors

SUGGESTIONS FOR WRITING

1. Write an unsolicited letter of application with a résumé using the strategy and format exemplified by Robert Eddy's letter and résumé.
2. Write a letter to Eddy from Carlson thanking him for his application but explaining there won't be an opening for 2–3 months. Provide an explanation that shows Carlson's enthusiastic interest in Eddy.

LETTERS OF INQUIRY AND REPLY

The first letter is an inquiry letter sent to a manufacturer from a purchasing agent employed by a junior college school district, which needed information to decide on which type of sewing machines to order for its home economics classes. The purchasing agent needed a prompt and complete reply in order to make her report to the board within a month of sending the letter. To assist the reader, she clearly identifies her reason for writing, lists specific questions for the reader to answer, and concludes with a specific date by which she must have the information. The writer makes it clear that it is in the reader's best interest to send a reply quickly. In the second letter the district manager at the Walton Company responds to Cassandra Burroughs's inquiry. Knowing that his reply could bring the company a large contract, the manager had two goals to achieve: (1) to answer all of Burroughs's questions completely and at the same time (2) to convince her, and ultimately the board, to buy his company's sewing machines. His letter is clear, polite, and timely.

Inquiry and Reply Letters about Purchasing Sewing Machines

HENDERSON JUNIOR COLLEGE DISTRICT
Henderson, PA 19342

May 11, 1984

Mr. Tom Zappata
District Manager
Walton Sewing Machine Company
2300 South Clancy Blvd.
Columbus, OH 46321

Dear Mr. Zappata:

I am writing to request information about your sewing machines and storage cabinets, since we are thinking of purchasing these items for our home economics classes for the upcoming school year. We require a model that is both efficient and equipped with a variety of attachments to meet the needs of our students. 1

We intend to purchase 25 sewing machines that will occupy no more than 50 sq. feet. The machines will be used by approximately 100 students each school day for at least 8 to 10 years. The machine must offer a straight stitch, zigzag stitch, automatic winding bobbin, and an attachable buttonhole device. Furthermore, each machine must be enclosed in an individual wood cabinet that will function as a desk when closed and provide storage space. To comply with our school board's standards, the cabinet must be at least 30 inches high and 40 inches wide. 2

Our Junior College Board has allocated $11,000 for the purchase of these machines, and we cannot exceed this sum. 3

Would you please answer the following questions to help us make a decision: 4

1. What type of sewing machine and cabinet would you suggest for our use?

2

2. Can you supply 25 appropriate machines and stay within our budget?

3. Could you supply the machines by September 1, 1984, in time for the start of the school year?

4. What are the terms of your warranty? How would it meet our special needs?

5. What kinds of maintenance contracts does your firm offer after the expiration of the warranty?

I look forward to hearing from you and would appreciate your answer to my questions before June 10, the date the Board meets to decide on a vendor. I will notify you about the Board's decision soon after this date. 5

Sincerely yours,

Cassandra Burroughs

Cassandra Burroughs
Purchasing Agent

Walton Sewing Machine Company

2300 South Clancy Boulevard
Columbus, OH 46321

19 May 1984

Ms. Cassandra Burroughs
Purchasing Agent
Henderson Junior College District
Henderson, PA 19342

Dear Ms. Burroughs:

Thank you for your inquiry about Walton sewing machines. To 1
meet your needs, I would suggest that you consider purchasing our free-arm convertible machine, the most durable and reliable model we make. It is equipped with all the functions you are looking for plus it has the additional benefit of the free arm. This machine can be installed in our wooden drop-down cabinet, custom-made to any dimension you choose at no extra cost.

We are always pleased to service organizations, and as a way 2
of showing our appreciation we offer a 15% discount on each machine, when 20 or more are purchased. Since you will be purchasing 25 machines, the price for each would be $385—a total cost of $9625, which is well within your budget.

All of our machines are backed by a 3-year warranty, which 3
provides maintenance at no additional charge. With the expiration of your warranty, our company will continue to provide maintenance, if you make arrangements with the service department. The terms of our service contract are especially attractive. I am enclosing a sample agreement for your inspection. Additional parts for machines can be ordered directly through our office and delivered to you within 1 to 3 days.

Since you request a cabinet that is not standard size, there 4
would be additional time required for the ordering. To have the machines ready and delivered by September 1, we need to receive your order by June 20.

2

I hope that I have answered all of your questions and that this information will assist you and the Board in reaching a decision. Please call me if I can provide any further assistance. Thank you again for writing to the Walton Company.

Sincerely,

WALTON SEWING MACHINE COMPANY

Tom Zappata
District Manager

Encl.

QUESTIONS FOR DISCUSSION

1. Evaluate the effectiveness of the first paragraph of the inquiry letter as a summary of the entire letter. What questions does it answer for the reader?
2. What types of information does the second paragraph of the inquiry letter provide for the reader? If this paragraph were omitted, could Zappata answer all of Burroughs's questions?
3. Why is the information about the college board budget in the inquiry letter listed in a single brief paragraph?
4. Are the questions that Burroughs asks relevant to her purposes? Why? How has she made these questions easy to answer for the reader?
5. Why does Burroughs wait until the last paragraph to mention the date by which she needs the information from the Walton Company?
6. Describe the tone of Tom Zappata's letter. In what ways does it reflect the writer's "You Attitude" toward the reader? Identify examples of this attitude and explain their effectiveness.
7. In what order does Zappata answer Ms. Burroughs's questions? Why does he follow that order?
8. How does Zappata make the cost of the machines an attractive feature for his reader?
9. Why does he include a service contract with his letter as opposed to summarizing its features?
10. Why does Zappata wait until the second to the last paragraph to discuss the delivery date? Why didn't he put this information in his first or second paragraph?
11. Discuss the style of Zappata's letter (word choice, sentence length and variety). Is his letter easy to read? Explain your answer by analyzing specific sentences.
12. How does Zappata's closing paragraph reaffirm his interest in the Henderson Junior College District and encourage a favorable reply?

SUGGESTIONS FOR WRITING

1. Write a letter of inquiry to an appropriate company or agency about one of the following subjects; identify a special problem you or your company has:
   a. meals on wheels for the elderly or handicapped
   b. fire extinguishers for a factory
   c. microcomputers for a college freshman dormitory
   d. centrifuges for a laboratory
   e. convention facilities at a hotel or motel
   f. a printer's procedures and capabilities in producing a company newsletter
   g. lasers for a plant's security

2. Write a reply to the letter of inquiry you wrote on one of the topics for question 1.
3. Write a letter from the Chair of the Henderson Board accepting Zappata's offer or write a letter from the Chair of the Henderson Board rejecting Zappata's offer and supply reasons for your refusal.

The following inquiry letter is not sales-related. It is written by a student to the director of research and development at a large company asking for information to help the student complete a paper on computerized blackboards used in teleconferencing. Realizing that he is asking for a favor from the writer, the student states why he needs the information, what he will do with it, and how he will acknowledge the reader's help. In the second letter Loretta M. D'Orsay replies to Ben Hutos's inquiry politely and helpfully. The director has taken time away from a busy schedule to help the student write his paper. Although she gives him most of the information he asks for, she does not answer all of his questions. Her reply will show you how to respond when you must include a refusal as part of your letter.

Inquiry and Reply Letters about Information on Computerized Blackboards

6409 Stewart Ave., Apt. 15
Springfield, CT 02134
September 10, 1984

Ms. Loretta M. D'Orsay
Research and Development Director
Landmark Industries
P.O. Box 8456
Mannington, CT 02365

Dear Ms. D'Orsay:

I am a first-semester senior at the University of 1
Northeastern Connecticut majoring in computer science and minoring in marketing. This semester I am writing a long paper for my business technology class on teleconferencing. My instructor, Dr. Maureen Eihler-Hayes, has told me of your company's use of computerized blackboards. I am writing to ask your assistance as I prepare my paper.

I plan to devote a major section of the paper to 2
computerized blackboards as teleconferencing tools and I would appreciate your answers to the following questions.

1. What is the cost of a computerized blackboard?

2. What is the standard computer communication link?

3. What are the major teleconferencing advantages in using a computerized blackboard?

4. What are the drawbacks in using the computerized blackboard in teleconferencing?

Dr. Eihler-Hayes also mentioned that your firm is 3
experimenting with modifications in the design of computerized blackboards you presently use. Would it be possible for you to tell me about these changes?

I will be grateful for your assistance and will acknowledge 4
your help in my paper. I will gladly send you a copy of it so that you might see how I used the information you send.

Sincerely yours,

Ben T. Hutos

Ben T. Hutos

LANDMARK INDUSTRIES
P. O. Box 8456
MANNINGTON, CONNECTICUT 02365

September 19, 1984

Office of Research and Development

Mr. Ben T. Hutos
6409 Stewart Ave., Apt. 15
Springfield, CT 02134

Dear Mr. Hutos:

I was pleased to learn of your interest in computerized 1
blackboards. They represent one of the most fruitful
applications of computers to business communications. For
the last year we have used these tools with a relatively
high degree of satisfaction.

To answer your specific questions, the cost of leasing the 2
computerized blackboard is modest, $500.00 per month. But do
not forget that there is an installation fee of about
$450.00. The standard communication link is two dial-up
telephones; the data link operates at 1300 bits/second.

In addition to its reasonable cost, the computerized 3
blackboard has offered us a number of advantages, including
a savings of thousands of dollars in travel expenses. Using
this technology, we are able to transmit both visual and
audio messages to remote sites. (The standard tele-
conferencing equipment from the phone company carries sound
only.) Moreover, the computerized blackboard allows us to
reproduce our training sessions at all four of our plants,
from Connecticut to Texas.

The blackboards do have some drawbacks. The message trans- 4
mitted is limited to what can be written on the blackboard.
Elaborate charts and diagrams cannot be transmitted; nor
does the computerized blackboard allow the use of color.

Regarding your request for information about our modifica- 5
tion of the system, I am sorry to tell you that this

2

information is confidential. However, you might be interested in some recent articles on computerized blackboards that have appeared within the last three or four months in *The Office*, *Computer Systems*, and *Engineering and Educational Technology*.

I hope this information helps you. By all means, send me a 6
copy of your paper. I am looking forward to reading it.

Sincerely yours,

Loretta M. D'Orsay

Loretta M. D'Orsay
Director

QUESTIONS FOR DISCUSSION

1. What kinds of research did the student have to complete before he could write this letter?
2. Why does the reader need the information included in the first paragraph of the inquiry letter? Could this information follow the four questions Hutos asks?
3. Discuss the function of the first sentence in paragraph two of the inquiry letter as both a transitional and a forecasting sentence.
4. In what order does Hutos ask his questions? Why doesn't he include the information contained in paragraph 3 as question 5?
5. How will the closing paragraph of the inquiry letter help the writer to receive a favorable reply from D'Orsay?
6. Is the information in the first paragraph of D'Orsay's letter helpful to the reader or could the letter simply start with paragraph 2? What effect would the letter have on the reader if it started immediately with the second paragraph?
7. Discuss the way D'Orsay moves from one question to another. Identify the transitional phrases she uses and comment on their effectiveness. Evaluate her topic sentences.
8. How does D'Orsay inform Hutos that she will not give him all the information he has requested? Should she have begun her letter with this news? How does her response reflect a "You Attitude"?
9. Why does D'Orsay refer the student to recent articles on the computerized blackboard?
10. Evaluate the closing paragraph of her letter, especially its tone. What impact is it intended to have on the reader?

SUGGESTIONS FOR WRITING

1. Write an inquiry letter to a professor at another school or to a local business for information that will help you to complete a project—a paper, survey, laboratory report, demonstration.
2. Write an inquiry letter to a foreign college or university at which you would like to study next summer. Indicate that you need information on the type of transferable credit, tuition, room, board, and local transportation. Specify why you want to attend.
3. Write a reply letter to the letter you wrote for questions 1 or 2. Include a refusal of part of the requested information in your reply.
4. Write a thank-you note to Loretta D'Orsay from Ben Hutos.

CLAIM AND ADJUSTMENT LETTERS

The claim letter printed below describes problems experienced by a delivery company that had a service contract with an automotive center. The writer gives his reader detailed information in order to show that the problem was serious and that the automotive center was responsible for it. He directly, though diplomatically, explains what he wants the automotive center to do. The adjustment letter that follows responds favorably to A. T. Eldred's claim letter. Written by the Vice President of the automotive center, the letter explains why the problem occurred and gives the reader the things he asked for. The writer has two goals—to rectify a wrong and to keep the reader as a customer.

Claim and Adjustment Letters about Service at an Automotive Center

NORTHWEST ILLINOIS DELIVERY SERVICE
Des Plaines, Illinois 60656

July 2, 1984

Mr. Robert L. Pierce
Vice President
Industrial Automotive Center
8800 Milwaukee Road
Niles, IL 60648

Dear Mr. Pierce:

Over the last two years, we have been pleased with our service contract with Industrial Automotive Center. However, a recent event has caused us a great deal of inconvenience, and I was sure that you would want to be informed about it. 1

Early yesterday morning (about 7:00 a.m.), one of our refrigerated trucks (a 1983 Badger) developed problems with its cooling system. This is one of the busiest times of the year for us, just before the 4th of July holiday. One of our drivers, Karen Biondella, took the truck to your shop for repair. 2

Shortly after Ms. Biondella arrived at your shop (about 7:40 a.m.), your Service Manager, Brent Grange, told her that a single hose had to be replaced and that the repair would take no more than 30 minutes. After waiting for more than 3 1/2 hours, Ms. Biondella was informed that the truck could not be repaired for at least four days--until a compressor would arrive from your Indianapolis distributor. 3

Because of your delay in reporting the correct information to our driver, we have experienced considerable trouble. Had we known earlier in the morning that the truck would be out of commission for four days, we would have had time to assign the driver's deliveries to other trucks still here in the warehouse. But because we found out about the repairs so late in the day, all of the other trucks had already left. Consequently, we were placed in the embarrassing position of having to call our customers to explain a delay that was all the more intolerable because of the approaching holiday. 4

2

Please let us know why our driver was kept waiting so long 5
before she received the correct information. Also, we wish
to be reimbursed for the time Ms. Biondella spent waiting in
your garage--$52.50 ($15.00 per hour × 3 1/2 hrs).

I look forward to hearing from you within the next four or 6
five days.

Sincerely yours,

NORTHWEST ILLINOIS DELIVERY
SERVICE

A.T. Eldred

A. T. Eldred
Manager

INDUSTRIAL AUTOMOTIVE CENTER
8800 Milwaukee Road
Niles, Illinois 60648
Phone: 345-6780

July 7, 1984

Mr. A. T. Eldred
Manager
Northwest Illinois Delivery Service
Des Plaines, IL 60656

Dear Mr. Eldred:

Thank you for bringing to my attention the problem you 1
experienced with the servicing of your 1983 Badger. I am sorry for the inconvenience you suffered and the embarrassment the delay caused you with your customers. The enclosed check for $52.50 will help to compensate for the time your driver lost in waiting in our shop.

As soon as I received your letter I spoke with Brent Grange 2
to find out why there was a delay and to make sure that the problem will not recur. Mr. Grange told me that the morning Ms. Biondella brought in your vehicle there was another Badger in the shop for repairs. The tickets for the two vehicles were mistakenly exchanged in our Parts Department, thus causing Mr. Grange to think that your job was relatively minor and could be handled quickly. To prevent this problem from happening again, I have established a new coding system whereby the Parts Department will have to verify order numbers with the service manager before customers are notified of the kinds of repairs that their vehicles require.

Your continued satisfaction with the Industrial Automotive 3
Center is important to us. If there is anything we can do in the future to improve our service, please give me a call.

Sincerely yours,

Robert L. Pierce

Robert L. Pierce
Vice President

Encl. 1

QUESTIONS FOR DISCUSSION

1. Before writing his letter, Eldred had to gather a number of facts. Point to specific details he uses to prove his claim against Industrial Automotive Center.
2. Why does Eldred begin his letter by saying that his company has been pleased with the Industrial Automotive Center? Why doesn't he begin with his complaint?
3. What is Eldred's purpose in including paragraph four? Why does his reader need this information?
4. Describe the tone of Eldred's letter, especially the types of verbs he uses to portray actions. Are any of his words loaded, e.g., angry, vindictive? Does the tone of his letter help to convince the reader that there is a problem and that Northwest Illinois is not to blame?
5. Eldred asks for a check and an explanation. Are both of his requests reasonable and necessary? Explain.
6. Pierce's first paragraph soothes the reader in a number of important ways. What are they and in what order are they taken up?
7. Find examples of a "You Attitude" in Pierce's letter that exhibit a sincere respect for the reader.
8. Why does Pierce tell Eldred about the new code system? Did Eldred ask for this information? How would such information be used given the two firms' business relationship?
9. Why does Pierce wait until the closing paragraph to acknowledge Eldred's comment about his satisfaction with the previous service at the Center? Shouldn't Pierce have begun his letter with this information? Explain.

SUGGESTIONS FOR WRITING

1. Write a claim letter about a faulty service you or your company has received recently. In your letter, include the specific circumstances surrounding the service and what inconvenience or loss resulted because of the poor service.
2. Write a letter refusing the request made in the letter for question 1 above.
3. Write a letter granting the request made in the letter for question 1.
4. Your company has been paying for time sharing on a computer system for the last year. You have just received a statement that billed you for time you did not use. Write an appropriate claim letter.

The following claim letter is about a defective product. It describes the problems the writer encountered and states specifically what action he wants the manufacturer to take. The writer's tone is objective and polite. The adjustment letter following favorably replies to the ALCO Industries claim letter. Notice how Carol Schwartz, the Customer Relations Director at Metro Media Products, provides an explanation that, while not going into unnecessary detail or blaming specific individuals, satisfies the reader's request for information.

Claim and Adjustment Letters about a Defective Product

ALCO INDUSTRIES OF AMERICA
3400 Ross Boulevard
Reno, Nevada 87650

November 10, 1983

Ms. Carol Schwartz
Customer Relations Director
Metro Media Products
P.O. Box 4356
Dallas, Texas 74563

Dear Ms. Schwartz:

In early April of this year, our Training Center purchased 1
two study carrels from one of your dealers, Sanderson's Electronics located on Pinewood Drive, Reno. One of these carrels, model #320a-700, has worked fine. But the tape player in the other carrel, model #320a-698, has destroyed the beginnings of two of our training tapes within the last week. This morning I returned the defective player to you by UPS.

We have followed all instructions contained in the 2
manufacturer's guide to owners but the player still malfunctions. Each time we try to use it, a tape gets caught. It is impossible to remove the tape from the machine without ruining it. Since much of our in-house classroom instruction is supplemented by slide-tape presentations, we have been seriously inconvenienced and have had to postpone three programs because we had only one carrel in operation where we have needed two.

Please repair the defective tape player. Since more than 3
eight months remain on our warranty, we think the repair should be made without cost to ALCO Industries. We would like to have the carrel #320a-698 in operating condition by the end of next week. If it will not be possible to repair the tape by this time, please give us a new tape player.

As we need to make plans for a program presentation next 4
week, I would appreciate hearing from you by the 19th.

Sincerely yours,

ALCO INDUSTRIES

Randall Higgins

Randall Higgins
Training Coordinator

Metro Media Products

P.O. Box 4536
Dallas, Texas 74563

November 14, 1983

Mr. Randall Higgins
Training Coordinator
ALCO Industries
3400 Ross Boulevard
Reno, NV 87650

Dear Mr. Higgins:

We have received the tape player you sent to us on November 10. I am sorry that the player from one of our carrels has destroyed your tapes and that your training schedule was disrupted. Because the study carrel is still under its one-year warranty, it will be repaired without any cost to ALCO Industries. 1

After examining your tape player, we found that failure of the take-up reel to turn caused your tapes to wrap around the push roller. The problem was the result of a faulty belt, which we are going to replace free of charge. However, I am sorry to say that it will not be possible to repair your tape player by the end of the week. We will make every effort to ship the repaired player to you by November 27. But to spare you any further inconvenience, I am sending you a temporary replacement today by express mail. You should receive this replacement player the same day this letter reaches you. 2

I do hope that you experience no further problems with our players. If you have any additional questions, please call me--collect. 3

Sincerely yours,

Carol Schwartz

Carol Schwartz
Customer Relations Director

QUESTIONS FOR DISCUSSION

1. Why does Higgins tell Metro Media Products that model #320a-700 works fine when his purpose is to complain about the other model?
2. Why does Higgins begin paragraph two with the statement that ALCO has followed the manufacturer's instructions?
3. Why does Higgins mention how the tape player is used?
4. Comment on the last two paragraphs of the claim letter. What is their function and how do they accomplish it?
5. Where does Schwartz announce the good news the reader wants to hear? Why in this place?
6. How does Schwartz's explanation of what went wrong help to keep the reader as a customer?
7. Why doesn't Schwartz tell her reader in the first paragraph that she cannot repair the tape player within a week? How does she break this news to Higgins without angering him?
8. Schwartz has to discuss a number of cause-and-effect relationships. How do her sentences effectively provide this type of information?

SUGGESTIONS FOR WRITING

1. Write a claim letter about a product you or your company has had difficulty with recently.
2. Write an adjustment letter in reply to the complaint letter you wrote for question 1. Either grant or refuse the request.
3. Write a reply to Mr. Higgins's letter in which you deny his request for one of the following reasons:
   a. ALCO did something wrong with the tape player (not cleaning the heads, using wrinkled tape).
   b. ALCO's warranty has run out.
   c. Another company manufactures the tape players Metro Media Products uses in its study carrels.

SALES LETTER

The sales letter below was mailed to engineers who work for utility companies. This audience, therefore, is carefully restricted—a highly trained group of readers (experts). The writer can safely assume that his readers know a great deal of information about distribution transformers. To convince them to buy or at least to consider buying his test equipment, the writer uses a rational, not emotional, approach by emphasizing the readers' need for his product and the variety of services his company offers.

A Multi-Amp Corporation Sales Letter

Multi-Amp Corporation

An Indian Head Company

4271 Bronze Way
Dallas, Texas 75237 USA
Telephone (214) 333-3201
TWX 910-861-9052

April 29, 1983

Mr. Lewis Ross
Davidson Power Company
2300 South Main Street
Davidson, NM 84076

Dear Mr. Ross:

Do you know that up to 20% of the system losses incurred by many utilities are caused by distribution transformers? 1

That's why utilities today are reevaluating their distribution transformers--new as well as older ones that were purchased before the losses were a significant factor. 2

That's where Multi-Amp can help. 3

An expanded line of distribution transformer test equipment now offered by Multi-Amp can help you to determine transformer losses per ANSI standards . . . accurately, quickly, and easily. 4

Whether your test program is targeted at initial acceptance of new transformers, or at evaluating rebuilt or reconditioned units, Multi-Amp can assist you. We have portable test sets for use in the transformer yard, and we also offer console-style test tables for your shop use. Plus, we offer you our application expertise to guide you in selecting the test equipment that is best suited for your requirements. 5

For more information about the benefits Multi-Amp distribution transformer test equipment can make in evaluating your transformers, please return the enclosed card. You will receive a complimentary copy of our new bulletin "Distribution Transformer Testing." If you would like to have one of our sales representatives in your area explain our equipment, please call us. 6

Very truly yours,

Robert D. Libbey

R. D. Libbey
Vice President

QUESTIONS FOR DISCUSSION

1. Libbey begins his letter with two well-known ways to arouse a reader's interest. What are these two techniques and how appropriate are they for the audience he wants to reach?
2. Identify the problem(s) the audience has that Multi-Amp wants to solve. Are these problems described in terms that are appropriate for the audience?
3. What specific advantages or benefits does the letter promise to the reader? How are these benefits tied to the reader's problems? That is, how does Libbey describe these benefits so that the reader will want to use the test equipment?
4. Where is the call to action part of the letter? Could this call be placed anywhere else in the letter? What does Libbey want his readers to do? How does he make it easy and enjoyable for them to do it?

SUGGESTIONS FOR WRITING

1. Write a letter to Libbey one month after your company has used his product. Refer to his sales letter in yours.
2. Write a sales letter to specialists in your field asking them to purchase a piece of equipment your company manufactures and services. Use the techniques found in Libbey's letter.

MEMOS

The federal government frequently prepares long reports on proposed changes in the environment. Called environmental impact statements or environmental assessments, these reports are sent to state agencies for comment and review. The following memo is the response of a state water-resources engineer to one environmental assessment. Originally sent to a state planning coordinator, the memo was subsequently attached to comments from the governor's office, which were then forwarded to the appropriate federal agency.

A Memo from the Wyoming State Engineer's Office

THE STATE OF WYOMING

ED HERSCHLER
GOVERNOR

State Engineer's Office

BARRETT BUILDING CHEYENNE, WYOMING 82002

MEMORANDUM

January 11, 1982

TO: Dick Hartman, State Planning Coordinator

FROM: Louis E. Allen, Water Resources Engineer L.E.A.

SUBJECT: State Identifier No. 81-156, Big Sandy/Salt Wells Oil and Gas Draft Environmental Assessment

After examining the subject Draft EA, I have no adverse comments. Water-related matters are addressed, and the necessary State Engineer permits are noted. Several water concerns could have been covered more completely, however, and I emphasize the following points: 1

The State insists that the options be kept open for development of the State's water for beneficial uses. 2

We are concerned about the protection of water wells as to both quantity (water levels) and quality. Owners of wells that are adversely affected by mining or other activities must be compensated with an alternate water supply or monetarily. The protection of aquifers from intrusion of poor quality water must be vigorously pursued, particularly in those areas where saline shales are common. 3

We feel that even more than the usual care must be exercised in assuring that all seismic exploration holes and any abandoned holes in these two areas should be adequately plugged and sealed to prevent leakage between aquifers, and to avoid the emergence of saline flows that could enter the surface water system. Most of the drainage in the subject areas is to the Colorado River, and we would not want to see any increase in salt loading to the system because this would tend to negate the efforts being made to reduce salinity in the Colorado River. 4

Thank you for the opportunity to review this Draft EA. Your referral memorandum is being returned as you requested. 5

LEA/ht

cc: George L. Christopulos
State Engineer

QUESTIONS FOR DISCUSSION

1. How does the format of this memo differ from that of the memo in Figure 12–3?
2. How is this memo organized? What are its parts or sections?
3. Based on Allen's language and the level of his technical description, to what audience type does Hartman belong? Evaluate the ways Allen meets Hartman's needs, paying special attention to the first sentence of the memo.
4. What rhetorical techniques discussed in Chapters 2–5 does Allen use? Provide some examples and comment on the appropriateness of these techniques to Allen's purpose.

SUGGESTIONS FOR WRITING

1. Write an appropriate reply to Allen from Hartman in which you ask for additional information or clarification.
2. Write a memo to a decision maker on any report or budget you have been asked to analyze. Make sure you specify why you read the report or budget, how you evaluated it, and what you think the reader should now do after having your assessment.

The following memo provides instructions for changing the interior configuration of a C-130 used in Aeromedical missions to Turkey. Two days each week the aircraft was modified to accommodate these flights. The other days of the week it served as a cargo carrier. Issued by the colonel, a wing commander, the memo was sent to two different Air Force detachments—the flight crew responsible for actually making the changes inside the aircraft and an Airevac section from the base hospital that accompanied the patients on the flight.

A U.S. Air Force Memo with Instructions for Modifying a C–130

DEPARTMENT OF THE AIR FORCE

TO: Tactical Airlift Detachment 12/CC
Airevac Detachment 41

FROM: Col. James G. Rodriguez
Wing Commander, Det. 25

SUBJECT: C-130 Aeromedical Configuration

DATE: 1/8/84

1. The C-130 routine cargo mission must be modified on Tuesday and Thursday flights to facilitate the on- and off-loading of medical passengers destined for Karamursel and Izmir. The attached schematic depicts the proper cargo compartment configuration for Det. 41 medical technicians and C-130 loadmasters to employ for the Tuesday and Thursday Airevac flights. This memo gives instructions for setting up the proper configuration. 1

2. The following steps must be taken to insure that the C-130 Airevac flight uses the appropriate configuration. Refer to the attached schematic as you read these steps. 2

   a. The line duty medical technician will advise the loadmaster at the show time about the final configuration. There might be some slight variations as each medical mission may dictate, e.g., the right side litter area may not be required and six seats may be rigged in its place. But there must always be an emergency left litter tier. In addition, the left center aisle must be left open for easy access by both the air and medical crews.

   b. The Comfort Pallet should be positioned at fuselage station 342 with the front end placed at station 298. This pallet requires AC power from the right forward galley outlet and DC power from the DC winch receptacle on bulkhead 245. The pallet comes supplied with an AC cable. The winch DC cable must be used for all DC hookups.

   c. Loadmasters must store all unused intermediate cargo floor and ramp conveyor sections (aft of the Comfort

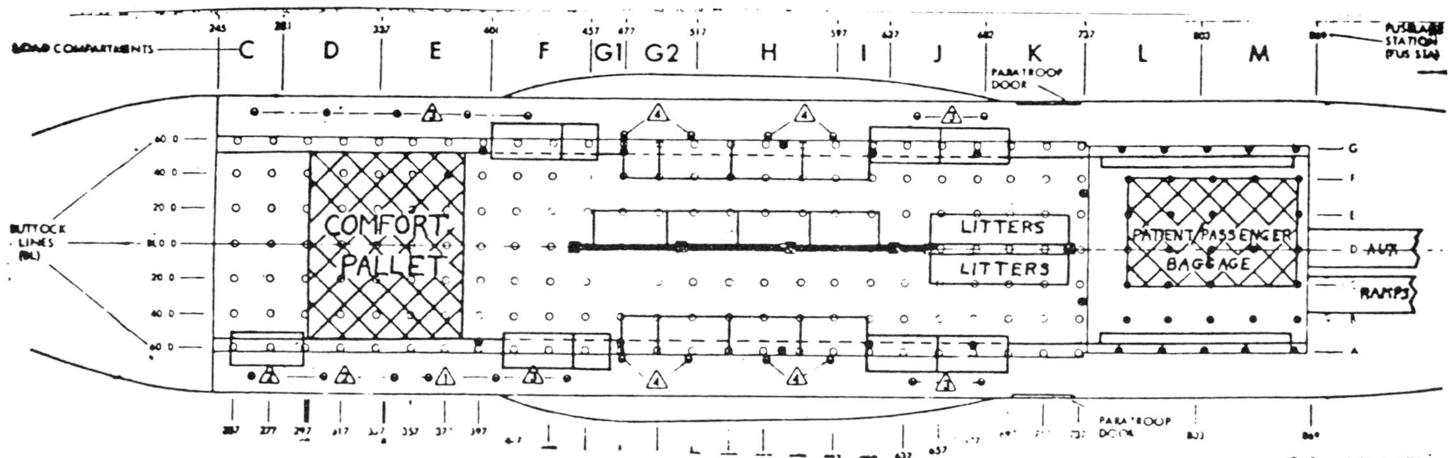

Figure 1. Schematic for the interior of a C-130.

Pallet) on the top of the left and right outboard rollers on the cargo ramp.

d. Four seat stanchions are installed at stations 452, 512, 572, and 732. If seats are to be installed in place of the right litter tier, an additional stanchion must be placed at station 692. The ladder should be installed at the normal location--station 642.

e. Loadmasters will include the following auxiliary equipment on each Airevac flight:

1. Three #6 upper seat back supports and #12, #13, and #14 lower seat back supports if seats are installed in place of the right litter tier

2. Four 2-person seats (or seven if the right litter tier is not used) for the center seat arrangement

3. All wheel-well seats

4. All side seats aft of the wheel wells

5. One 2-person seat, but <u>NO 1-PERSON SEAT FORWARD OF EACH WHEEL WELL</u>

6. Four operable litter straps

7. Two auxiliary loading ramps

8. Six 5000 lb. straps

f. All baggage for the Karamursel stop must be stowed on the right side of the ramp and the baggage for Izmir on the left. The auxiliary ramps should be positioned side-by-side to align with the aisle. These ramps will be folded back and strapped in place during the flight to enhance safety.

g. The aircraft winch, crew baggage, and life support equipment will be stowed forward of the Comfort Pallet in compartments "C" and "D."

3. Questions concerning the aircraft configurations for the Airevac missions should be directed to the Chief Flight Nurse (Det. 41_SGR, Airevac Section, Ext. 4732). The Flight Nurse will be my representative to resolve technical matters. 3

James G. Rodriguez

James G. Rodriguez
Col., USAF
Commander

1 Attach.
Airevac Schematic

QUESTIONS FOR DISCUSSION

1. Why is a memo the best means of conveying the colonel's information? His memo uses a standard format employed by the U.S. Air Force. Comment on the advantages this format offers to both the reader and the writer.
2. Why does the colonel send the memo to both the medical and flight crews? What are the specific needs of each audience and how do their needs overlap?
3. The memo gives detailed technical instructions. Explain how the colonel's tone and language are appropriate for his and the reader's purpose.
4. Why is the schematic an important part of the memo? Where and how effectively does the colonel refer to it?

SUGGESTIONS FOR WRITING

1. Using the format of this Air Force memo, write instructions to employees at a plant about one of the following subjects:
   a. parking
   b. priorities for using word processing equipment
   c. cash registers
   d. travel plans
   e. security clearances
   f. safety policies for a particular job
   g. catering services
   h. training programs
2. Write a memo informing employees about some change in company policy—e.g., parking, vanpooling, vacation time, fringe benefits, overtime.

Chapter 3 introduction. Illustration of typical incandescent lamp reprinted courtesy of GTE Lighting Products, Danvers, MA 01923.

"Speech Disorders." From *Encyclopedia of Science and Technology,* v. 12, pp. 873–875. Reprinted by permission of McGraw-Hill Book Company.

Chapter 4 introduction. Classification of fire doors from *Illustrated Encyclopedic Dictionary of Building and Construction Terms* by Hugh Brooks. © 1976 by Prentice-Hall, Inc.

"The Importance of pH Measurements." Courtesy of GTE Lighting Products, Danvers, MA 01923.

"A Chemist's Definition of pH." From *The Condensed Chemical Dictionary,* p. 666. Copyright 1977 by Van Nostrand Reinhold. Reprinted by permission.

"A Letter to Patients of a Pain Clinic." Reprinted by permission of Carla Waddell.

"How the Electrostatic Copying Process Works." Published with permission from *Word Processing & Information Systems.* Copyright © 1979 by Geyer-McAllister Publications, Inc., New York.

"A Life Support Decision Tree." © American Heart Association. Reproduced with permission.

"How to Install a Sensor into the Ball Valve of a Water Chiller System." Reprinted by permission of the Foxboro Company.

"Individual Lipids and Proximate Analysis of Various Foods." Reprinted from the *Journal of Agricultural and Food Chemistry* (1977, *25,* 406–407). Published 1977 by the American Chemical Society.

"Hand-Washing Patterns in Medical Intensive-Care Units." Reprinted by permission of the *New England Journal of Medicine,* vol. 304, pp. 1465–1466; 1981.

"Women in Engineering: Attitudes, Motivation, and Experiences." From *Engineering Education,* November 1981. Reprinted by permission of the author and *Engineering Education.*

"A Proposal to Management by the Lockheed Employees' Recreation Club." Reprinted by permission of the Lockheed Employees' Recreation Club.

"A Proposal to Supply Maintenance Service for National Home Life's Office Equipment." Reprinted by permission of the author.

"An Unsolicited Proposal to Provide Assistance in Upgrading a City Snow Control Program." Reprinted by permission of the author.

"A Feasibility Study on Establishing a Central Repair Facility at Midcontinent Power Company." Reprinted by permission of the author.

"The Progress Report for December 1982 on the Construction of a Pulp Mill for the Leaf River Corporation." Reprinted by permission of the author.

"A News Release Announcing a New Product from Abbott Laboratories." Reprinted by permission of Abbott Laboratories.

"A News Release Announcing a New Sylvania Outdoor Lighting System." Reprinted by permission of GTE Products Corporation.

"Two Advertisements from a Telephone Company." Reprinted by permission of AT&T Communications.

"An Advertisement on Beer Production from the PQ Corporation." Reprinted by permission of PQ Corporation.

Figure 12.2 reprinted by permission of Janis Monk.

"A Solicited Application Letter and Résumé." Reprinted by permission of Tina Bond.

"An Unsolicited Application Letter and Résumé." Reprinted by permission of Robert Eddy.

"Inquiry Letter about Purchasing Sewing Machines." Reprinted by permission of Cassandra Burroughs.

"A Multi-Amp Corporation Sales Letter." Reprinted by permission of Multi-Amp Corporation.

Audience Index to Models

The following index identifies the audience type(s) to which readers of each model belong. Keep in mind that a model may be addressed to multiple audiences and that even one reader can belong to several audience types.

CHAPTER 1

Living with the Kudzu	**Consumers**
Memo about Landscaping an Industrial Park with Kudzu	**Decision Maker**
How to Grow Kudzu	**Operators**
Influence of Time of Herbicide Application on Control of Kudzu	**Experts**
Kudzu	**Consumers**
Word Processors for Attorneys	**Consumers**
Memo about Word Processing Needs at a Corporation	**Decision Maker**
Apple II Super Serial Card Instructions	**Operators**
Exploring Technical Aspects of Formatting Floppies	**Experts/Decision Makers**

CHAPTER 2

New Floating Trap for Capturing and Preserving Emerging Aquatic Insects	**Experts/Operators**
The Freedom 200	**Experts/Decision Makers**
Jack Pine . . . An American Wood	**Consumers/Experts**
Shock Absorbers	**Operators**
Geologist's Memo about the Groundwater Potential at an Industrial Site	**Decision Maker**

CHAPTER 3

NIOSH Fact Sheet	**Operators**
Kinds of Maps and Kinds of Map Data	**Consumers**

Speech Disorders	**Consumers/Experts**
Radio Frequencies	**Consumers**
The Colony	**Consumers**
Keeping Your Balloon in Shape	**Operators**

CHAPTER 4

The Importance of pH Measurements	**Consumers**
A Chemist's Definition of pH	**Experts**
A Definition of Computer-Related Crime	**Experts**
A Letter to Patients of a Pain Clinic	**Consumers**
The Preservation of Historic Glazed Architectural Terra Cotta	**Operators/Experts/ Decision Makers**

CHAPTER 5

The Brewing Process	**Consumers**
How Your Body Deals with Drugs and Other Chemicals	**Consumers**
How the Electrostatic Copying Process Works	**Decision Makers**
EPA Air Pollution Inspection Procedures for Cement Plants	**Operators/Experts**

CHAPTER 6

Instructions for Installation and Maintenance of Steelflex Couplings	**Operators**
A Life Support Decision Tree	**Experts**
Starting an Engine with Outside Power	**Operators**
How to Make a Thermostat Control Timer	**Consumers**
How to Install a Sensor into the Ball Valve of a Water Chiller System	**Operators**
Troubleshooting Hints for the Apple II	**Operators**

CHAPTER 7

Individual Lipids and Proximate Analysis of Various Foods	**Experts**

A Chemist's Laboratory Report for the General Aluminum Corporation	**Decision Maker**
A Field Inspection Report of 1A Mill Motor	**Decision Makers**
Hand-Washing Patterns in Medical Intensive-Care Units	**Experts**
Polymeric Sealant Used to Stop Shale Degradation in Coal Mines	**Experts/Decision Makers**
Women in Engineering: Attitudes, Motivations, and Experiences	**Experts**

CHAPTER 8

A Plan to Reduce Supply Loss at Willis Machinery	**Decision Maker**
A Proposal to Management by the Lockheed Employees' Recreation Club	**Decision Makers**
A Proposal to Supply Maintenance Service for National Home Life's Office Equipment	**Decision Maker**
An Unsolicited Proposal to Provide Assistance in Upgrading a City Snow Control Program	**Decision Makers/Experts**

CHAPTER 9

A Feasibility Study on Storage Techniques for Tractor-Rite Inc.	**Decision Maker**
A Feasibility Study on Establishing a Central Repair Facility at Midcontinent Power Company	**Decision Maker**
A Feasibility Study of Alternative Uses for the Federal Correctional Institution at Alderson, West Virginia	**Decision Makers**

CHAPTER 10

Progress Report #1 of Phase III of a Cardiac Rehabilitation Program	**Decision Maker**
Memo to a Physics Instructor on the Progress of a Student Research Project	**Expert**

Index